Biological Museum Methods
Volume 2

Biological Museum Methods

Volume 2
Plants, Invertebrates and Techniques

George Hangay
Michael Dingley

The Australian Museum, Sydney, Australia

ACADEMIC PRESS
(Harcourt Brace Jovanovich, Publishers)
Sydney • Orlando • San Diego • New York
London • Toronto • Montreal • Tokyo

ACADEMIC PRESS AUSTRALIA
Centrecourt, 25–27 Paul Street North
North Ryde, N.S.W. 2113

United States Edition published by
ACADEMIC PRESS INC.
Orlando, Florida 32887

United Kingdom Edition published by
ACADEMIC PRESS, INC. (LONDON) LTD.
24/28 Oval Road, London NW1 7DX

Printed in Australia

National Library of Australia Cataloguing-in-Publication Data

---

Hangay, George.
Biological museum methods. Volume 2.

Bibliography.
Includes index.
ISBN 0 12 323302 X.

1. Museum conservation methods. 2. Museum techniques.
3. Natural history—Technique. I. Dingley, Mike.
II. Title.

069.5'3

Library of Congress Catalog Card Number: 84-73526

In this book, where drawings are not given a credit in the legend, they are the work of K. Khoo. The authors are indebted to K. Khoo for his assistance.

CONTENTS

Foreword ix
Preface xi
Acknowledgements xiii
Contents of Volume 1 xv

1 **Introduction: The methods and procedures of preservation** 1
Definition of preservation 2
Purpose 2
Methods in preservation 3
References 10
Source material 10

Plants

2 **Plants** 13
Dry preservation 13
Wet preservation 15
Display 22
Artificial plants 23
References 26
Source material 27

Invertebrates

3 **Sponges and coelenterates** 31
Sponges 31
Coelenterates 33
References 38
Source material 38

4 **Echinoderms** 40
Collecting 40
Killing, fixing and preserving 42
Dry specimens 43
Display 43
Source material 44

5 Worms and worm-like animals 45

Acanthocephala (spiny-headed worms) 45
Annelida (segmented worms) 47
Aschelminthes (Nemathelminthes) 50
Echiura (spoon worms) 51
Nemertines (ribbon worms) 51
Pentastomides (tongue worms) 52
Phoronides 52
Platyhelminthes (flat worms) 52
Pogonophora 53
Priapulida 54
Sipuncula (peanut worms) 54
Display 54
References 55
Source material 55

6 Molluscs 57

Terrestrial molluscs 57
Aquatic molluscs 59
Display 63
References 65
Source material 65

7 Arthropods (crustaceans) and chordates 67

Collecting 67
Killing and fixing 69
Preserving and storage 70
Display 70
Chordates 74
Reference 75
Source material 75

8 Arthropods: Insects and others 77

Collecting terrestrial insects 77
Collecting aquatic insects 104
Preservation and storage in the field 119
Laboratory preservation and mounting 125
Microscopic preparations 146
Storage of dried specimens 148
Synopsis of treatments for each insect order 152
Display 160
Arachnids 168
Diplopoda and chilopoda 173
Symphyla and pauropoda 173
Onychophora 173
References 173
Source material 174

Techniques for display and safety

9 Freeze drying 177

Freeze drying plant and invertebrate specimens 177
References 186

10 Plastic display box manufacture 187

Acrylic plastics 187
Glues and cements 189
Machinery 189
Making the box 198
Special heat formed boxes 207
Special considerations 210
References 214
Source material 214

11 Plastics workshop, embedding and workshop safety 216

Plastics workshop 217
Embedding biological specimens in clear plastic 223
Workshop safety 244
References 249
Source material 249

12 Modelmaking 251

Sculpting 251
Carving 259
Moulding 264
Flexible moulds 270
Casting 276
Vacuum forming 286
Surface finishing of models 287
Scale models 290
Landscape surface finishing and painting techniques 296
References 299
Source material 299

13 Dioramas 300

Displays 300
Special foreground effects 310
References 314
Source material 314

Formulae 315
Index 319

FOREWORD

NO OTHER class of artisans has ever better served the interests of science and education, or taught the public more about nature in such an exciting and interesting way, than the museum plant and animal preparator. Museum cabinets throughout the world are full of tray after tray of carefully catalogued specimens, that, if put on display that way, would bore the most casual visitor to distraction. No matter how knowledgable or skilled professional scientists or educators happen to be, they have never possessed the skills of the museum artist in preparing specimens for habitat display. Therefore, while scientists were providing the carefully assembled facts, museum artists were just as faithfully presenting them to the public by means of their unique skills.

Yet, though a unique class of artists and craftsmen, museum workers spend entire careers largely unknown and unrecognized. Their numbers are so small that few ever bother to record their skills and techniques for posterity. Consequently, new books on museum display techniques are as rare as annual sightings of endangered species from a single nation. Much of what they learn of technique is passed verbally from generation to generation or gleaned from a small body of out of print books and antiquated classics. It is largely from these books that amateur and commercial taxidermists learn most of what they know about mounting animals, making artificial plants, and constructing habitat exhibits. Without their skills and dedication, our natural history museums would be dry warehouses of cold facts instead of centres of public education and entertainment.

While almost any book on museum preparation techniques is welcomed by everyone interested in the subject, few attempt to cover the subject thoroughly. Many are published for a single discipline, or describe an author's favourite method. So the aspiring artist entering one of the fields of museum work must sort through an already thin literature and experiment for years before discovering what could be learned from a single good book on the topic. Having had the privilege of reading the manuscript for *Biological Museum Methods*, I can truthfully say that this work is one of those stout timbers that bridges the gap between those who know how and those who want to learn; between one generation of classics and another.

I met George Hangay during the period he spent researching in the United States and can testify to his knowledge and experience. He is one of those professional museum workers who enable an Australian to see the flora and fauna of America in an Australian museum, or an American to see a rare Australian specimen in an American one. On behalf of museum workers everywhere, I would like to thank George and Michael for sharing their knowledge with us.

Joe Kish

Editor,
Taxidermy Review

PREFACE

THE PRIMARY AIMS of this book are to highlight the main aspects of the preparator's work, describe as many techniques as practical, and guide the reader to the relevant literature, from which more information can be gained.

To prepare a book of this kind, one must draw heavily on the published and unpublished research of many other workers, present and past. As it is not possible to describe every technique which may be practised in a biological museum, lists of useful publications are provided at the end of the chapters.

In most cases when animal preparation is discussed, the grouping is not governed by any of the accepted scientific systems but rather by preparational requirements.

The preparation of specimens for natural history and building exhibitions has unlimited possibilities. The advances in technology are strongly felt in the museum field. We hope that this book will help the reader to explore ideas, learn techniques and understand the relationship between science and craft involved in the creative process that is the work of the preparator.

ACKNOWLEDGEMENTS

MANY PEOPLE assisted in the preparation of this book. We are grateful to all. First we would like to thank all our major contributors, whose names and institutions are mentioned at the appropriate sections of the text or in the captions of the illustrations. We are indebted to the Director, Dr Des Griffin, and the Trustees of The Australian Museum, Sydney, Australia, for permission to reproduce the many photographs and selections from the writings of museum workers. We are grateful to all those who helped us by giving advice through personal communication, namely (in no particular order): to K. Küng, Naturhistorisches Museum, Bern, Switzerland; D. Oppliger, Naturhistorisches Museum, Basel, Switzerland; H. Inchumuk, Natural History Museum, Denver, U.S.A.; A. James, British Columbia Provincial Museum, Victoria, Canada; R. Séguin, National Museum of Natural Sciences, Ottawa, Canada; R. Hursey, Provincial Museum of Alberta, Edmonton, Canada; E. Parkman, The Cleveland Museum of Natural History, Cleveland, U.S.A.; L. Jensen, Bergen Museum, Bergen, Norway; B. Jensen, Naturhistorisk Museum Aarhus, Denmark; R. Angst, Landessammlungen für Naturkunde, Karlsruhe, Germany; R. Buob, Staatliches Museum für Naturkunde, Stuttgart, Germany; G. Pucka, Niedersachsisches Landesmuseum, Hannover, Germany; K. Wechsler, Übersee Museum, Bremen, Germany; O. Post and H. Hjorta, Zoological Museum, Copenhagen, Denmark; R. Gajeelee, The Mauritius Institute, Port Louis, Mauritius; I. Chisselet and S. Endrödy-Younga, Transvaal Museum, Pretoria, Republic of South Africa; P. Tempest, Natal Museum, Pietermaritzburg, Republic of South Africa; W. Schaefer, Port Elizabeth Museum, Humewood, Republic of South Africa; R. Bullock, University of Sydney, Sydney, Australia; Z. Catalan, University of Los Banos, Republic of the Philippines; E. Diehl, Dolok Merangir, Indonesia; D. Horning, MacLeay Museum, Sydney, Australia; A. Vojnits, Natural History Museum, Budapest, Hungary.

Our special thanks go to the experts who have read parts of the manuscript and, with their constructive criticism, helped to make it better. The readers were: J. Kish, *Taxidermy Review*, Montana, U.S.A.; and L. Vail, F. Rowe, P. Hutchings, J. Paxton, R. Strahan, J. Lowry, R. Springthorpe, M. Gray, C. Smithers, W. Boles, L. Gibson, H. Cogger, O. Keywan and P. Colman (who not only read but also rewrote parts of Volume 1, Chapter 6) all of The Australian Museum, Sydney, Australia.

We are indebted to: K. Khoo for his drawings; J. Fields, Head, Photography Department, The Australian Museum, Sydney, for the photographs; J. Dingley, I. Toohey and K. Hangay for typing and proofreading the manuscript; J. Miller and Ivo Docking for sorting the manuscript; and J. Giffen for translating from Afrikaans to English.

Last but not least we thank Jeremy Fisher, Senior Editor, Academic Press Australia for all his patience and help.

CONTENTS OF VOLUME 1

1 **Introduction: A short history of taxidermy**

Vertebrates

2 **Fishes**
3 **Amphibians**
4 **Reptiles**
5 **Birds**
6 **Mammals**
7 **Embalming**
8 **Anatomical preparation**
9 **Freeze drying**
10 **Skeletal preparation**

1

Introduction: The methods and procedures of preservation

ROBERT BOYLE (1627–1691) proved that air was needed to sustain life; he also found that biological and anatomical specimens would not decompose if they were placed in spirits of wine. This was very significant for the understanding and teaching of anatomy as, until this time, anatomists had to work quickly on their subjects before decomposition made their task unbearable. The first specimens Boyle placed in spirits of wine were a linnet and a snake, which were placed in Gresham College in 1663. Due to the high cost of spirits of wine, use in preservation was limited and did not radically change the style of wet preservation for a long time.

In 1666, Frederick Ruysch (1638–1731) preserved the whole body of Vice Admiral Sir William Berkeley by infiltrating the embalming fluid into the tissues through the vascular system. As embalming fluid, Ruysch used diluted alcohol distilled from fermented barley mash, and about 67% alcohol by volume. Although the body was perfect in tone and lustre, and had the freshness of youth, the alcohol solution was not sufficient for permanent conservation, although no deterioration was observed during Ruysch's lifetime.

From the seventeenth century to the middle of the nineteenth century, museums evolved from "curious" and "interesting" places into more specialised education establishments where the presentation of specimens was of the greatest importance. In 1863, August Wilhelm von Hofmann, then chemist at the British Royal Mint, discovered the gas formic aldehyde. It was not until the 1890s that Ferdinand Cohn (1828–1898), a botanist, experimented with formaldehyde (formic aldehyde dissolved in water) and found that it preserved plants. This was another great breakthrough; it did not take long for zoologists and anatomists to use this preservative. It was far cheaper than spirits of wine and, when mixed properly, did not cause excessive hardening or shrinkage of tissues.

Up until this time, anatomical museums were the most widely used and best documented of all museums as the study of the human body was of paramount importance. The greatest of all anatomical collections was that of John Hunter (1728–1793), who deposited nearly 14 000 preserved specimens, all of which he had personally prepared, in the Royal College of Surgeons.

From the last part of the nineteenth century to the present, preservation techniques have improved tremendously allowing workers to experiment, and perfect techniques to preserve colour as well as form.

DEFINITION OF PRESERVATION

Good preservation is the immortalization of a specimen in a natural form, without losing its character. It should also retain its original colouration, markings and size.

PURPOSE

There are three main reasons for collecting and preserving specimens.

The first is to build and update collections, and to try and store as many species as possible with proper written data so that they can be used as references in the identification of unknown species.

The second is so that researchers in morphology and anatomy can assemble enough data to allow taxonomists to arrange animals and plants into the hierarchies generally used for classification. This purpose is also very useful in the process of creating reference collections.

The third purpose is display. This area of collecting and preserving is important. More and more people are becoming aware of natural history. This interest stimulates them into learning about and becoming familiar with each animal and plant.

Workers in natural history are stimulated to study thoroughly, as each display is different. Their interest and dedication produces a high standard of exhibits. Any visitor to a museum expects a well made display.

The rest of this chapter deals in detail only with gross or macroscopic specimens to be collected for museum collections, museum displays, or as teaching aids in educational establishments. It will not try to describe methods applicable to histology or slide mounting as this is a very complex task which has already been well documented in other books.

METHODS IN PRESERVATION

Parts of this section were taken from Kannemeyer (1982); the authors are indebted to Kannemeyer for allowing us to use his information.

LABELS

Solitary specimens in a wet collection should have a stainless steel tag which bears their catalogue number attached to them. Where there are many specimens in a jar, a single stainless steel tag can be used.

When labelling specimens in the field, matt-surfaced plastic labels or robust parchment paper labels, which are impervious to preservatives, should be utilized. Both of these materials can be written on with hard pencil or waterproof India ink. All field labels should be retained with the specimens in the jars in which they are permanently stored. This is to prevent the mistakes which often occur when transcribing data into catalogues.

For permanent storage, standard printed parchment labels should be placed inside jars. A printed label should contain the same information as a catalogue index card. Information is written on it in waterproof ink and, when thoroughly dry (about five minutes drying time should be allowed), the label is placed inside the jar so that its entire surface is visible and all information easily read. Labels on the outside of jars are often thought to be superfluous. However, these should be used as an added safeguard, particularly on jars containing type specimens.

The minimum information needed for each label is: locality where found; date when found; how caught; habitat where found (for example, in rocks, under bark); name (if possible); a number or identifying letter; and sex. More detailed information (for example, temperature, climatic conditions, compass bearings, whether the environment is wet or dry, collector's name and time collected) should be written down in a note book so that later it can be collated in the laboratory. Colour notes should be recorded as soon as the animal is caught, but a problem here is that people see colours differently; for example, someone may record a colour as green but there are many shades of green, or someone may record a colour as pale green whereas someone else may record the same colour as yellow green. Therefore, there is room for a universal colour chart with strict guidelines on how it is to be used. There are colour charts available, but these are not comprehensive enough for exact colour shades. Listing all colours and shades on a chart, and keeping those colours stable over many years, would produce a book thick enough to warrant a special vehicle to transport it. Maybe the answer lies in a portable electronic colorimeter which records the colours in nanometers (Ångstroms). Some collectors have used water colours to paint those colours seen on the specimen in their note book. The colours are built up until the correct colour and shade is achieved; experience made the task easier and quicker. Today, the collector does not have time to spend painting colours into note books; also some will say that for some specimens colour notes are not altogether important. Those labels which are to be included with the specimen, and are to be stored in alcohol or formalin, should have the data written with a waterproof India ink, and allowed to dry, before immersion. When collecting in the field, pencil can be used to write data on labels, but new labels in India ink should be made back in the laboratory.

Marine animals

The following information is required when registering marine animals collected offshore: position co-ordinates, in terms of latitude and longitude; the depth at which specimens were captured; station number; and the name of the vessel and the type of gear employed in capture. The quarter degree grid reference system is often used to pinpoint intertidal or terrestrial localities.

CATALOGUES

Catalogues should be durable and, preferably, leather bound. Pages should be quality paper and ruled vertically and horizontally to accommodate the various headings. All information should be written with waterproof India ink.

There should be no ambiguity about recording dates. The date should always be written in full; that is July 22 1985. Transcribing should be meticulous. Scientific names should be followed by author and date. Long pages of ditto marks should be avoided.

All catalogued data must be cross referenced to a card index system. These index cards should be arranged systematically by family according to modern taxonomic classification. Within each family, cards are arranged alphabetically by genus and species. Cards within species follow numerically, according to catalogue number. An alphabetically arranged index list, numerically coded to each family, simplifies the task of locating a particular group within the system. Index cards contain the following information: catalogue number; number of specimens; scientific name; nature of the specimen (for example, holotype); station number (if any); locality and position; depth; date of collection; collector; preserving fluid; and the name of the person who identified the specimen.

All of this data can be included in a computer based system, if one is available. However, computer cataloguing does not do away with the need to label specimens.

NARCOTIZING AND KILLING

Narcotizing certain creatures before killing them produces fully relaxed specimens. If snails, for example, are killed outright, they will retract into their shells; worms will shrink, warp, and, in some cases, break up. These animals require narcotizing. The methods for narcotizing and killing will be found in the sections that discuss the various animal groups.

FIXATION

Fixation is a process that is quite distinct from that of preservation. It is, often, a process that is poorly understood. A good fixative should inactivate enzyme systems and prevent changes due to autolytic, bacterial, or fungal activity. The aim of fixation should be the preservation of cells and tissues in a state identical to that existing during life. It is, of course, impossible to achieve this aim completely. Any method of fixation is at best a compromise. Fixation in the sense of inactivation of enzyme systems and stabilization and preservation of the structural components of tissues is, usually, accomplished by chemical means. A variety of chemicals may be employed. For general purposes, formaldehyde is, perhaps, the best. At correct concentration, and buffered to neutrality, it is a rapidly penetrating and effective fixative with excellent protein stabilizing properties. Classified as a denaturing fixative, it reacts with the nitrogen–hydrogen groups of protein molecules to form an additive compound. The resultant molecule is larger and very stable.

Formaldehyde

Formaldehyde is a gas, formic aldehyde, of which 40 g is dissolved in 100 mL of water, equivalent to 37% by weight. Formaldehyde is an all round preservative with good penetrative properties. It is acidic, but it has a pungent vapour which attacks the mucous membranes of the nose and eyes. It is best to use it in a well ventilated area. Its acidity causes decalcification of the calcareous structures in animals and it should therefore be neutralized before dilution. To neutralize it, add borax (sodium borate), calcium carbonate, or sodium carbonate until neutral pH is attained (this can be checked by using pH paper). Another method is to add 4% aqueous sodium hydroxide to the stock solution until it is neutral to phenol red. Alternatively, the stock solution can be buffered using 200 g hexamine (methenamine) per litre of formaldehyde. A buffer is one that allows the solution in which it is dissolved to resist changes in pH due to the dilution and addition of acids and alkalis. Most animals are placed into formaldehyde diluted with tap or distilled water but, when preserving marine animals, sea water is used. The sea water prevents shrinkage due to osmotic changes which occur if tap water or distilled water is used. Formaldehyde should be stored in a warm area: a white precipitate of paraformaldehyde forms when the solution is cold.

Formalin

The term formalin is used here to denote diluted formaldehyde in water. Thus 10% formalin is 10 parts of formaldehyde mixed with 90 parts of water. In reality, this is 4% formaldehyde. Formalin is most often used in a concentration of between 5%–10%. To calculate the concentrations of any solutions the following formula will be helpful.

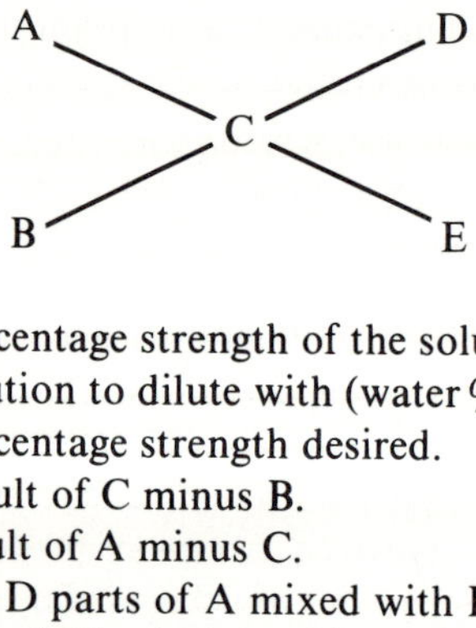

Where A is the percentage strength of the solution to be diluted.
Where B is the solution to dilute with (water %).
Where C is the percentage strength desired.
Where D is the result of C minus B.
Where E is the result of A minus C.
The answer will be D parts of A mixed with E parts of B.
EXAMPLE: To make a 10% solution of formaldehyde, using the formula and substituting:

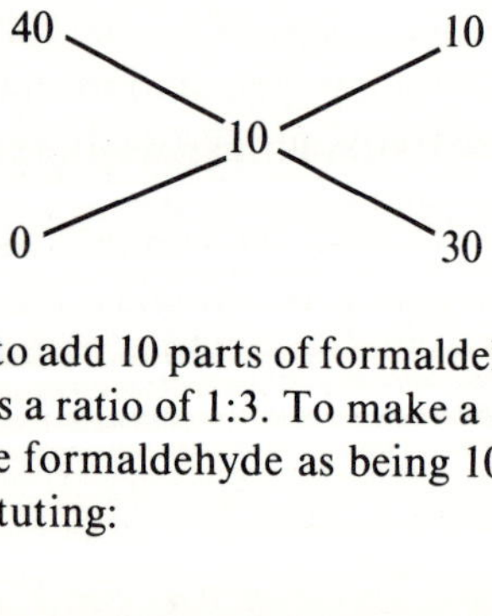

The answer will be to add 10 parts of formaldehyde to 30 parts of water, which is the same as a ratio of 1:3. To make a 10% solution of formalin you should take the formaldehyde as being 100%. Therefore using the formula and substituting:

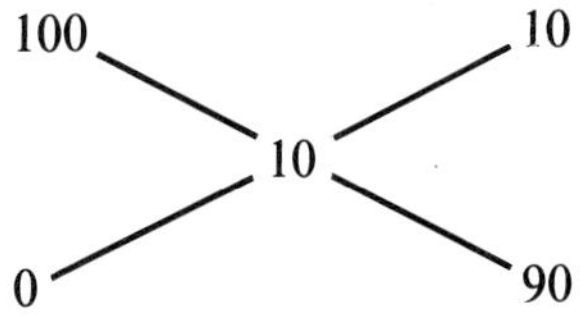

You will take one part formaldehyde and mix with nine parts of water.

Paraformaldehyde

This is a white crystalline powder having an odour of formaldehyde. It is obtained by concentrating formaldehyde solution. It is slowly soluble in cold water, but easier to dissolve in warm water. Some collectors use paraformaldehyde on collecting trips instead of the liquid formaldehyde. It is easier to handle and transport. There is no risk of broken bottles, which can allow the liquid to seep into the collecting gear. The herpetology department at The Australian Museum, Sydney, Australia, uses paraformaldehyde in the field on some trips and makes an aqueous solution of 10% formalin from two litres of water in which 80 g paraformaldehyde and 20 g anhydrous sodium carbonate have been dissolved. Dissolving time is decreased if frequent stirring takes place or warm water is used. When the chemicals have completely dissolved, 21 g citric acid is added and the solution stirred until no more carbon dioxide gas is given off. The solution is then ready for use.

Dowicil

Dowicil (1–(cis–3–Chloroallyl)–3,5,7–Triaza–1–azoniaadamantane chloride) is a yellow-white solid, highly soluble in water. It can be used as a fixative, and is normally made up as a 10% solution with either fresh or sea water, depending on the medium in which the specimens were caught. According to Lincoln and Sheals (1979), Dowicil releases formaldehyde in the presence of proteins. Specimens retain some flexibility as opposed to normal formalin preserving and the chemical does not give off the pungent and irritating vapour as does formalin, and does not have the flammable and volatile properties that alcohol possesses.

Alcohol (ethanol)

When diluted, alcohol is used as the storage medium for most museum collections. It is easier to handle as its smell is much more bearable than formalin. Most specimens are stored in a concentration of between 70%–85%, but never less than 50%. Alcohol is a strong dehydrating agent which extracts the water from specimens and thus becomes diluted itself. If the mass of a specimen exceeds the volume of alcohol in which it is immersed, the alcohol will become so dilute as to become a breeding ground for bacteria and moulds.

Some soft bodied animals shrink excessively if placed into a high concentration so it is prudent to place them into 40% alcohol and then, by successive increases in concentration, to 70%.

Most institutions purchase their alcohol in bulk in the form of a mixture of 95 parts ethanol and five parts methanol. The small amount of methanol does not affect the preserving qualities of gross specimens.

Propylene phenoxytol

This is really a postfixation preservative for specimens that have been fixed in formalin. Fixed specimens are transferred to a 1%–2% solution of propylene phenoxytol in water. It is very difficult to dissolve in cold water, but easier in boiling water. After adding it to boiling water, shake vigorously until it has dissolved; once it has cooled it is ready for use. Propylene phenoxytol may be sold under the name 2–phenoxyethanol. Compared to other preservatives is quite expensive, but, when diluted, the cost is not prohibitive.

STORAGE

Penrith (1971) has emphasized the great fire risk of a spirit collection and has drawn attention to the need for using fireproof materials in museum building design. Because of this particular fire risk, and because of their relatively greater scientific importance, type specimens should be segregated from the rest of the spirit collection. Types should be stored in locked cupboards, which, if at all possible, should be made of fireproof materials.

Jars and bottles containing type material should be marked conspicuously: A red 'type' label can be attached to the exterior label or the jar marked in some other manner to show that it contains type material.

Where wet specimens are stored in closed cupboards, they are protected from harmful natural and artificial light. When open shelving is used for storage, specimens should, preferably, be stored in the dark.

The spacing, width and length of shelving will, normally, be dictated by the specific nature of individual collections. Usually these factors will depend upon the height, weight and width of the largest container which must be accommodated.

Kannemeyer (1973) and Penrith (1971) have discussed the problem facing southern African museums in ensuring an adequate supply of glass containers suitable for wet biological use. Museum specimens, preserved in fluid, require well sealed containers. The Grathwol jar described by Kannemeyer (1973) is an ideal container.

Wooden tanks, lined with glass fibre, may be used to house large wet specimens (Kannemeyer, 1973). Additionally, plastic buckets, of the type commonly used for paint, may be utilized for larger specimens. These buckets have a press-on lid which forms an airtight seal. Alcohol does not seem to harden the plastic from which these containers are manufactured. They are marketed in five, 10 and 20 litre sizes.

All buckets and large bins should be numbered. The number of the bin or bucket in which specimens are stored should appear in the catalogue and card index system.

Specimens are stored in systematic order upon shelves. The same taxonomic order as that employed for the card index system must be followed. Shelf separators should be used to segregate families. Each shelf must bear a label indicating the particular family it contains. Within families, the specimens must be arranged alphabetically by genus and species. However, within species, jars do not follow in numerical sequence. Instead, they are placed in order of size so that the smaller museum jars are easily visible. The outside of the jar indicates the contents in terms of the catalogue entry and family, genus, or species. Cupboards and shelves are numbered. A list bearing a numerical key to families, and cross referenced to the catalogue system, indicates the location of each family by cupboard and shelf number. This list should be prominently displayed.

MAINTENANCE

Once specimens are safely stored in an airtight container containing the appropriate preservative, the only maintenance necessary is the periodic checking of the fluid levels in the jars. The frequency with which a collection must be checked will, of course, vary with climate, storage conditions, and the airtight abilities of particular jars. Where collections are in active use, this task is normally done annually usually just before the onset of summer. Where collections are inactive, more frequent checks are advisable. The addition of one part glycerine to 19 parts ethanol will ensure that, where collections are not used, specimens do not dessicate completely. Specimens stored in this manner, and from which most of the fluid may evaporate, need only have the preservative replaced to restore them to their original state. Pantin (1959) recommends the use of equal parts of alcohol, glycerine, and distilled water for reclaiming dessicated specimens. Van Cleave and Ross (1947) have suggested the use of a weak solution of trisodium phosphate for the same purpose. Specimens should be left in these solutions only as long as it is necessary to restore them, after which they should be transferred to preservative. Where fluid levels have drastically dropped, and particularly with jars containing many specimens, old fluid should be discarded and replaced with fresh preservative. To determine the strength of alcohol from time to time, a volume alcoholmeter is employed; if the strength is too low, preservative can be added until the solution reaches normal strength. A volume alcoholmeter gives a direct reading, but it is necessary to apply temperature corrections to the reading obtained. A table of adjustments for temperature is usually supplied with the instrument.

REFERENCES

Kannemeyer, S.X.: The storage of a wet collection. *Bulletin of the South African Museum Association.* 10/7 1973: 274–277.

Kannemeyer, S.X.: The curating of a wet collection. Paper delivered at the South African Museum Association Technical Conference, Pretoria, July 1982.

Lincoln, R.J. and J.G. Sheals: *Invertebrate animals: Collection and preservation.* Cambridge University Press, Cambridge 1979.

Pantin, C.F.A.: *Notes on microscopical technique for zoologists.* Cambridge University Press, Cambridge 1959.

Penrith, M.J.: Notes on new products of interest to curators of wet biological collections. *Bulletin of the South African Museum Association.* 9/16 1971: 499–507.

Van Cleave, H.J. and J.A. Ross: A method for reclaiming dried zoological specimens. *Science* 105/1947: 318.

SOURCE MATERIAL

Atkins, W.R.G.: The preparation of permanently non-acid formalin for preserving calcareous specimens. *Journal of the Marine Biologists' Association U.K.* 12/1922: 792.

Bianco, S.L.: The methods employed at the Naples Zoological Station for the preservation of marine animals. *United States National Museum.* 39/1899 (Part M).

Gohar, H.A.F.: The preservation of contractile marine animals in an expanded condition. *Journal of the Marine Biologists' Association U.K.* 22/1932.

Mahoney, R.: *Laboratory techniques in zoology.* Butterworths, London, 1973.

Owen, G., and H.F. Steedman: Preservation of animal tissues with a note on staining solutions. *Quarterly Journal Microbiological Science.* 97/1956: 319–321.

——. Preservation of molluscs. *Proceedings of the Maloc. Society of London.* 33/3 1958: 101–103.

Rau, R. The preparation of specimens for wet collections. *Bulletin of The South African Museum Association.* 9/14 1971: 589–592.

Smith, J.B.L.: A neutral solution of formaldehyde for biological purposes. *Transactions of the Royal Society of South Africa.* 31/3 1947: 279–282.

Wagstaffe, R. and J.H. Fidler: *The preservation of natural history specimens.* H. & G. Witherby, London 1957.

Waller, R.A. and W.N. Eschmeyer: A method for preserving colour in biological specimens. *Bioscience* 15/5 1965: 361.

Williamson, D.I., and G. Russell. Ethylene glycol as a preservative for marine organisms. *Nature* 206/1965: 1370.

PLANTS

2
Plants

PLANTS are an important part of any natural history museum, and their collection and preservation must be undertaken with the same care and diligence given to all other areas of the preparator's craft. The methods described below will ensure that specimens are presented at their best.

Field notes should be recorded as soon as plants have been collected and should include data such as collector's name, date of collection, locality (using latitudes and longitudes where possible), habitat, shape, size and colours of the flower parts and, lastly, altitude. The notes must also indicate whether the plants were growing in natural, cultivated or pasture areas and, if possible, rock and soil types.

DRY PRESERVATION

Small herbaceous plants should be dug from the ground leaving the roots intact. Specimens are washed to remove soil and other extraneous matter and blotted dry using paper towelling.

Plants that have been dug from the ground, or have been cut across stems or branches, quickly lose water from the cells and vessels. This makes the leaves and herbaceous stems wilt and curl and increases the risk of bacterial and fungal attack. To overcome these problems, plants in herbariums are

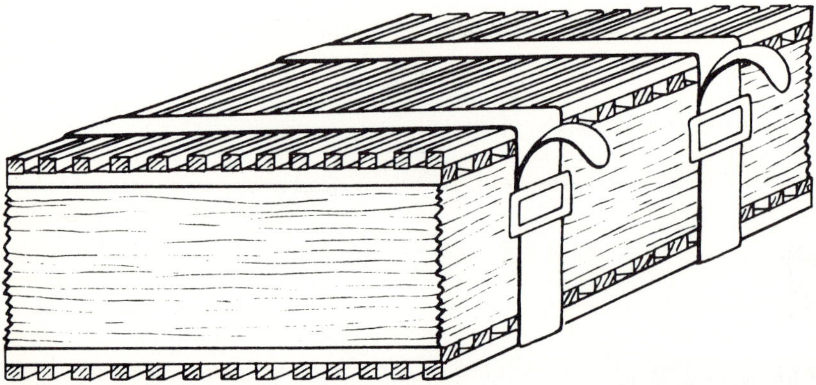

Fig. 2.1. A plant press in operation showing the straps that keep the press closed whilst pressing the material.

pressed flat and dried between sheets of semi-absorbent paper before they have time to curl and shrivel. The most successful piece of apparatus that is employed to flatten and dry plants is called the plant press. This can be made very easily in the workshop using few tools and materials.

Fig. 2.1 shows a finished press. It is made as follows: strips of wood 18 mm square are cut to a length of 600 mm, and enough are made so that, when placed every 18 mm apart, they will make a top and base having a width of 430 mm. More strips of wood 18 mm square are cut 430 mm in length and are placed across the others and screwed, or nailed and glued, in position. When the press is in use the top and base are employed to squash the plant material flat; this is achieved using straps and lock ties. The straps make it easier to add more plant material quickly than the more conventional way of using long threaded bolts and wing nuts. To start using the press, a sheet of corrugated packing paper is laid onto the base; on top of this is laid a thin sheet of cardboard. A sheet of newspaper is unfolded, laid on top of the cardboard, and the plant is arranged, with its identification tag, in position. The plant is covered by folding the newspaper over it, then laying another sheet of cardboard on top, followed by another piece of corrugated paper. The process is repeated for every plant. When arranging the material, the size of the plant determines the final appearance; long stemmed plants will have to be bent into a V, N or M shape. Dry material will probably break if attempts are made to bend it. If a specimen is uneven in thickness (say, a thick and bulbous base tapering to narrow twigs) a sheet of polystyrene foam can be included between the layers of newspaper to distribute the weight more evenly. Unless you work in a very humid climate, the drying time of most plants should be less than three weeks; succulent plants will take a little longer, and the newspaper will have to be changed more frequently.

Drying times can be speeded up by placing the press (with specimens) in drying cabinets for one to three days at 35° C. The time will vary depending on the water content of the plants.

Pine cones, large inflorescences, fruits, and seeds should be placed into plastic bags and kept with the main part of the plant or labelled so that they may be brought together once drying has been completed.

Plants such as mistletoe and fig loose their leaves, or remain alive for long periods, in the press. They first have to be killed by freezing, placing them into boiling water for a few minutes, or by soaking them in 100 mL glycerine, 100 mL 50% ethyl alcohol, 5 mL formaldehyde, 5 mL glacial acetic acid, and one lump of copper sulphate for 10 days (or until normal colour prevails).

Thick plants such as cacti should have their segments cut lengthways in half; the fleshy material can then be scooped out, leaving the outside skin to be pressed.

MOUNTING ON HERBARIUM SHEETS

The dry material should be removed from the press and stuck onto herbarium sheets prior to storing. Each herbarium sheet should have the following information stamped on it: common name; scientific name; habitat; exact locality; collector's name; altitude; date of collection; date of mounting; and a space for any other relevant data.

The plants are stuck in a position which allows easy observation of important characteristics using herbarium tape, good quality adhesive paper, or linen binding tape (Fig. 2.2)

Dry specimens can be stored for many years as long as insect pests do not come in contact with them. Plant specimens can be insectproofed by being placed into boxes which have tight fitting lids and a few grams of naphthalene spread on the bottom. The naphthalene should be checked every year or so and must be replenished when necessary.

WET PRESERVATION

Specimens which are intended for classroom study should be preserved in formalin. A 4% solution suits small plants and a 6% solution is necessary for fleshy specimens. Plants which have a waxy coating can be preserved in 4% formalin to which has been added, 25% by volume, 95% alcohol. It must be remembered that bleaching of the colours will occur and in most cases they will bleach entirely.

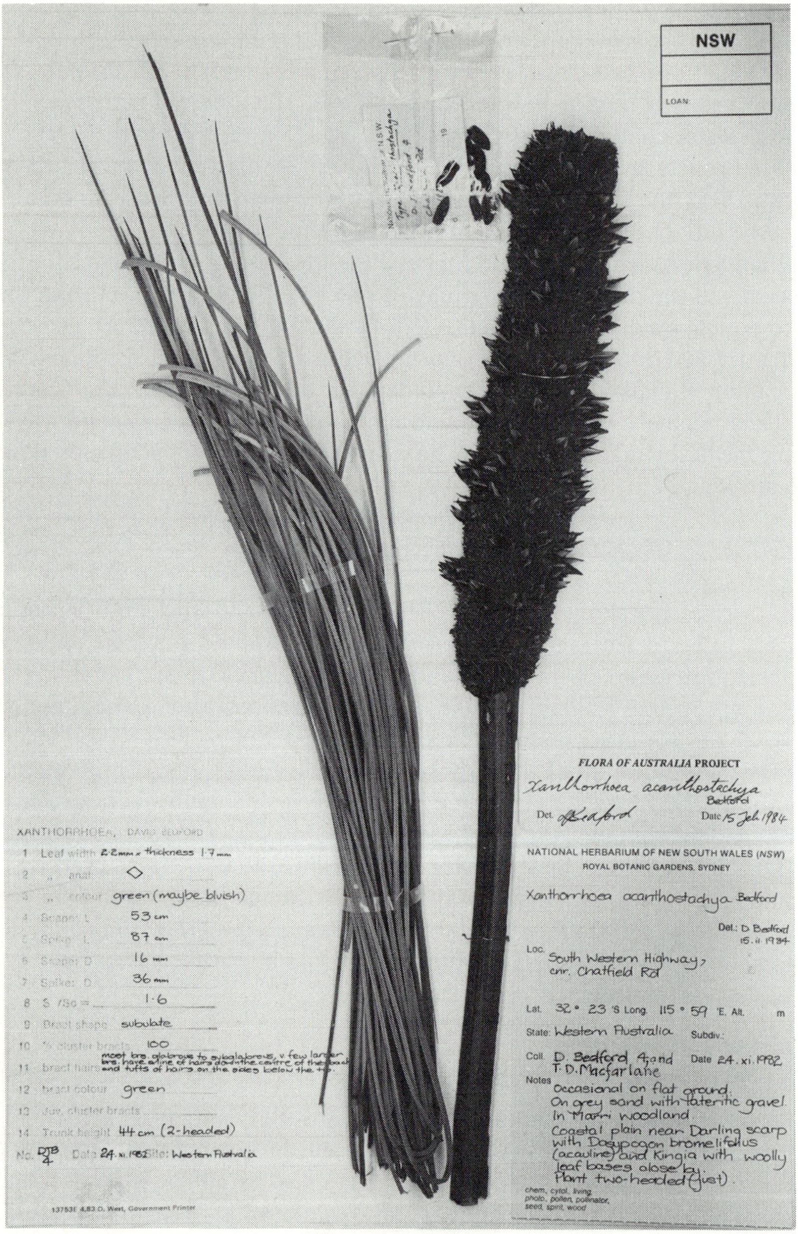

Fig. 2.2. An herbarium sheet showing a pressed specimen in position and data. Photo courtesy of D. Bedford, Royal Botanic Gardens, Sydney, Australia.

The following list of preservatives is included to enable the reader to be able to choose a formula that best suits the plant that is to be preserved. They are intended as storage media and must not be allowed to evaporate and cause specimens to dry out as this renders the plants useless where examination is concerned.

ALGAE

For general purposes, where specimens are to be examined within a few days of collecting, the following formula is suitable for algae; water 60 mL, ethyl alcohol 30 mL, and formaldehyde 10 mL. For all marine forms, seawater should be substituted for tap or distilled water. Microscopic single celled algae may be preserved indefinitely in the following: 0.5 g iodine; 1 g potassium iodide; 4 mL glacial acetic acid; 24 mL formaldehyde; 400 mL water. All components are mixed and added to an equal volume of water containing the specimens. Macroscopic algae may be preserved and stored in the following: 10 g potassium chrome alum; 5 mL 4% formalin; 500 mL water. This is said to preserve the green colour.

Formol acetic alcohol (FAA)

This solution, in its various forms, is widely used for preservation. Three formulae are given —

A.	50% alcohol	100 mL
	formaldehyde	6.5 mL
	glacial acetic acid	2.5 mL
B.	absolute alcohol	90 mL
	formaldehyde	5 mL
	glacial acetic acid	5 mL
C.	70% alcohol	85 mL
	formaldehyde	10 mL
	glacial acetic acid	5 mL

Formulae A, B and C are variations of the same constituents and are given as a guide only. Formula B is the most widely used.

Other preserving solutions for algae

Some other preserving solutions are described below:

1.	water	72 mL
	formaldehyde	5 mL

glycerine	20 mL
glacial acetic acid	3 mL

2. *Transeau's algal preservative*

water	12 parts
95% alcohol	6 parts
formaldehyde	2 parts
glycerine	1 part

To this solution add copper sulphate crystals to make a saturated solution to retain the green colour of chlorophyll.

3.

chromic acid	1 g
glacial acetic acid	3 mL
water	100 mL

4. Mix 0.5%–1% copper acetate in 2% formalin and keep specimens in this for 24 hours. Store specimens in 5% formalin.

5. *Acid Lugol's solution*

iodine	10 g
potassium iodide	20 g
glacial acetic acid	20 mL
distilled water	200 mL

For use, dilute 1 mL of this stock solution with 100 mL of the sample solution.

6.

formaldehyde	5 parts
propionic acid	5 parts
50% alcohol	90 parts

This solution is especially designed for use in staining and slide mounting.

7.

copper acetate	1 g
camphor water	227 mL
glacial acetic acid	5 mL
glycerine	180 g
corrosive sublimate	1 mL

8.

cupric sulphate	10 g
boiling water	90 mL

9a. *Montagu Brown*

borax	14 g
alum	14 g
potassium bichromate	2 g
sodium chloride	56 g
water (warm)	9 L

	saturated solution of bichloride mercury	4 drops
9b.	potassium nitrate	56 g
	water	1137 mL
	saturated solution of bichloride mercury	1 drop

Marine algae may also be preserved in the following solutions:

formaldehyde	25 mL
saturated solution of cupric acetate	25 mL
sea water	350 mL

ethyl alcohol	70 mL
sea water	30 mL

formaldehyde	3 mL
sea water	97 mL

For large specimens such as kelp use 5 mL formaldehyde mixed with 95 mL of sea water.

FERNS

Dig the fern from the ground, wash the root system to remove soil, and store in formal acetic alcohol (FAA; *see* Formulae). Dissolve a crystal of copper sulphate in the FAA to maintain the green colour of chlorophyll.

SEED PLANTS

Store specimens in FAA (*see* Formulae) which contains a dissolved crystal of copper sulphate.

GREEN PLANTS

Plants are placed directly into the following solution and stored until needed. Complete preservation of material takes from two to 14 days, depending on the bulk of the specimen.

phenol crystals	200 g
lactic acid	200 g
glycerine	400 g
cupric chloride	2 g
cupric acetate	2 g
distilled water	2 L

Alternatively, take the green plant and remove all of the air from the intercellular spaces using a vacuum chamber or by immersing it in 90%–95% alcohol. The plant can then be placed into 5% glycerine to which has been added enough copper sulphate or copper acetate to impart a bluish tint. The copper combines with the chlorophyll to produce copper phyllocyanate (Hopkins, 1923) which is insoluble except in strong alcohol and is not affected by light.

Fresh green plants may be fixed for 24 hours in a mixture of 10 mL sodium silicate (waterglass) and 990 mL water.
They should then be rinsed in clean water and placed into the following solution for storage:

zinc sulphate	50 g
copper sulphate	5 g
hydrochloric acid	12 mL
sodium sulphite	16 g
water	950 mL

To mix this, dissolve the zinc sulphate in 700 mL water, then dissolve the copper sulphate. Add the hydrochloric acid while vigorously stirring the mixture. After adding the acid, stir in the sodium sulphite and add the rest of the water. This solution may become cloudy, but will clear on standing.

Using a sulphur dioxide solution is a rather smelly procedure. The solution should be made up in a fume extractor. Bubble sulphur dioxide gas through 750 mL of water until the volume increases to 755 mL. Add 15 g copper sulphate and stir until dissolved. Place the specimens in this solution for three to five days (depending on the thickness of the specimens) and store in 2%–5% formalin.

Place specimens into the following solution for 10 days or until the leaves show a normal green colour which is permanent.

glycerine	100 mL
100% ethyl alcohol	100 mL
formaldehyde	5 mL
glacial acetic acid	5 mL
copper sulphate	0.5 g

The specimens can then be rinsed in water, placed between driers and pressed as described for dry herbarium specimens.

For retaining colours, Knudsen (1972) suggests that specimens be placed in a 5% solution of copper sulphate for 24 hours. They should then be washed several times in water and placed in: 16 mL sulphuric acid; 21 g sodium sulphite; and 1 L water.

COLOURED FLOWERS

There are times when cut flowers are required for a temporary exhibit and must still retain their colour. The following solution should be prepared and the cut stems should be immersed in it: 22 mL salicylic acid; 20 mL formaldehyde; 57 mL ethyl alcohol; and 1137 mL distilled water. For other colour preserving solutions refer to Volume 2, Chapter 11.

Flowers can be artificially coloured to obtain special effects where, for example, different colours are required from the same species of flower. This can be seen in certain types of commercial flowers such as carnations which have been coloured blue or green. Such colouring may be achieved by standing freshly cut flowers in vessels containing aniline dyes. The petals quickly take the dyes up via the stems.

Alternatively, freeze dried specimens may be sprayed with alcohol soluble dyes to either enhance their colours or give them totally new colours, depending on the required effects.

FUNGI

Fungi may be stored in FAA (*see* Formulae). If green parts are attached to the roots, add a small amount of copper sulphate crystals to the preservative to impart a bluish tint. Other fungi preservatives include: A. 10 mL glacial acetic acid; 1 g mercuric acetate; 10 g neutral lead acetate; and 1 L 90% alcohol. B. 5 mL glacial acetic acid; 10 g mercuric acetate; and 1 L distilled water. C. 5 mL formaldehyde; 90 mL alcohol (ethyl or isopropyl); and 5 mL glacial acetic acid.

LICHENS AND MOSSES

These are best preserved dry. If they are to be studied in the classroom, soak them in water for half an hour to make them soft enough for study. If they are required to be stored in liquid use FAA (*see* Formulae).

FRUIT

The following formula is useful for preserving red and green apples: 40 mL distilled water; 2 g zinc chloride; 2 mL formaldehyde; and 2 mL glycerine. The fruit should be stored in the same solution. To make up the solution, dissolve the zinc chloride in hot distilled water and filter when hot before adding the formaldehyde and glycerine. If a sediment appears upon cooling, decant the clear fluid and use this liquid as the preserving and storage medium. Yellow and red apples may be preserved in 750 mL white paraffin oil mixed with 10 mL formaldehyde. Place the apples in this fluid and seal the container for two weeks. After the two weeks, transfer the apples into a sulphur dioxide solution (see above) for storage.

DISPLAY

The simplest and quickest way to produce ground cover for a display is to collect leaf litter, pieces of bark, and stones and rocks from the habitat that is to be shown. There is no need for preparation of the material, but fumigation against insect attack is recommended. Other vegetation that can be used without preparation includes vines, mosses and lichens, dead branches and tree trunks. Botanical specimens that can be dried in air and later spray painted include spinifex bushes, tall grasses, sedges, and spikey leafed bushes that have hard leaves. Trees that have smooth bark can be sprayed with a matt laquer to revitalize the patterns.

Specimens with fine needle like leaves, or processes which dry and are prone to loss if knocked, can have a very thin layer of latex sprayed on as a fine mist so that they stick in position. Alternatively, a fine spray of water-thinned glue can be applied; if large petioles are to be stuck, small blobs of glue may be dabbed on with a small artist's brush.

Some grasses can be dug from the ground, have their roots washed and bathed in ethylene glycol (Davitt, 1978). When the diorama is ready to house the plants, plaster of Paris is poured around the roots to act as a base and prevent the plants falling over.

Ironing sometimes prevents or rejuvenates curled up leaves.

Sometimes, botanical specimens are needed to enhance a showcase; unfortunately, the production of dry specimens from succulent and broad leafed plants has not been too successful. This is an area that is open for experimentation. The following text is a summary which may not be correct for all plants. Plants may be dried using a 50:50 borax–sand mixture. The mixture should be laid in the bottom of a suitable container so that the specimen can be pushed into it. More of the mixture should then be poured in very carefully and slowly until the specimen is covered completely. The container should then be placed in a warm, dry atmosphere for many weeks. The mixture can then be slowly removed through a hole in the base or side. Plants fade slightly with this treatment, but can be revitalized with paint.

Silica gel may replace the borax–sand mixture. If kept at a temperature of 40° C, flowers such as roses take about 4 days to dry (Steedman, 1975).

Usually, parts of trees (branches and trunks) and shrubs are required for inclusion in a diorama. Many leaves curl and become brittle if allowed to dry in air. If stems or branches of trees are required, then the wood has to be treated to resist insect attack, but must remain lifelike. Stems and branches should be sawn, then the cut surface placed upright in a tank containing three to four kilograms of copper sulphate per hectolitre of water. The liquid will be absorbed into the leaves and extremities through the capillaries.

Small branches with leaves have been successfully treated in 100 mL toluene mixed with 100 mL polyethylene glycol 600. Specimens were soaked for a week, then removed and drained. Several days later the leaves were sprayed with a laquer based paint tinted to the correct shade of green. Examples were still supple and lifelike two years later. Tests have shown that no insects attack the foliage. At this stage, only small twigs have been treated with this mixture as the cost of materials is quite high.

The freeze drying of flowers has not produced consistent results. Much more experimentation is needed to solve the problems that exist.

ARTIFICIAL PLANTS

Many museums rely on manufactured plant material to supplement the lack of lifelike dried specimens. There are some firms that sell artificial plants and flowers; their preparators may find a plant similar to that which is needed with only minor alteration required to complete its appearance.

Sometimes, there is a need for miniature, or highly enlarged, models of plant material in certain dioramas. These types of models require much time and material. Tebble (1973) used lino cut methods to make a mould to produce fibreglass leaves seven times their natural size. An enlarged picture was superimposed onto a piece of lino; the outline drawn on the lino with a pencil. All details, such as the leaf margin, veins and cellular structures, were cut in using chisels and cutting tools. The leaf shape was fastened in its required position and coated with a layer of shellac, followed by a wax and finally a separator. When all the layers had dried, a gel coat of polyester resin was applied and thin strands of glass fibre were run down the main stem. Glass fibre mat was positioned on top of the gel coat and allowed to set. Trimming was completed before the resin had time to cure properly.

Frazier (1973) produced 300 leaves, individually shaped and varying in size from small to very large, which were fixed to a twisting vine to give a Jack-and-the-beanstalk effect. Each leaf had to be translucent and flexible. They were modelled to the desired shape, including the placement of veins, with Plasticine and sheet lead. One half was moulded using an epoxy–sand mix and the other half was moulded directly from the first. The two halves had a coating of release agent applied before they were finished in epoxy resin. This method of reproducing a large number of leaves proved very successful for, after 10 years of constant display, they are still in good shape.

Petersen (1958) described several methods of producing artificial plants.

WAX LEAVES

These are made by first making two piece plaster moulds of natural leaves. Molten wax is poured into one half of the mould and the other half is brought into contact, thereby squeezing the wax into the shape of the leaf. When the wax has had sufficient time to solidify, the moulds are released and the wax leaf is removed and trimmed.

PAPER LEAVES

Using different kinds of paper to give the required effect of translucency, lightness, texture, and strength, it is possible to make hundreds of leaves in a day using a jigsaw or band saw. The method is to lay many loose sheets of paper on top of one another to form a thick wad. Outlines of leaves are drawn onto a sheet of cardboard or Masonite which is then placed on top of the pile. A similar material is laid under the pile. The pile is compressed and held tightly while it is pushed through the band saw, which cuts around the leaf outlines. Wires are glued to the undersurface to give strength and support to the leaf when it is mounted on its branch.

PLASTIC LEAVES

Acetate sheets can be softened in a tray of acetone for a predetermined length of time and then laid in one half of a plaster mould and secured into shape by resting the other half of the mould on top, allowing the acetate sheet to become rigid before it is removed from the mould. It is trimmed, and wires are glued to the undersurface, before it is painted.

Another method, again using acetate sheet, is to vacuum form leaves. For this technique, a one piece plaster mould is made upon which the top side of the leaf is in relief. Many different sizes and shapes can be made at one time. The mould is placed into a vacuum apparatus (Fig. 2.3) and an acetate sheet is laid across the top of the apparatus. The acetate is heated via the heating elements positioned above, and a vacuum source (in this case a vacuum cleaner) is switched on. As the air is sucked down and out, the acetate is pulled onto the plaster mould. The vacuum source should be switched off after a few seconds when the acetate is seen to stop moving and there is no more air being sucked out. The leaves are cut from the acetate and tapered wires used to reinforce them.

Leaf impressions may be produced in the same manner. The following three methods are taken from the British Columbia Provincial Museum's publication *Wild Harvest Exhibit 1983.* The publication details methods used to produce many plants included in an exhibit.

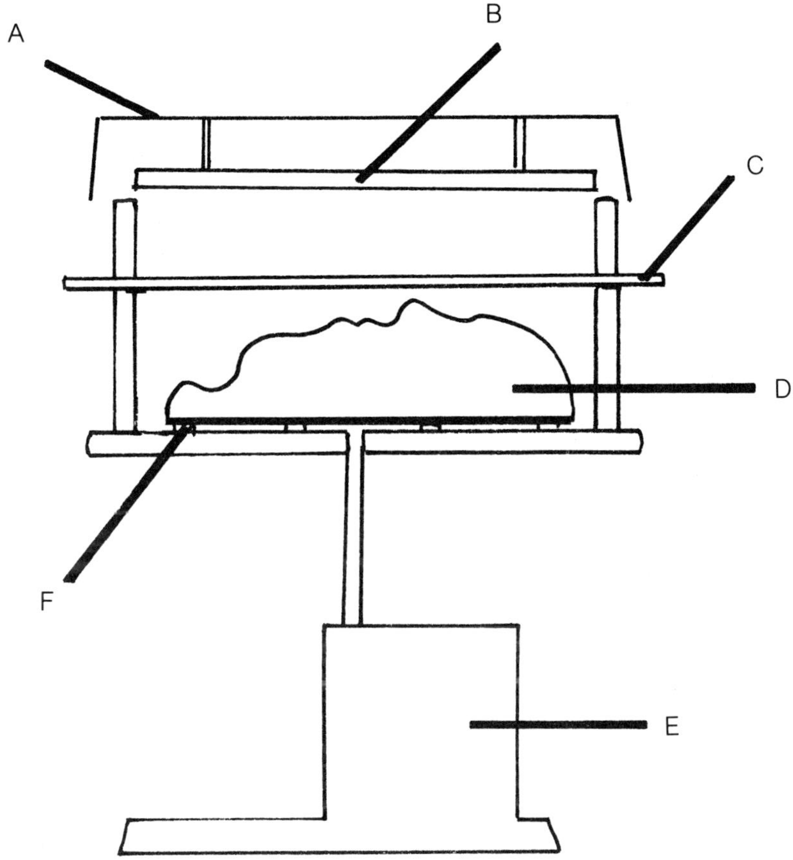

Fig. 2.3. An apparatus that produces vacuum formed shapes from thin plastic sheet. A. Reflective shield. B. Heating element. C. Acetate sheet. D. The mould. E. Vacuum pump. F. Perforated platform, raised above air outlet.

Impressions can be made by choosing appropriately sized fresh or preserved leaves and placing them *face up* on smoothed, individual porcelain clay mounds, 1 cm–2 cm high. It is necessary to very carefully lay the fresh leaf onto the clay mound. It is especially difficult to achieve this when dealing with young, unopened or halfopened leaves. Then, scalpel cuts are made half way to three quarters into the veins on the front surface of the leaves, to allow the leaf to open, before they are placed on the clay. When the leaf is on the mound, gently press the veins into the clay with a round-tip, wooden modelling tool. Next, gently press the fleshy area and the leaf edge so that the leaf is approximately 1 mm below the surface of the clay. Allow the clay to dry for two to six hours, then very carefully peel the leaf away from the clay. Start by catching the petiole with a pair of tweezers, then slowly peel the leaf back. The detail in the clay can be augmented with carving.

Another method is to place fresh or preserved leaves *face down* on shallow clay mounds which have been preformed to approximate the shape and size of the leaf perimeter and undulations. It is always advisable to slope the sides of the mould plates and to keep the mean height of them consistent. These are considerations made for the vacuum former. The more dips and variations on the mound plate, the less chance of consistent detail in the product. Vacuum forming can take place as soon as there are enough impressions to cover the mound plate. This method is limited by the fact that most impressions will have to be discarded when the leaves dry.

The third method is to place fresh or preserved (if wet, dry between paper towelling) leaves *face down,* directly onto the felt platten of the vacuum former. This is the most successful method for reproducing detail. The resultant plastic leaves are flat, but they can be shaped after the cutting and glueing of the petioles, or by folding the midrib and main veins, as one would fold a fan or in half for ovate leaves, after painting. The leaf can be folded in position and the correct shape if it is run over the nozzle of a heat gun kept stationary with the nozzle perpendicular to the work surface.

Wires are used to attach the leaves to the stem. They become the leaf vein and petiole. Copper wires are suitable, but they need to be tapered in order to be realistic. This is achieved by dipping the wire into a beaker containing concentrated nitric acid. The wire is dipped into and out of the acid to facilitate oxidation. Repeat the procedure until the correct taper has been achieved. The wire is washed in water and scrubbed using steel wool. Be careful if dipping many wires as the reaction could be quite violent (the work should be done in a fume cupboard), but incomplete or rough tapering may result if a tight bunch is dipped, or if wires are crossed, preventing the acid reacting with the copper.

Petersen (1958) has described the processes involved in making fruit and vegetables using plaster, rubber and latex moulds.

REFERENCES

British Columbia Privincial Museum: *Wild Harvest Exhibit 1983* (program of exhibit). British Columbia Provincial Museum, Vancouver 1983.

Davitt, L.M.: A simple approach to botanical accuracy. *Curator* 21/4 1978: 311–313.

Frazier, J.: Plastics and workshop techniques. *Kalori* 45/February 1973.

Hopkins, A.A.: *Scientific American cyclopedia of formulas.* Scientific American cyclopedia of formulas. Scientific American Publishing Co., San Francisco 1923.
Knudsen, J.W.: *Biological techniques: Collecting preserving and illustrating plants and animals.* Harper and Row, New York 1966.
Petersen, G.E.: Artificial plants. *Curator* 2/3 1958: 12–35.
Steedman, H.F.: Flowers in plastics. *Museum Journal* 75/4 1975: 142–144.
Tebble, T.P.: Making fibreglass leaves using lino-cut as a mould. *Kalori* 46/April 1973.

SOURCE MATERIAL

Burns, W.A.: A process for making artificial foliage for museum dioramas. *Curator* 13/1 1970: 27–31.
Collection, preparation and preservation of plant specimens. National Herbarium of New South Wales, Royal Botanic Gardens, Sydney.
Gillett, J.M.: Plants: From the field to the herbarium. *Syllogeus* 44/1983: 161–167.
Miller, D.F., and G.W. Blaydes: *Methods and materials for teaching biological sciences.* McGraw-Hill, New York 1938.
Plants. Instructions for collectors, no. 10. British Museum of Natural History, London 1965.
Turtox service leaflet no. 3. *Preserving botanical specimens.* General Biological Supply House, Chicago n.d.

INVERTEBRATES

3 Sponges and coelenterates

THE invertebrate sponges and coelenterates have been collected throughout history, for varying reasons. They are very simple life forms, but require a good deal of attention from collectors and preparators.

SPONGES

Before the advent of synthetic material, sponges (sub-kingdom *Parazoa;* phylum Porifera) were collected extensively and used as bathing accessories. Today, their chief commercial interest lies in the fact that they are a source of biochemical products which can be tested for any pharmaceutical activity.

COLLECTING

Fresh water sponges may be collected by hand from lakes, ponds and streams, where they grow on hard substrate. Marine sponges also live on hard substrate, either natural or synthetic. They are generally found at depths from 10 m (some live many metres deep), although many species can commonly be collected from depths of less than 30 m. When collecting sponges, a knife may be necessary to remove them from the substrate. Once collected, specimens can be placed in buckets or bags of water until the time for preparation.

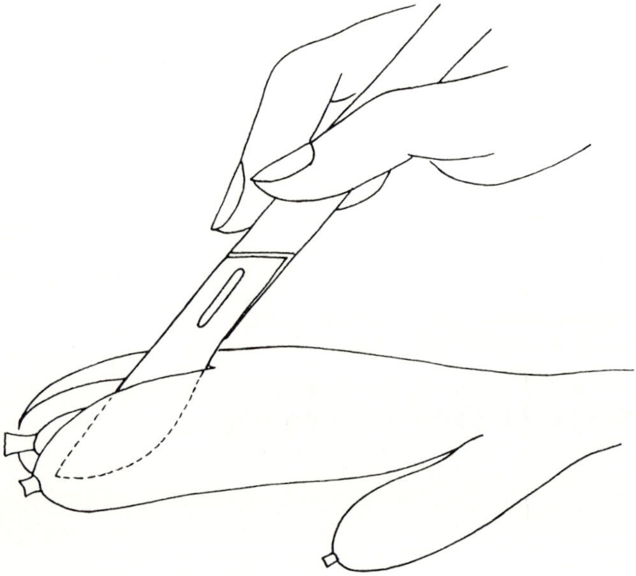

Fig. 3.1. Slicing a sponge using a scalpel.

KILLING, PRESERVING AND STORAGE

Sponges are killed by being placed directly into 70% alcohol. This also acts as a storage medium.

In many collections, sponges are stored in a dry state. The sponges are dried by being left in a warm, dry place with good air circulation for a day or so. They are then stored in cardboard boxes with a few flakes of naphthalene to keep out unwanted insect pests.

DISPLAY

Sponges are best displayed dry. Freeze drying is not necessary as they dry very well in air. If the specimen is attached to a piece of wood, or if there are other marine invertebrates connected to it, it is best preserved in 70% alcohol. Any soft bodied or fleshy animals attached to it should be injected. It should be left to soak for a week in the alcohol then allowed to dry in air. When dry, sponges can be cut to show the spongocoel and excurrent canals (Figs. 3.1 and 3.2). A simple but effective method of attaching sponges to a wall panel is to cut a hole in the back of the sponge using a cork borer or scalpel, and then insert a bolt. Epoxy glue is used to fix the bolt permanently.

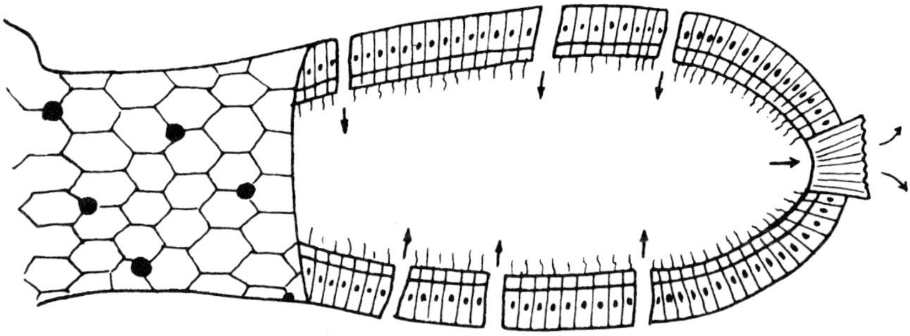

Fig. 3.2. This figure shows the cut section of the sponge and the incurrent and excurrent canals.

COELENTERATES

JELLYFISH

Jellyfish (Fig. 3.3C) are commonly seen in the medusa stage of their life cycle when they show a bell like structure with tentacles hanging beneath. Most people know the marine and estuarine forms and it will be those that will be referred to here.

Collecting, killing and preserving

Jellyfish float on or near the surface. They can be collected using dip nets or, better, a plastic bucket. If a net is used, the jellyfish should then be tipped into a bucket. The bell should be contained in the bucket and overhanging tentacles can be flipped in with long forceps or by hand, provided the collector wears rubber gloves to prevent stings. The specimens are killed by pouring enough neutral formalin into the sea water to make a 5% solution. The specimens are usually stored in 70% alcohol.

Display

Jellyfish make good displays if mounted in an acrylic box (Fig. 3.4); therefore, they should be preserved in formalin instead of alcohol as alcohol attacks the acrylic. The authors have successfully mounted a jellyfish (*Catostylus mosaicus*) in an acrylic box measuring 270 mm × 270 mm × 350 mm without losing the delicate colour markings of the specimen. It was fixed

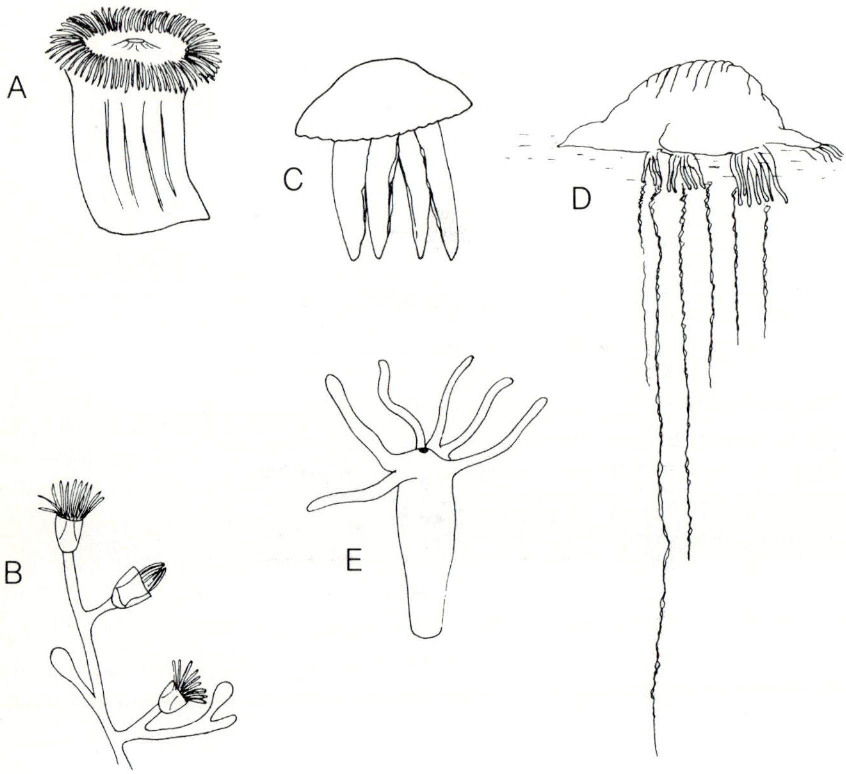

Fig. 3.3. A variety of coelenterates: A. Sea anemone. B. Hydrozoa. C. Jellyfish. D. Physalia. E. Hydra.

using 5% formalin in tap water and later stored in Wentworth's storage solution 1953, omitting the sodium hydrosulphite (see Volume 1, Chapter 7). The main problem associated with a large jellyfish is that it cannot be removed from the liquid medium or handled often, as the weight of the tentacles tends to break their connection to the bell. Therefore, in the instance referred to above, the size of the specimen made preliminary fixation possible only in a plastic tub measuring 1 m × 1 m × 0.8 m. All procedures until final mounting in the acrylic box were carried out in this container.

A method by which the specimen could be attached to the container without impeding the view was carefully thought out. Two strips of 3 mm thick acrylic sheet 30 mm wide and 180 mm long were bent slightly using a Bunsen burner flame. These were inserted under the bell and above the tentacles to form a cross. The strips were cemented together using acrylic cement and clamped together with a very small G-clamp until rigid. Then a 280 mm length of acrylic rod (diameter 18 mm) was placed in a lathe where one end was turned down to a diameter of 8 mm. A predetermined point on the backing plate had a corresponding hole drilled. The rod was bent using a low Bunsen burner flame. The rod was rotated once or twice in the flame, then withdrawn from the heat and rotated again, this time in the air. This was repeated many times until the rod became soft. The rod should not stay in the flame too long or the surface will bubble and burn. Taking the rod out of the flame allows the heat to penetrate deeper, preventing bubbles forming and producing an even heat throughout. The rod was bent to a 90° angle and cooled under running tap water for a minute. Then it was placed flat in a container and cold tap water was poured onto it for 15 minutes. The end to be attached to the specimen was cut and sanded to the required length; the other end was inserted into the hole on the finished backing plate and permanently glued in position using acrylic cement. The rod was positioned correctly; a small force was exerted on the backing plate to achieve a strong bond. Next, the mouth of the specimen was cut (using scalpel and forceps) to produce a larger hole approximately 25 mm in diameter. To fix the specimen permanently, the backing plate and rod were supported vertically in the tub of formalin with burrette clamps and the jellyfish was very carefully positioned so that the cross — already inserted between the bell and tentacles — balanced on the surface of the rod (Fig. 3.4). Acrylic cement was pipetted between the two surfaces and allowed to harden. After 24 hours, the acrylic box, minus lid, was submerged into the tub, and, with the aid of a helper, the jellyfish and backing plate were slid very carefully into the box. The container and specimen were removed from the tub. Every day for two weeks, about a quarter of the volume was siphoned off and fresh 5% neutral formalin was added to refill the container. This was to replace the tap water with distilled water which was used to make up the formalin. Just prior to the lid being cemented into position, 2% of the volume was siphoned off and replaced with glycerine. The lid was cut and polished to fit the box exactly and a hole drilled and tapped to take a nylon screw. The lid was cemented onto the box using acrylic glue (see Volume 2, Chapter 11).

SEA ANEMONES

These animals (Fig. 3.3A) are very easy to locate as they are generally sessile creatures, although a few are able to burrow in sand. The common forms are attached by means of their pedal disc to shells, rocks and other submerged substrate. During low tide, rock pools are a good place in which to start looking. A knife slipped between the pedal disc and substrate will remove the specimen. The specimens should be placed in a container of sea water to prevent dehydration.

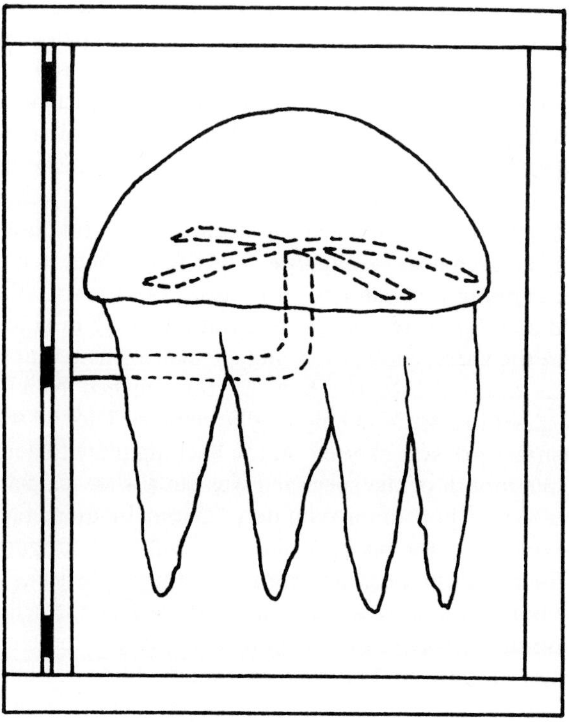

Fig. 3.4. The side view of a jellyfish mounted in a fluid-filled acrylic box. The mechanisms by which the specimen has been mounted are shown.

Killing and preserving

The sea water which contains the specimens should be aerated continuously during transportation. Aeration should continue back in the laboratory until the specimens have been narcotized. If specimens are fully expanded before the narcotizing agent is added, it makes the job of relaxing and killing much easier. Sea anemones should be placed alone in containers during treatment to achieve best results. Using the sea water in which the specimens were transported, two drops of clove oil are added to each litre of sea water every hour for three or four hours. This is best done in the afternoon so that the specimens fully relax overnight. Test their sensitivity to touch by probing them with a long instrument. If there is no reaction, neutral formalin should be added to the sea water to make a 5% solution.

Another method for relaxing specimens is to add a saturated solution of magnesium chloride drop by drop, as with clove oil, to the sea water. A slower method, and one which could cause the specimen to macerate, is to leave the anemone in the sea water at room temperature for a day or more. The dissolved oxygen is gradually used up and, combined with its own excreta, slowly kills the animal. Finally, menthol crystals can be sprinkled onto the surface of the sea water every half hour for three or four hours or until the specimen is insensitive to touch. Specimens should be stored in 70% alcohol. To display anemones, a liquid preservative is essential to show all important morphological features. Embedding in clear plastic is probably the best method of display, especially if it is to be used in a classroom situation where it will be handled by students. Harden (1963) has embedded anemones in clear plastic with success, but does not say whether colours are retained.

CORALS AND SEA PENS

These groups of animals are found in warm tropical waters. Most are brilliant in colour with many being luminescent. The best way to collect these animals is by scuba diving. A geology hammer or pick is very useful for chipping away the rocky substrate from under specimens. Plastic bags are the most efficient method of carrying specimens while the collector is scuba diving. Some specimens are too large to be placed in plastic bags. They have to be carried in the hands until very soft and floppy specimens can be carefully removed from the sea without damaging the fronds. Some collectors go out in boats, and use garbage bins to hold specimens.

Narcotizing and preserving

The corals and sea fans, according to Bianco (1899), should be left in sea water until the polyps fully expand. They can then be killed by adding chromacetic mixture no. 1 in a volume twice as great as that of the water they are in. As soon as possible after death, wash the specimen in tap water and transfer it to 70% alcohol. If the polyps are not required to be in an expanded state, the specimens can be killed directly in absolute or 90% alcohol and allowed to dry in air. They should be stored in lightproof boxes or in plastic bags.

After being relaxed, sea pens are placed into a measuring cylinder containing chromacetic mixture no. 1. This kills them rapidly. After only a few seconds, transfer the specimen to 50% alcohol and inject the alcohol into the internal cavities. Injection is best carried out through the base of the specimen until the body is distended. The polyps will also distend, but, when pressure is released, they will become limp. Tie cotton thread around the needle in the specimen, tightening the thread when the needle is withdrawn to prevent a back flow of alcohol. Sea pens, corals, and sea fans can also be narcotized by adding 5 g menthol crystals every half hour for three or four hours; they can then be killed in 90% alcohol, and injected as previously described.

Display

Corals and sea fans are displayed dry if the polyps are not required to be in an open position. There are various methods of fixing specimens. Which one is to be used depends on whether holes can be drilled into the specimens, or whether they must be stuck to their base. If they are to be fixed to a panel, sea fans can have a hole drilled into the back of the holdfast and a bolt attached with epoxy cement. If a hole cannot be made (as is the case with valuable specimens), three 6 mm acrylic rods can be used. The ends of the rods must be polished. Three very accurately placed holes, 6 mm in diameter, should be drilled into the panel and the rods inserted a short way. The sea fan is then very carefully pushed onto the rods until they protrude through the many branches, thus giving support. If the rods are pushed into the panel until they are flush with the front of the specimen, they are almost invisible.

Large corals and sea fans can be put in a diorama after they have been dried.

On the other hand sea pens may be set, frozen, and later freeze dried. They can then be painted. They can also be displayed in an acrylic wet box or embedded in clear plastic. Small sea pens are best embedded in plastic while large specimens are more easily prepared in a liquid preservative and mounted in an acrylic box.

REFERENCES

Bianco, S.L.: The methods employed at the Naples Zoological station for the preservation of marine animals. *Bulletin of the U.S. National Museum* 39/1899: 1–37.

Harden, C.E.: *How to preserve animals and other specimens in clear plastic.* Naturegraph Publications, Happy Camp CA 1963.

SOURCE MATERIAL

Entz, G.: A véglények kikészitésének és állandó eltartásanak a módja. *Term. Tud. Kozlony.* 13/1881: 381.

Friedlander, B.: Notizen zur Konservierungstechnik pelagischer Seetiere. *Biol. Zbl.* 10/1891: 483–491.

Frohlich, G.: Schlammgerät zur Untersuchung von Bodenproben und deren Entwicklungsstadien. *Vschr. Dtsch. Pflanzenschutzdienst.* new series 14/1960: 41–43.

Gohar, H.A.F.: The preservation of contractile marine animals in an expanded condition. *Journal of the Marine Biology Association.* 22/1 1937: 295–299.

Knudsen, J.W.: *Collecting and preserving plants and animals.* Harper and Row, New York 1972.

Lo Bianco, S.: Metodi usati nella Stazione zoologica per la conservazione degli animali marini. *Mitt. Zool. stat. Neapel.* 9/1890: 435–474.

Mocza, L., I. Haraszti and E. Nagy: *A biológiai szertárak fejlesztése.* Budapest, 1949.

Okland, J.: Notes on methods for collecting and persevering fresh-water invertebrates. *Fauna* 15/1962: 69–92.

Ostergren, H.: Aether als Beteubungsmittel fur Wassertiere. *Zeitschr. Wiss. Mikrosk.* 19/1902: 300–308.
Plate, L.: Beitrage zur Technik des Sammelns, der Konservierung und Aufstellung biologischer Gruppen mariner Tiere. *Vsch. Dtsch. Zool. Ges.* 1903: 143–158.
Randolph, H.: Chloretone (acetonchloroform), an anaesthetic and macerating agent for lower animals. *Zool. Anz.* 23/1900: 436–439.
Remiet (n.i.): Fixation et extension du pedicule des vorticellides. *Bulletin Societe Zoologique Francais* 63/1938: 222.

4

Echinoderms

ECHINODERMS comprise such invertebrates as starfish, sea urchins, sea cucumbers, brittle stars and feather stars (Fig. 4.1). They are a very interesting group, and often quite beautiful. The means of collecting and preserving them are outlined below.

COLLECTING

Echinoderms may be collected by several methods.

Shoreline collecting in rock pools and under boulders will result in catches of several species of star fish, urchins, brittle stars and sea cucumbers. It is the simplest method of collecting.

Skin diving to relatively shallow depths will allow the collection of additional species found in deeper waters.

Scuba diving will be necessary in order to collect species which are only found in waters at depths greater than 10 m, such as feather stars. Sea urchins are frequently found burrowed into rocks or in crevices, and only emerge at night. Sand dollars burrow into the sand, and leave rings of sand to mark their presence. Star fish may be found on rocks, sand, or under boulders, depending on the species and the size of the individual. Feather stars are frequently found in the open on sea fans, sponges or coral in the tropics; but some tropical, and most temperate, species are cryptic. Sea cucumbers may be found on sandy patches, with tentacles exposed, feeding on the sediment; or they may be found under, or disguised as, a rock.

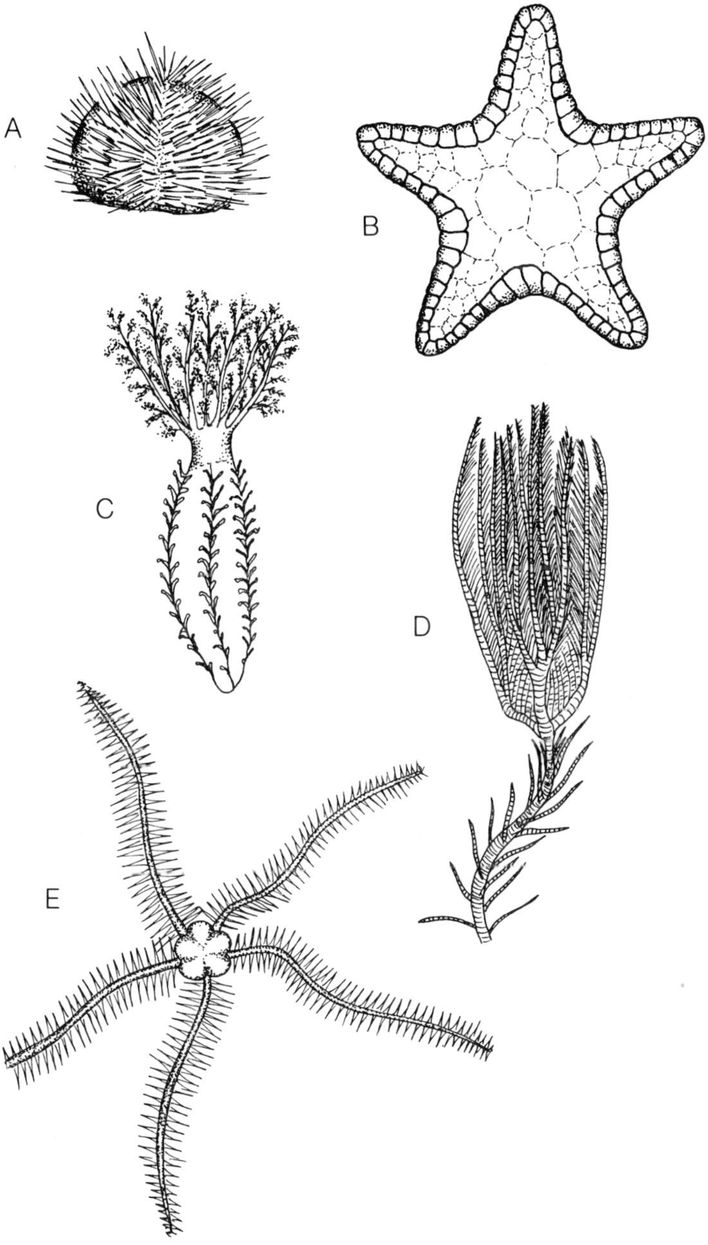

Fig. 4.1. Several varieties of echinoderms: A. Sea urchin. B. Sea star. C. Sea cucumber. D. Sea lily. E. Feather star.

Once found, the specimens should be placed in buckets containing sea water, which should be replaced every 15 minutes or so. This prevents deaeration and premature death of the specimens. Most echinoderms can be left in sea water for up to half an hour before the water needs changing; however, sea cucumbers need special care so that they can be killed in a relaxed state. These should be taken back to the laboratory alive, and the water should be aerated either by pumping air with a bicycle pump or with a battery operated fish tank air pump.

KILLING, FIXING AND PRESERVING

Echinoderms can be killed and fixed by being placed directly into 70% alcohol or 10% neutral formalin. If magnesium sulphate or menthol crystals are sprinkled onto the surface of the sea water every half an hour, the animal soon becomes insensitive to the touch and relaxed; it can then be killed by being placed into 70% alcohol. Star fish should be placed into 20% alcohol until the ambulacral feet are distended, and then placed into 70% alcohol. The ambulacral feet of star fish can also be fixed into their extended position if the star fish is placed in liquid nitrogen for immediate freezing. It need only take 15 seconds to freeze; then it can be preserved either dry or wet. Brittle stars and feather stars can be positioned by placing them in the palm of the hand, and then inverting them into a tray containing 70% alcohol. The hand should be left in position for a few minutes, after which the specimen can be gently prised off the hand and left to soak in the alcohol. The specimen should be completely covered by the alcohol. Sea cucumbers, after being taken back to the laboratory alive, should be in a relaxed position. Aeration of the water (if taking place) can cease, and magnesium sulphate should be sprinkled onto the surface. About 2.5 g of crystals per four litres of sea water should be added every half an hour. This procedure needs to be continued for four or five hours, or until the animal is insensitive to touch. If this procedure begins in the afternoon, the specimens may be left overnight without too much trouble.

Alternatively, specimens are allowed to expand in the sea water and then plunged into hot water to kill them. Other techniques used to narcotize echinoderms include immersing specimens in soda water, 1% 2-phenoxyethanol, or nicotine water. Nicotine water is a slow, but useful, narcotic in the field. It is prepared by placing four or five used cigarette butts into about 500 mL of water and shaking them until the yellow nicotine has been extracted and colours the water. Specimens are immersed in this solution until they become insensitive to touch.

After narcotizing, specimens are killed and fixed by adding enough neutral formaldehyde to the sea water containing the specimens to make a 10% formalin solution. Large sea cucumbers are split open to facilitate entry of this fixative and hence better preservation. Large echinoderms are pierced near the mouth, to allow the fixative to penetrate into the body cavity.

All echinoderms should be fixed for at least 24 hours, washed in tap water for 10 minutes, and then immersed in 70% alcohol for preservation and storage. Alcohol should be injected into the body cavity of specimens for complete preservation. If the alcohol storage medium discolours due to the leaching out of body fluids, it should be discarded and replaced with fresh 70% alcohol.

DRY SPECIMENS

All classes of echinoderms except sea cucumbers are suitable for drying. Some retain most of their colour if they are not subjected to continuous or long periods in a well lit room. Those specimens which have been preserved in alcohol make the best dry preparations; however, dry specimens are very fragile. The spines of sea urchins and the arms of brittle stars and feather stars may break if mistreated. Drying is achieved quickly if the specimen is left in a dry, draughty place, or placed in a clothes drying cabinet which has a fan to circulate the air. Drying time depends on the size of the animal, but should take no more than one or two days for larger specimens if the temperature is relatively low. If the temperature is too high the specimens will become brittle, or they will warp, causing them to crack.

DISPLAY

Sea cucumbers are the only echinoderms which need to be displayed in a liquid medium. The best specimens should be housed in an acrylic wet box. The holothurians need to be fixed in 10% neutral formalin, then preserved in the general preserving solution used for biological material, before being placed in the final mounting solution. The remaining classes can be displayed in a dry state, as described above. If the specimens are to be placed on a panel, a hole should be drilled into the mouth parts and a suitably sized bolt glued in with epoxy resin. This can be bolted onto the panel, and a washer and nut fixed on from behind. Those specimens which are meant to be seen on a sandy bottom in a display case can simply be placed in position without the need for a fixing support. To prevent specimens slipping if they are positioned at a slight incline, a silicone sealant can be applied to the mouth parts which should be pressed down lightly to achieve a good bond.

The very fragile brittle and feather stars are suitable for embedding in clear plastic. They then become immune to attack by insect pests and damage from being dropped.

SOURCE MATERIAL

Bianco, S.L.: The methods employed at the Naples Zoological Station for the preservation of marine animals. *Bulletin of the U.S. National Museum* 39/1899: 1–37.

Dingley, M.O.: A note on the surface finish of a plastic embedment. *Abstracts of Papers Presented at the Conference of Museum Preparators October 9th and 10th 1980.* Australian Museum, Sydney 1980.

Grafe, M.: Uber die Konservierung von Seegurken (Holothuroidea) und marinen Naktschnecken (Nudibranchia) in lebensnäher Stellung. *Der Präparator* 29/1 1983.

Kannemeyer, S.X.: The preservation of natural colour in biological specimen preserved in fluid. *Sama* 8/1983.

Knudsen, J.W.: *Collecting and preserving plants and animals.* Harper and Row, New York 1972.

MacGilivray, J.: Hints on the preservation of specimens of natural history, intended for country residents. Pamphlets 8vo series. Vol. CXVII. Australian Museum, Sydney n.d. (but before 1916).

Szüts, A.: Ungarische Adriaforschung. *Zool. Anz.* 45/1915: 422–432.

Tullberg, T.: Über Konservierung von Evertebraten in ausgedehten Zustand. *Verk. Biol. Ver. Stockholm.* 4/1891: 4–9.

5

Worms and worm-like animals

As MANY individuals of the group of worms and worm-like animals are very small and usually prepared for microscopic studies, the following description of methods for killing and preserving and storing will be directed towards gross morphological studies used in museum collections and display areas. If microscope slide mounts are required, readers are directed to Knudsen (1972) and Mahoney (1966).

The animals are listed below in their particular phyla which are in turn listed in alphabetical order and not in ascending order within the animal kingdom as this will enable the preparator, not skilled in taxonomy, to locate a particular phylum easily and hence treat specimens quickly.

Details of external morphological characteristics have been omited. Identification can usually be obtained using standard zoological textbooks (see Fig. 5.1).

ACANTHOCEPHALA (SPINY-HEADED WORMS)

These worms are endoparasites living in the digestive tracts of birds, fish, and mammals and require an arthropod as an intermediate host for the larvae to flourish.

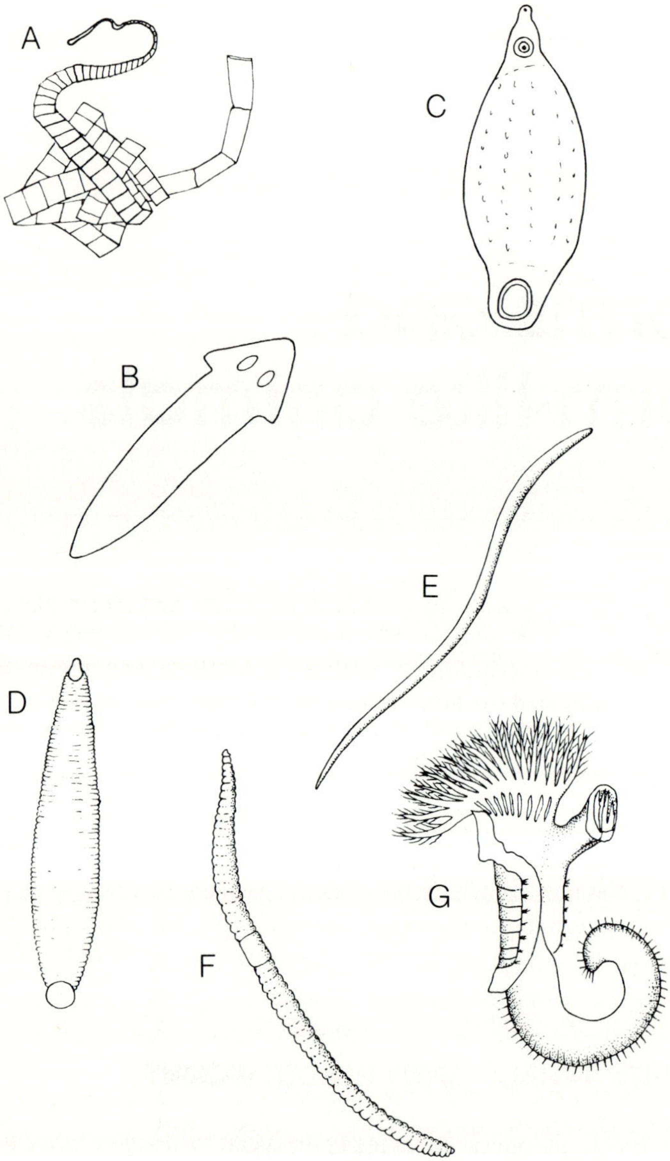

Fig. 5.1. Several varieties of worms: A. Tapeworm. B. Planarian. C. Fluke. D. Leech. E. Roundworm. F. Earthworm. G. Feather-duster worm.

KILLING AND PRESERVING

Acanthocephala should be killed by gently squeezing the animal between two glass plates until the proboscis is extended. The plates are secured with rubber bands, then plunged into hot 3%–5% formalin, hot 70%–90% alcohol (Lincoln and Sheals, 1979), or hot corrosive acetic (Mahoney, 1966). If hot, corrosive acetic is used, specimens must be washed afterwards in iodized alcohol. Knudsen (1966) gives several methods by which specimens can be chilled before fixing. Specimens are stored in 70% alcohol.

ANNELIDA (SEGMENTED WORMS)

The three main classes of annelid worms are polychaeta, oligochaeta, and hirudinea. There are other classes, but mention of these is omitted as the preserving techniques may be extracted from the three classes that are defined.

OLIGOCHAETA

Members of this class are found in fresh water and terrestrial habitats. The most common is the earthworm (see below). Others, such as *Tubifex*, are collected from their fresh water habitat by sweeping a dip net through submerged vegetation. If aquatic mud samples are placed in a bucket containing clear water, the oligochaeta will be found wriggling in the water and can be scooped up.

Oligochaeta may be found amongst detritus or masses of filamentous algae, and this may be brought to the laboratory and rinsed in a sieve. Very small specimens can be trapped using a sieve with mesh size of 0.3 mm. Dredge, grab or core samples may also be washed using the same sieve. Mud containing silt is usually harder to wash than mud containing sand. Washing is carried out by adding water to the sandy sample and stirring, so that organisms are suspended in the water. Pass the stirred water through the sieve to concentrate the animals into a smaller volume. Each sample may be washed eight to 10 times, depending on the volume of the sample.

Depending on the volume of sediment retained, sieved samples require 10% formalin to ensure adequate preservation. Preservation is achieved by approximating the total volume of the sieved sample and adding concentrated formaldehyde to make a 10% formalin solution. Bulk quantities can be stored in small glass phials using 5% formalin. All phials should be labelled to show the locality, date, number of individuals, replicate number, and station number (if used).

For taxonomic purposes, oligochaeta are cleared and mounted on glass microscope slides using Amman's lactophenol (*see* Formulae).

Earthworms

Earthworms can be dug from moist soil. If the soil is very dry, soak the area with water of a 0.5% solution of formalin, which will make the worms surface. Other sites harbouring a good supply of earthworms are garden compost heaps, manure heaps, camp areas with a rich source of vegetation, and under rotting logs.

Killing and preserving: Best results are achieved if earthworms are first relaxed by being placed in water, then adding absolute alcohol drop by drop to make a 10% solution. Other methods of relaxing specimens are to add a few drops of chloroform to the water, to place specimens in chlorobutanol, or to immerse specimens in 2-phenoxyethanol. After narcotizing, specimens may be killed by being stretched out for one hour in a shallow dish containing 5% formalin or FAA. They should then be washed in tap water to remove debris. They can be stored in 70% alcohol. If specimens are permitted to expand on a glass plate, and then killed using hot FAA or Bouin's fixative, they will also be in a relaxed state.

POLYCHAETA

These marine animals live in mud and sand, or on rocks, vegetation, and some marine animals. A screen will separate these worms from the sand and mud. It should be filled and then washed in sea water to enable the sand and mud to filter through the mesh, leaving the worms to be picked off and placed into containers. Some polychaeta are free living and may be found in oyster beds, seaweed, and other algae.

Narcotizing

Various agents have been used to relax specimens, but most are inconsistent in their effects. Further experimentation is needed before agents such as chloroform (drop by drop), chlorobutanol, Epsom salts, clove oil, and nicotine (from soaking used cigarette butts in water) can be used with consistency. The best method (but one that does not always work) is to place specimens in a quantity of sea water until they are just covered, and then slowly add 80% alcohol drop by drop until a 10% volume of alcohol to sea water is achieved.

Killing and preserving

Kill relaxed specimens by stretching them on a board in a shallow dish, and flooding them with FAA. Preserve specimens with 5% formalin of 70% alcohol, and store in 70% alcohol or 50%–70% isopropyl alcohol.

Some collectors place specimens directly into 7% formalin, then transfer them into 70% alcohol for storage.

HIRUDINEA (LEECHES)

If a collector walks around rainforest areas or moist vegetation, leeches will usually attach themselves. They can be removed from the skin by holding the point of contact between the thumb and forefinger and pulling. The application of a flame or a lighted cigarette will cause them to drop off.

Aquatic leeches are found in lakes, ponds and slow moving water courses where they adhere to the underside of submerged rocks, structures, and vegetation. They have also been found attached to aquatic beetles, frogs, turtles and fishes.

Narcotizing, killing and preserving

Aquatic leeches are placed in a small jar half filled with water. A few drops of 5% ethyl alcohol should be added until the specimen does not respond; add more drops of alcohol and repeat until there is no reaction when the specimens are touched with a probe. Leave the specimens in this solution for a further hour before removing them. Place them on a piece of blotting paper dampened with 5% formalin to check for movement. Extend the leech using a fine probe or paintbrush, and transfer to a Petri dish containing 5% formalin. After 24 hours, transfer to a phial containing fresh 5% formalin for storage.

Terrestrial leeches can be placed into a glass tube containing three quarters water. If the leech floats, invert the tube until it finally sinks. After this, the tube should be rested top uppermost. Add either so that it forms a thin layer on top of the water and leave for several hours until the leech is fully relaxed. They can then be fixed and stored in 5% formalin.

Madill (1983) suggests relaxing individual specimens in a shallow dish containing enough water to just cover them; 95% alcohol can then be added drop by drop until the specimen ceases all movement and is insensitive to touch. Specimens are fixed in 5% formalin for 24 hours, then preserved in 75%–85% ethyl alcohol or 5% formalin.

ASCHELMINTHES (NEMATHELMINTHES)

Several classes are found in this phylum. Mostly, ashchelminthes inhabit fresh and marine waters, but the class nematoda are terrestrial.

The classes are: rotifera; gastrotricha; nematoda; nematomorpha; and kynorhyncha.

ROTIFERA, GASTROTRICHA AND KYNORHYNCHA

These classes are microscopic and are of no importance as far as museum exhibition work is concerned. They also require special equipment and methods. For further information, refer to Hanley (1949), Knudsen (1972), Mahoney (1966), and Wagstaffe and Fidler (1957).

NEMATODES (ROUND WORMS)

Round worms are a very successful group, inhabiting the sea, fresh water, soil, and vertebrate animals (the latter being parasitic). All vertebrates are potential hosts for parasitic round worms; therefore, they can be collected in prolific numbers from fish and reptiles. To collect the worms perform the same dissection as described for cestodes, but also look in the stomach. Keep a record of the locality in which they are found within the animal as this is important. The lungs, liver, and urinary tract should also be examined for likely specimens. Round worms are also found in damp and wet areas, such as the mud from the bottom of rivers and lakes, seashores, rainforest soil, and under vegetation lining streams and creeks. To acquire specimens, soil or mud should be washed in a shallow dish to break up lumps and dilute the sample. With light shining over the dish, specimens can be located and pipetted, or picked up with forceps; they can then be transferred to collecting bottles. Soil samples can also be washed through descending mesh sizes until specimens are trapped.

Parasitic round worms require washing in several changes of isotonic saline solution, and should be shaken vigorously. They can be killed and fixed by being dropped into hot 70% alcohol (WARNING: fumes are highly flammable) or hot 3%–5% formalin. Allow the fixative to cool, and transfer the specimen to fresh fixative for storage. Free living specimens are fixed and preserved as described for trematodes, or are narcotized as described for cestodes, then fixed in 7% neutral formalin, and stored in 70% alcohol containing a trace of glycerine.

NEMATOMORPHA (HAIR WORMS)

These are very long and slender worms which are free living as adults and parasitic in arthropods during their juvenile stage. The adults are predominantly aquatic (marine and fresh water), but some inhabit damp soil.

Killing and preserving

Specimens for general taxonomic purposes are killed and fixed in 3%–5% formalin and stored in 5% formalin.

ECHIURA (SPOON WORMS)

Animals belonging to this phylum live in a marine environment or brackish water. They range in length from a few millimetres to 600 mm. They live in burrows, or in crevices in coral and rocks.

KILLING AND PRESERVING

Specimens should be narcotized to avoid contraction. This can be done by adding 70% alcohol, drop by drop, or crystals of menthol, to the sea water they are in. Alternatively specimens may be immersed in 7% magnesium chloride solution. Kill and fix specimens in 5% neutral formalin and store in 70% alcohol.

NEMERTINES (RIBBON WORMS)

The nemertines are mostly free living animals. They are generally found amongst seaweeds, in sand, mud and gravel, and under stones, though some are commensal. Some species are found in fresh water and terrestrial habitats.

KILLING AND PRESERVING

Specimens must be narcotized before fixing. This can be done by gradually adding crystals of magnesium sulphate or chloral hydrate to the water containing them, or by placing specimens in a 7% solution of magnesium chloride. When the specimens have been relaxed, small specimens can be placed into lengths of glass tubing, the diameter of which must be slightly larger than the diameter of the worm, and placed horizontally in the fixative bath so that the specimens are fixed reasonably straight. Fixatives suitable for nemertines are 10% formalin and 70%–90% alcohol. Storage solution should be 3%–5% formalin or 70% alcohol.

PENTASTOMIDES (TONGUE WORMS)

The adult worms of this phylum inhabit the respiratory systems of certain reptiles, birds and mammals; herbivorous animals are usually used as intermediate hosts by the larvae, although one species is known to live in the cockroach *(Periplaneta americana).*

KILLING AND PRESERVING

Specimens are killed, fixed and stored in 70% alcohol, 3%–5% formalin, or formol–alcohol.

PHORONIDES

Animals belonging to this phylum are worm-like coelomates that live in a tube built from their own secretation in the marine upper littoral zone.

KILLING AND PRESERVING

Specimens should be first narcotized by being placed into 7% magnesium chloride solution, or by adding alcohol slowly into the water containing them. When they are insensitive to touch, the solution should be drawn off and replaced with either 5% formalin or formol–alcohol. The specimens should be kept straight for five minutes, or until suitably fixed. After 24 hours, specimens should be removed from the fixative, and can be stored in fresh fixative indefinitely.

PLATYHELMINTHES (FLAT WORMS)

There are three classes of flat worms.

CESTODES (TAPE WORMS)

These worms inhabit the digestive tract and are common in the large and small intestines of vertebrate animals. When a bird or mammal has been acquired, remove the large and small intestines from the body, and lay them in a shallow tray. Using scissors, cut along the length of the intestines, sorting through the organic matter. When a worm has been located, the head or scolex should be held with forceps and the point of attachment to the intestinal wall cut, removing the scolex and a small portion of the wall. The worm is relaxed by being placed in the freezer until death occurs. Alternatively, it can be narcotized by being placed in a dilute (1%) solution of magnesium chloride until it is ready for fixing in 7% neutral formalin.

Specimens housed in museum collections are usually stored in 70% alcohol with a trace of glycerine added to keep them supple and to prevent drying out in the event of container breakage.

TREMATODES (FLUKES)

Flukes are parasitic, as are the cestodes, and as adults they occur in vertebrate hosts or as ectoparasites. The ectoparasites only live on aquatic hosts, especially on fish fins and gills. These can be picked off using forceps. To locate endoparasitic flukes, the host must be dissected as described for cestoda.

Once the flukes have been collected, they should be washed in several rinses of isotonic saline, while being shaken vigorously to remove foreign matter. The flukes can be fixed if placed, for 24–48 hours, into any of the following fixatives: 10% formalin; Bouin's fluid; Helly's fluid; or a saturated aqueous solution of mercuric chloride (corrosive sublimate). If the latter fixative is used, all traces of the fixative must be removed by washing in tap water for three to five hours, numerous changes of 70% alcohol, or by placing the specimen in iodized alcohol overnight. Some workers prefer to place the specimen on a glass microscope slide, and then flatten it with another slide. The two slides sandwich the fluke, and rubber bands are used to hold the slides together. The flattened specimen can then be placed into a shallow dish containing fixative. Specimens are stored in 3%–5% formalin.

TURBELLARIANS

Turbellarians are free living flat worms with a few terrestrial species living in damp soil, but the majority are aquatic and, mainly, marine. Terrestrial species can be picked up with forceps, but aquatic species require sweeping underwater plants or dredging (see volume 2, Chapter 8; Aquatic insects).

When caught, specimens should be allowed to extend in water (sea water for marine species) for a few minutes. Then, most of the water can be poured off and hot 70% alcohol (WARNING: the fumes are highly flammable) poured over them. They can also be fixed rapidly if dropped into a beaker containing boiling 70% alcohol, before being preserved and stored in cold 70% alcohol.

POGONOPHORA

These are marine, benthic animals without a digestive tract that live in tubes within the sea floor.

KILLING AND PRESERVING

Specimens may be narcotised using 7% magnesium chloride or specimens may be killed directly by fixing in 70% alcohol or 2%–3% formalin. The fixative should be changed after 24 hours and specimens stored in 70% alcohol.

PRIAPULIDA

This is a small group of marine worms which often exceed 70 mm in length and live buried in the soft sea floor at depths of up to 500 metres.

KILLING AND PRESERVING

Before being killed, specimens should first be narcotized by immersion in 7% magnesium chloride solution. Larger specimens will take longer to narcotize than small specimens. The animals can then be fixed using 5% formalin in sea water for 48–72 hours. They should be stored in 70% alcohol.

SIPUNCULA (PEANUT WORMS)

These are sedentary, marine animals found from the intertidal zone down to considerable depths. They are found living in burrows in the sea floor, as well as in crevices in rocks and corals, and in seaweed.

KILLING AND PRESERVING

Specimens can be narcotized by adding crystals of magnesium chloride or alcohol, drop by drop, to the sea water containing them until they are insensitive to touch. Place the specimens in a shallow dish and lay a glass plate or slip on top to make the specimen rigid. Then flood the specimen with 3%–5% formalin or 70%–90% alcohol. Fix for 12–24 hours and store in 70% alcohol or formol–alcohol.

DISPLAY

Worms have been rather neglected as far as display material is concerned; therefore, more research is needed to obtain consistent results. The following information is to be used only as a guide. It will be up to the preparator or technician to decide upon the proper method of display for worms.

The most important factor to bear in mind is that fresh specimens must be used in order to correctly position specimens. In some cases, specimens may need to be injected with a filler to prevent collapse. Gelatine (15% by weight dissolved in water) is suitable, but the gelatine must not be too hot, or it will take a longer time to cool down. If specimens are placed into 5%–10% formalin after injection, the gelatine will solidify. It cannot be remelted, but this will seal the needle hole quickly. Worms can be displayed in acrylic wet boxes, embedded in clear plastic (although this method may produce cleared or darkened specimens), freeze dried, or models may be made. Models of worms need little or no maintenance, and can be coloured to an almost natural appearance. These models can be placed into mini dioramas especially suitable for burrowers and tunnel dwellers.

REFERENCES

Hanley, J.: the narcotising and mounting of rotifera. *Microscope* 7/6 1949: 154–159.

Knudsen, J. W.: *Collecting and preserving plants and animals.* Harper and Row, New York 1972.

Lincoln, R. J. and J. G. Sheals: *Invertebrate animals: Collection and preservation.* Cambridge University Press, London 1979.

Madill, J.: The preparation of leech specimens: Relaxation, the key to preservation. In: Proceedings of the 1981 workshop on the care and maintenance of natural history collections (ed. D. J. Faber). *Syllogeus* 44/1983: 37–41.

Mahoney, R.: *Laboratory techniques in zoology.* Butterworths, London 1966.

Wagstaffe, R. and J. H. Fidler: *The preservation of natural history specimens.* H. & G. Witherby, London 1957.

SOURCE MATERIAL

Bragg, J. H.: Some observations on the staining of tapeworms. *Proceedings of the Oklahoma Academy of Science* 14/1934: 21–22.

British Museum of Natural History: Instructions for Collectors. *Invertebrate animals other than insects* No. 9A/1954.

Cable, R. M.: *An illustrated laboratory manual of parasitology.* Burgess, Minneapolis 1958.

Chubb, J.C.: Acetic acid as a diluent and dehydrant in the preparations of whole, stained helminths. *Stain Technology* 37/1962: 179–182.

Courtney, W. D., D. Polley, and V. L. Miller: TAF. An improved fixative in nematode technique. *Plant Disease Report* 39/1955: 570–571.

Gelei J.: Untersuchungsmethoden für Turbellarien. *Z. wiss. Mikrosk.* 46/1929: 45–88.

Goodey, J. B.: *Laboratory methods for work with plant and soil nematodes.* Ministry of Agriculture, Fisheries and Food, Technical Bulletin No. 2 (4th edition). HMSO, London, 1963.

Gower, W. C.: Technique on staining and mounting trematoda. *Stain Technology* 14/1939: 31–32.

Hargis, W. J.: Chloretone as a trematode relaxer and its use in mass-collecting techniques. *Journal of Parasitology* 39/1953: 224–225.

Hauser, I.: Zur Fixierung und Konservierung von Turbellarien. *Mikroskopie* 8/1953: 127–128.

Heynemann, B.: Cuticular peeling: A dissection technique for preparation, of cestode whole mounts. *Journal of Parasitology* 45/1959: 573–574.

Hope, I. G.: Experiments in the paraffin infiltration of marine worms. *Bulletin of the Wagner Free Institute, Philadelphia* 6/1931: 41–44.
Meyer, M. C.: Spring clips for mounting helminths. *Transactions of the American Microbiological Society* 76/1957: 344–345.
Pop, I.: A kevés sertéjü gyürüsférgek (Oligochaeta) gyüjtése és konzerválása. *Fragm. Faun. Hung.* 7/1944.
Seinhorst, J. W.: A rapid method for the transfer of nematodes from fixative to anhydrous glycerine. *Nematologica* 4/1959: 67–69.
——. On the killing and transferring to glycerine of nematodes. *Nematologica* 8/1962: 29–32.
Wordle, R. A.: On the technique of cestode study. *Journal of Parasitology* 24/1932: 241–252.

6
Molluscs

MOLLUSCS are one of the widest ranging invertebrate groups. The various molluscs are to be found on land, in the ocean, in fresh water, at high altitudes, and in deserts. Some examples of molluscs are squids, cuttlefishes, octopuses, bivalves, slugs, and snails.

TERRESTRIAL MOLLUSCS

COLLECTING

Land snails (Fig. 6.1A) are extremely adaptive environmentally. They have established habitats in rainforests, mountains, and deserts. In Australian deserts, it has been estimated that, in an average year, snails have only about 25 days in which to ensure the survival of their species. On such days, when rain is either falling or has presented a suitable environment, the snails must feed, build shelter, copulate, lay eggs, and store enough food to allow them to survive until the rains come again (possibly for some years). It is necessary for the snails to seek shelter from the sun. This may mean digging 20 cm to 30 cm into the earth, or adhering firmly to the side of a rock. These sites should be considered when collecting. In dry weather, snails and slugs will harbour under the dead bark on the sunless side of trees, or several centimetres below the surface of the soil and the base of trees, or well down in rock

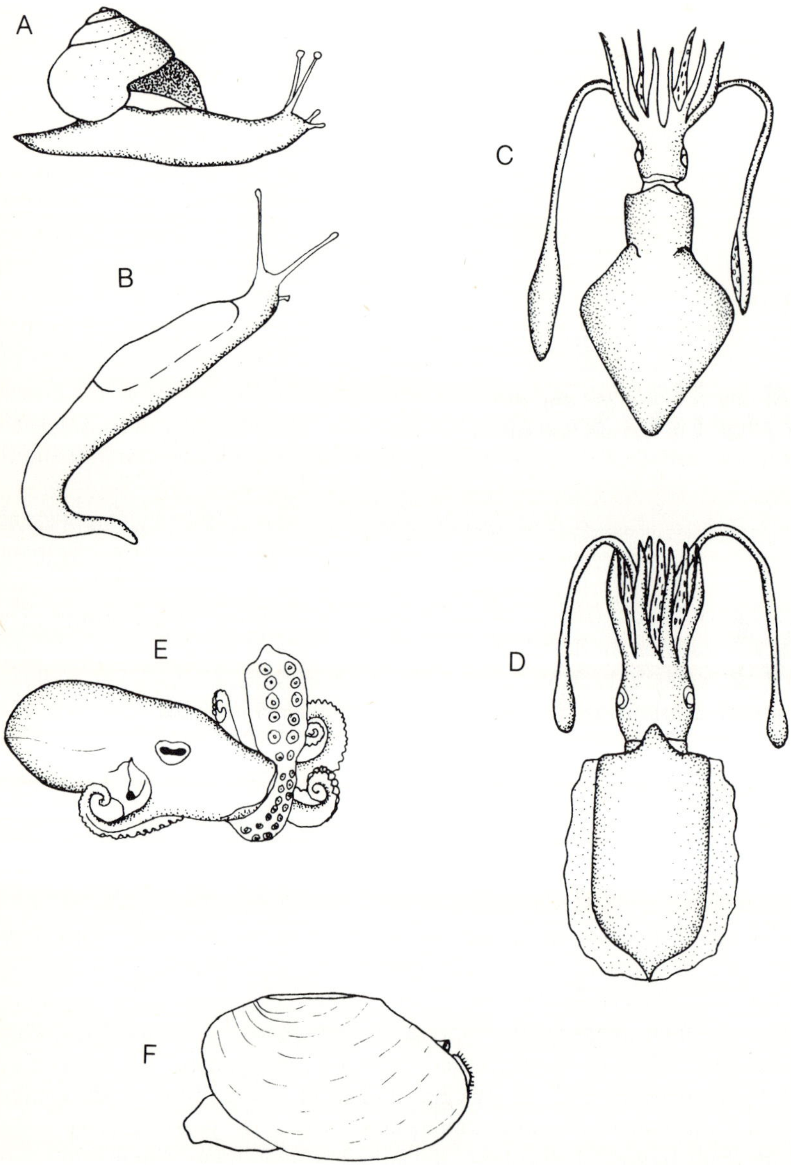

Fig. 6.1. Several varieties of molluscs: A. Snail. B. Slug. C. Squid. D. Cuttlefish. E. Octopus. F. A bivalve.

screes. Arboreal species are generally opereculates (that is, they have an operculum or door on their foot, which they pull in to seal themselves against the elements). Thus equipped, they can then seal the shell to a leaf or trunk, thus presenting a double barrier to evaporation. These species should be sought on tree trunks, branches and under leaves in those areas least reached by the rays of the sun.

Terrestrial molluscs are one of the most important agents for the decomposition of forest litter and, though most litter species are quite small, there are generally numerous specimens. It is not uncommon to find 15 species in litter samples. More than 30 have been recorded in litter samples in New Zealand. About 100 specimens per litre of litter have been recorded in rainforests of north eastern Australia. Once collected, litter should be thoroughly dried out. It can then be fraction sieved; the larger fractions should be scanned for snails. The smaller fraction (10 mm down) should be poured into a bucket of water where the soil and heavier matter will sink. Any snails and leaves will float, and so can be skimmed off. Skimmed material should be dried, then sorted under magnification.

In temperate forests, most species will lodge under bark, logs, and stones in dry weather, but in wet weather they normally forsake their hiding places and can be collected in the open.

KILLING AND PRESERVING

Snails and slugs

Snails and slugs (Fig. 6.1B) can be drowned in fresh water (boiled, if possible) in an airtight container. Add a crystal or so of chloral hydrate. Leave 24 hours or more, unless temperatures are extreme. Fix the specimens in 95% ethyl alcohol and preserve in 70% alcohol. Slugs, especially, should be photographed alive as colours are quickly lost in preservatives.

AQUATIC MOLLUSCS

COLLECTING

Snails, limpets and chitons

The most accessible place for collecting aquatic molluscs is the intertidal zone. It is an easy and pleasant task to wade through shallow waters picking out specimens from rock pools and under ledges. Shells are found under loose rocks, amongst seaweed, and around structures such as wooden piers. A blunt knife will help to prise off animals such as limpets and chitons, which can adhere very strongly to the substrate. Marine snails are mainly nocturnal; therefore, and this is especially true in the tropics, the best time to collect them is night.

A sample can be taken from sandy or muddy bottoms around estuaries at low tide. The sample should be graded through sieves to separate small animals. Rivers, streams, and lakes are home to a variety of snails and freshwater bivalves which can be collected by hand. Specimens will be found hiding under submerged tree limbs on rocks, in vegetable matter, and in the sand or mud bottom.

Squids, cuttlefishes and octopuses

Squids, cuttlefishes and octopuses are caught using different collecting techniques. Squids (Fig. 6.1C) and cuttlefishes (Fig. 6.1D) are most easily collected at night. A boat is normally needed. From this, a strong light can be beamed onto the surface of the sea. This attracts the animals so that they can be scooped up with a net. Many professional fishermen catch cuttlefish in two nets. Squid are caught with specially constructed jig lines. If the specimens are required for display only, and collection data are not required, the fish markets provide an excellent source of material.

Octopuses (Fig. 6.1E) are found in crevices, under rock ledges, and in discarded containers such as tin cans. Octopuses eat crustaceans and other molluscs. The parts of the exoskeleton they have not eaten, and the shells, can be found littered around the entrance of an octopus home. If hiding in a crevice, octopuses can be caught by hand by taking hold of the muscle on either side of the mantle and pulling. The octopus will release its grasp quite easily, and can then be placed in a collecting bag. Collectors must be very careful in waters that contain the highly venomous blue-ringed octopuses *(Octopus maculosus).* These occur in waters from Australia to the Philippines. Before venturing out on a collecting trip collectors should make doubly sure that they are familiar with any poisonous animals they may encounter.

Micromolluscs

Where access to the site is possible, by far the best technique for collecting marine or fresh water micromolluscs is to use simple gear such as a one litre light coloured ice cream container (round or rectangular), and a one millimetre mesh plastic kitchen sieve. Plastic is best, as metal sieves corrode.

For normal rock or algal sampling, wash the sample caught in the sieve vigorously in the container, let it stand 30 seconds, then pour off most of the liquid with any floating algal fragments. Any molluscs will remain in the residue, and generally will have adhered to the container walls, from which they are easy to remove.

For sampling of freshwater algae and sedges, heavily silted substrates, and similar salt water environments, scrape the sieve over or through the sampling area, then knock the contents into an empty container. Repeat this several times, then carefully swirl water into the container to stir up lighter matter. Decant this. Repeat the process several times, if necessary, to clean the sample. The sample can then be poured straight into a preserving vial for later sorting. This technique is so efficient that less than five minutes of effort will indicate whether even cryptic molluscs are present or not. This technique is also excellent for sampling water for such other organisms as ostracods, amphipods and other crustacea, water beetles and bugs, *Odonata* larvae, and so on.

KILLING AND PRESERVING

Shelled molluscs

Generally speaking, all shelled marine molluscs (Fig. 6.1F) can be killed and preserved by being dropped into a 10% neutral formalin solution made with salt water. There seems to be no standard way of fixing the animals in a relaxed position. Magnesium chloride sometimes will relax some species; other occasionally effective measures include menthol crystals, drowning in fresh water, chloral hydrate crystals, freezing, and leaving the specimen in a plastic bag in the sun.

Bourne (1984) experimented with the effect of various anaesthetic agents on the Moon snail *(Polinices lewisii)* so that surgery could be performed without death occurring. These agents could possibly be used in order to relax specimens before they are killed and fixed. The anaesthetic was added slowly to the water containing the animals and mixed until the correct concentration was reached. The best agents were found to be alphaxalone and alphadolone (Alfathesin; Glaxo) and Xylazine (N-(2,6-Dimethylphenyl)-5,6-dihydro-4H-1,3-thiazine-2-amine; Rompun-Cutter laboratories). If cost and availability of the anaesthetics prove excessive, ethanol can be used with acceptable results. A physical method (dropping the temperature of the sea water to 2° C over a one hour period) was tried successfully.

Carriker and Blake (1959) experimented on methods for anaesthetizing muricid gastropods and found an acceptable agent in 1-naphthyl N-methylcarbamate. They dissolved it in filtered sea water to obtain a concentration of 10 ppm. After one hour in this solution, the snails were placed in a fresh solution in one atmosphere of CO_2 at room temperature for three hours. The ventral, or lateral, part of the foot of the snail was then held quickly against a block of dry ice, and left in this position (in a freezer) for 24 hours. Fixation was achieved by dropping frozen specimens into a suitable fixative.

If the specimens to be preserved are large, and are required for a scientific collection, it is advisable to crack their shells to allow efficient penetration of the preservative; in any case, the shells will have to be cracked to extract the animals. For display, shells are usually kept intact and the animal is extracted, as it is normally the shells that are displayed. Fresh water specimens are drowned using boiled water plus a narcotic such as nicotine sulphate, or cigarette butts soaked in water, or chloral hydrate, or menthol in 10% neutral formalin. Specimens can be preserved in 5% neutral formalin.

To collect small to medium sized shells, the animal should be killed by being tossed into a container of cold water which should be brought to the boil for a few minutes, after which the animals can be removed and allowed to cool slowly. They should not be placed in cold water as cracking of the shell may occur. The animal can then be extracted using a bent pin or needle. The shell should be left to dry, out of the sun. Very small shells can be soaked in 5% neutral formalin or 90% alcohol for a day, and then dried.

Glossy shelled species, such as cowries *(Cypraea)*, and thin shelled land snails are placed in cold water, heated, simmered for 10 minutes, and then allowed to cool slowly before the animal is extracted. After initial fixing in 10% neutral formalin, the shells can be transferred to 5% neutral formalin for long term preserving. Several weeks should be sufficient for fixing, but if they are left for months or even years the specimens will not be ruined.

Photographs should be taken of nudibranchs (sea slugs) as they are collected. The print should be tied to each specimen. Colour is all important in the identification of most species. Other important data to collect are: state of the tide (low, high, coming in, going out); date; and time. To preserve specimens, place them in a container of salt water. Add menthol crystals (two or three are enough) and leave them for up to 24 hours by which time the specimens should be ready for fixing. Fix them in neutralized formalin, brought to 10% with salt water, and preserve them in 5% neutral salt water formalin.

Cephalopods

Octopuses, squid and cuttlefish are all best preserved in 5% neutral formalin after initial fixing in 10% neutral formalin. For large specimens, the preservative should be injected into the body cavity and tissues.

Certain species will cause the pH of the preservative to alter drastically over a very short space of time, so it is necessary to test the pH each day and refix in fresh formalin several times until a balance is reached.

Ideally, cephalopods are best preserved so that the body and arms are fixed in a straight line. This makes study later easier and to avoid entanglement of specimens. However, the lack of suitable long term storage vessels of sufficient size normally precludes this technique being used.

Most preserved molluscs can be posted or shipped in cotton wool soaked in formalin. However, with cephalopods the cotton wool fibres become entangled in the hooks and barbs on the suckers and are extremely difficult to remove. Use muslin or cheesecloth for wrapping.

DISPLAY

Those animals with an external shell can be displayed dry. They can be mounted using a bolt glued into the shell opening with epoxy cement. The animal can be modelled in clay or Plasticine, using the shell as a guide to size. Some workers prefer to model the animal from wax. The model can then be moulded and cast, and the shell glued onto the cast after the cast has been painted (Fig. 6.2). Most museum collections have dry shells arranged in rows, with labels nearby. The purpose is mainly for identification.

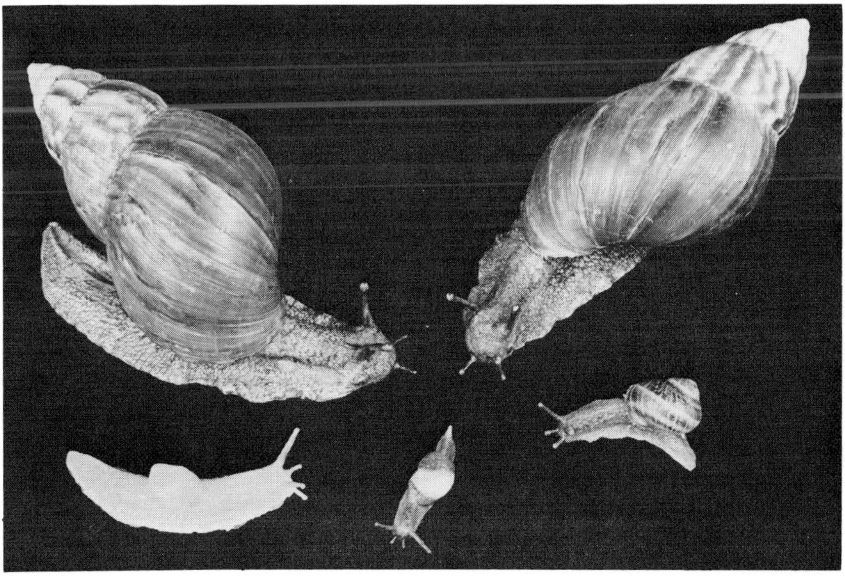

Fig. 6.2. Display techniques for snails include modelling. Plastic bodies together with real shells produce realistic looking snails. Photo A. Pitman. Courtesy of The Australian Museum, Sydney.

Cephalopods lend themselves to display in a wet state. During preliminary fixing the animals can be positioned in the manner in which they are to be displayed. Octopuses may be required to lie flat with their tentacles arranged around the head, whereas squids will have their tentacles laid straight out in front and the fins splayed out. Preliminary fixing is accomplished using 10% netural formalin injected into the body cavity and tentacles. The specimens are best arranged in a tray with a layer of hard wax in the base so that pins, if required, can be stuck into the wax to hold the specimen. A solution of 10% formalin is gently poured into the tray to cover a third of the animal and paper towelling or cloth is laid over the animal with the edges submerged in the fixative. This forms a capillary bridge which keeps the animal moist at all times. It can be left for two days, after which it should be placed into a vessel containing the general preserving liquid used in acrylic wet boxes.

The tentacles may not stay in position during the transfer but this is of no great concern as they will be fixed permanently in the desired pose when it is mounted in an acrylic box.

Squids have a long, narrow, semitransparent structure called a pen situated under the dorsal surface of the abdomen. This can be displayed dry; it should be dissected from the animal carefully as it can be damaged very easily. The quickest method of removal is to lay the squid on its dorsal surface and make an incision along the entire length of the ventral surface. Once the internal organs are removed the pen will be observed buried slightly in the tissue. Cut away the connective tissue until the pen can be removed, then cut away any tissue adhering to the pen. When it is clean, lay it out on a soft board. Cut a strip of stiff, clear acetate sheet and lay it over the pen, pinning it to the board. Do not place pins through the pen, but on either side of it. The board and specimen should then be submerged in a tray containing 5% neutral formalin and soaked for two or three days. After soaking, remove the board from the tray and allow the specimen to dry. When the pen is completely dry, the acetate sheet can be removed. The acetate allows the pen to be held in the correct shape throughout its preserving and drying times.

The beak of the squid also makes a good display if treated properly. The beak is dissected from the specimen and cleaned of all extraneous matter. It is composed of an upper and lower portion which should be fitted loosely together and placed into a beaker containing 5% neutral formalin for two days. After fixing, it is removed from the formalin and the two portions must be locked together. A small container is half filled with clean dry sand and the beak carefully placed on top, embedding it to a depth of 1 mm until it can hold its position without falling over. Clean, dry sand is then very carefully poured around the beak to give extra support in all crevices until it is completely covered. Watch for the sand distending any parts unevenly, and counteract this by applying more sand to the opposite side. Try and cover the beak as slowly and evenly as possible. Leave it undisturbed for two weeks, by which time it should have dried out completely. Pour off the sand carefully and remove the beak after which the two portions can be glued together in proper alignment.

REFERENCES

Bourne, G. B.: Anesthetic methods for the moon snail. *(Polinices lewisii). The Veliger* 26/4 1984: 327–329.

Carriker, M. R. and J. W. Blake: A method for full relaxation of muricids. *Nautilus* 73/1 1959: 16–21.

SOURCE MATERIAL

Baker, F. C.: Preparing collections of the mollusca for exhibition and study. *Transactions of the American Microbiological Society* 40/1921: 31–46.

Bauermeister, W. B.: Ein vereinfachtes Verfahren zur Herstellung durchsichtiger Totalpräparate. *Mikrokosmos.* 48/1959: 272–274.

Bethe, A.: Formaldehyd, nicht Formol oder Formalin. *Anat. Anz.* 11/1895: 358–359.

Bianco, S. L.: The methods employed at the Naples Zoological Station for the preservation of marine animals. *Bulletin of the U.S. National Museum* 39/1899: 1–37.

British Museum of Natural History: Instructions for Collectors *Invertebrate Animals Other than Insects* No. 9A/1954.

Dahl, F.: Die mechanische Methode im Sammels von Tieren. *Zool. Anz.* 31/1907: 917–919.

Gohar, H. A. F.: The preservation of contractile marine animals in an expanded condition. *Journal of the Marine Biology Association* 22/1 1937: 295–299.

Hope, I. G.: Experiments in the paraffin infiltration of marine worms. *Bulletin of the Wagner Free Institute, Philadelphia* 6/1931: 41–44.

Knudsen, J. W.: *Collecting and preserving plants and animals.* Harper and Row, New York 1972.

Lincoln, R. J. and J. G. Sheals: *Invertebrate animals: Collection and Preservation.* Cambridge University Press, London 1979.

Loos (n.i.): Zur Sammel und Konservierungstechnik von Helmichen. *Zool. Anz.* 24/1901.

MacGilivray, J.: Hints on the preservation of specimens of natural history, intended for country residents. Pamphlets 8vo series. Vol. CXVII. Australian Museum, Sydney, n.d. (but before 1916).

Mayer, F. K.: Röntgenographische Untersuchungen an Gastropodenschalen. *Jena. Z. Naturw.* 65/1931: 487–512.

Noll, F.C.: Präparieren der Schneckenzunge. *Neckr. Bll. Dtsch. Malak. Ges.* 14/1882: 186.

Owen, G.: Use of propylene phenoxetol as a relaxing agent. *Nature* 175/1955: 434.

Rosewater, J.: An effective anaesthetic for giant clams and other molluscs. *Turtox News* 41/12 1963.

Rotarides, M.: A puhatestűék (Mollusca) gyűjtése és kezelése. *Fragm. Faun. Hung.* 7/1944.

Runham, N. W., K. Isarankura and B. J. Smith: Methods for narcotising and anaesthetising gastropods. *Malacologia* 2/1965: 231–238.

Schmidt, G.: Blutgefassystem und Montalhohle der Weinbergschnecke (Helix pomatica). *Z. wiss. Zool.* 115/1916: 201–261.

Schmied, G.: Versuch einer trockenen Konservierung von Nacktschnecken. *Mitt. Dtsch. Malak. Ges.* 2/1962: 19–20.

Schwanecke, H.: Das Blutgefassystem von Anodonta Cellenis. *Schrt. Z. wiss. Zool.* 107/1913: 1–77.

Szilady, Z.: Eine Präpariermethode für Schnaken. *Ent. Z.* 56/1943: 190–191.

Wagner, L.: A csigák es kagylók gyújtése. *Búvár.* 5/1939.

Wachtler, I.: Zur Technik des Mollusken Sauelns. *Archiv fur Molluscenkunde* 57/1924.
Warburton, F. E.: Removing intact bivalves from their shells. *Pubbl. Staz. Zool. Napoli.* 30/1957. H. 1.
Williamson, M. H.: The separation of molluscs from woodland leaf litter. *Journal of Animal Ecology* 28/1959: 153–155.

7 Arthropods (crustaceans) and chordates

CRUSTACEANS have been a source of food for humans for countless years. Therefore, the techniques for catching them are many and varied. However, specimens for preservation need to be unmarked and undamaged, unless they are going to be preserved and displayed in part.

COLLECTING

The methods of collection vary depending on which particular specimen or group is required. For example, most intertidal crabs can be caught by hand, as can small fresh water crayfish. Large crabs and lobsters can be caught by a skin diver, or by using lobster pots or commercial funnel type traps with fish or meat baits inside. A dip net can be useful for scooping up small prawns and other small crustaceans.

Larval forms and the small free-swimming crustaceans that inhabit the uppermost part of the marine habitat are caught using tow nets (Fig. 7.1) or plankton nets. These nets are cone shaped and made from Terylene or other synthetic fibres. The larger open end is made from canvas; this is attached to a metal hoop with in turn is secured to ropes. The smaller end has a collecting jar attached which can be removed in order to release the sample. The net may be towed through the water behind a boat, or it can be secured to a pylon in a fast flowing current.

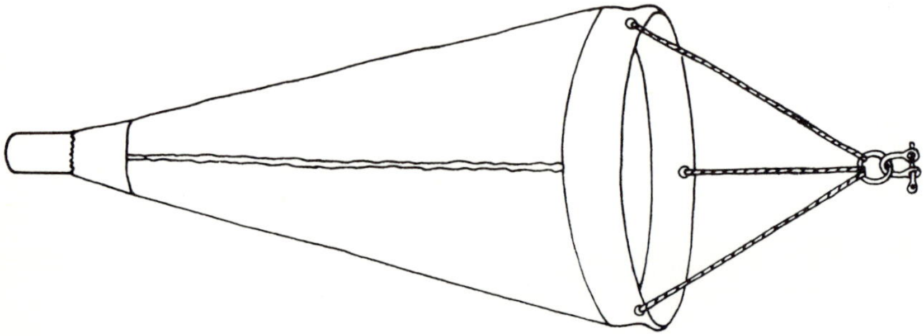

Fig. 7.1. A tow net.

Small tow nets can be used in fresh water habitats if they are dragged through open water, or along the sides of banks, or among the water vegetation. Two people are needed for this; one walks, with the net attached to a rope, to a position opposite the other who holds the other end of the rope. The net is placed or thrown into the water and pulled to the other side. The contents of the collecting jar can then be emptied into a storage container and the procedure repeated.

If collecting from small expanses of water, a small tow net, secured to a long handle, can be used. This can be made by sewing a stocking to a metal hoop attached to a handle. The stocking should be cut about 20 cm from the toe to produce an open end. This end can be attached to a small glass or plastic jar. It should be swept through vegetation and open water near the bank.

To collect those animals which live in mud or sand, a spade should be used to turn over the mud and sand. This is then sorted through by hand; it must be spread thinly to find specimens. Another efficient method of capture is the "yabbie pump", used by fishermen to acquire their bait (Fig. 7.2). It consists of a stainless steel tube with a handle projecting at a 90° angle near the top. A plunger slides up and down inside, similar to a bicycle pump. The bottom of the tube is placed onto the sand and pushed down while the inner plunger is pulled up. A core sample is sucked up. When the tube is removed from the sand, the plunger is depressed and the core expelled. The contents can then be searched for animals.

Terrestrial crustaceans such as woodlice can be collected from damp soil or under rotting logs, or from under stones and in vegetation. Sometimes they are collected in pit traps and in leaf litter which should be sorted using a Berlese funnel (see Volume 2, Chapter 8).

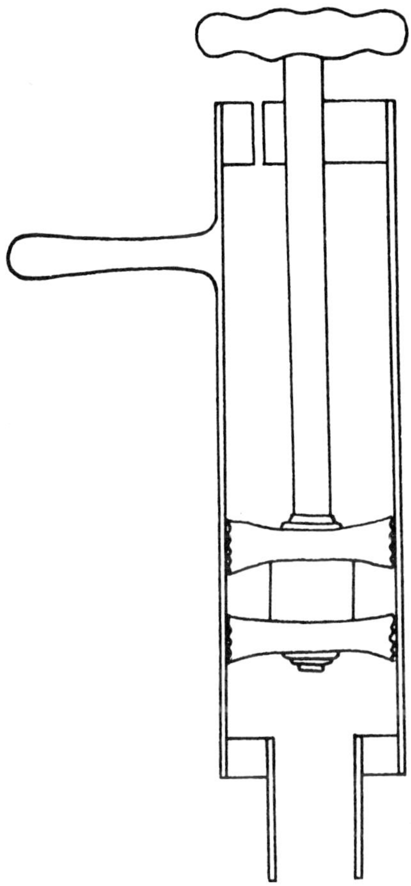

Fig. 7.2. A yabbie pump or slurp gun.

KILLING AND FIXING

Specimens can be placed in a freezer but care must be taken in thawing them. Thawing can be a problem with *Callianasa arenosa* in which, if completely frozen and then thawed, the thorax and abdomen separate very easily. Specimens should be thawed out in a bucket containing 10% neutral formalin, then transferred to the storage bottle containing 70% alcohol. If more than one specimen is to be killed, this should occur in separate jars as they may fight and parts of the body may be broken. Ostracods, isopods, and amphipods can be killed and preserved by being placed immediately into 10% neutral formalin. Cirripedia should be killed in 35% alcohol, which aids the distension of the cirri. Malacostraca (crabs, prawns, lobsters, and others) should be killed by being placed into 10% neutral formalin.

PRESERVING AND STORAGE

For museum collections, specimens should be washed in tap water after 24 hours fixation, then stored in 70% alcohol. Specimens can be housed in glass bottles with screw top lids or clip on plastic lids. The size of the specimen usually determines the size of the container. Five or six prawns may be stored in a 500 mL jar, but it will contain only one blue swimmer crab *(Portunus pelagicus).* Some institutions keep a dry collection of crabs and lobsters, but as internal organs are not retained this type of storage is becoming less popular. Very small crustaceans (ostracods, isopods, and amphipods) are stored in a different manner to larger members of the group. They should be placed into small glass phials (16 mm x 50 mm, or 7 mm x 50 mm) containing the storage medium. The phials can then be plugged with cotton wool, then placed into large tubes (25.4 mm x 80 mm) containing fluid. The identification label should be placed inside the larger tube with the small phial. Damage may occur if it is placed in with the specimen. If many examples of the same species are collected, but each has slightly different data (such as capture in a different locality), they may be stored in the same way. All the tubes can then be placed into a large jar containing 70% alcohol. A large label should be inserted to be read easily. The label should contain such data as the number of phials present, their catalogue or computer numbers, and the name of the family (if that is how specimens are catalogued for easy shelf access).

DISPLAY

Crabs, lobsters and prawns can be displayed in a dry state. This can be achieved in the following ways.

A. The specimens can be injected with 10% neutral formalin and arranged in a suitable display position with the use of strategically placed pins. They can be dried in the air or, more quickly, by using a drying cabinet (Fig. 7.3). Once it has been dried and painted the specimen will appear as the one in Fig. 7.4.

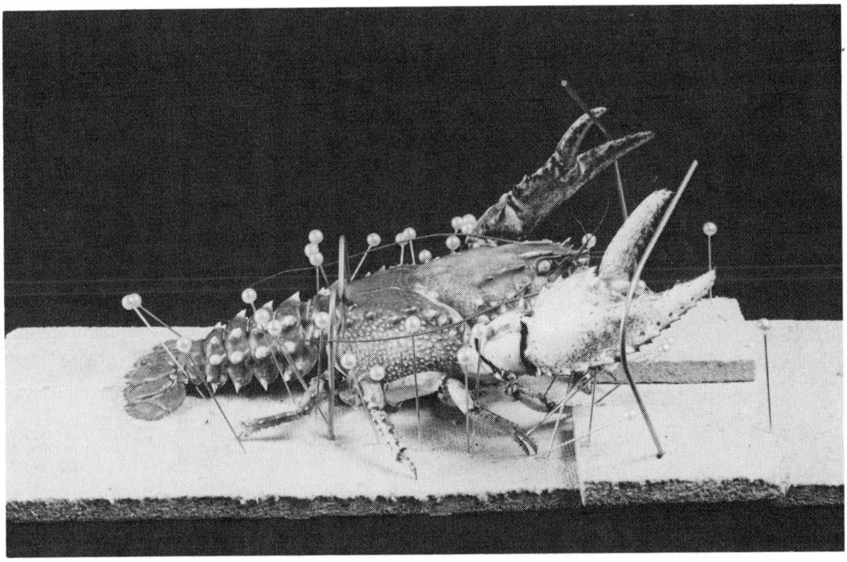

Fig. 7.3. A lobster is prepared for drying by pinning the appendages into their correct position. Photo K. Lowe. Courtesy of The Australian Museum, Sydney.

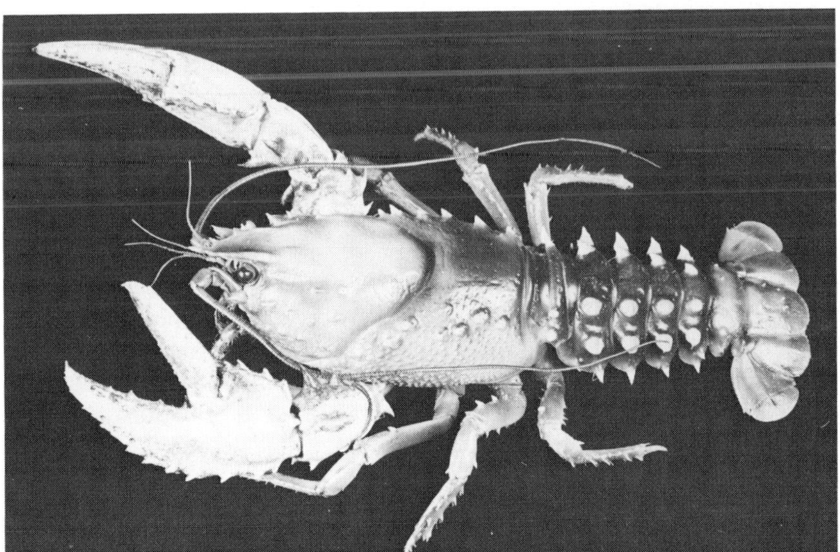

Fig. 7.4. A finished lobster after drying and painting. Photo A. Pitman. Courtesy of The Australian Museum, Sydney.

B. It is possible to remove the carapace of large crabs. With lobsters, it is possible to cut the abdomen across at the point where it joins the thorax, and then separate the two parts. In both cases, the flesh can then be removed using forceps and scrapers. The appendages can be removed at the first coxal joint and the flesh can be removed using a hooked piece of wire. The carapace can then be glued back on, the abdomen can be attached to the thorax, and the appendages must also be attached (if they were removed) using a quick setting epoxy resin. The specimen can then be allowed to dry in the desired position. Preservation will be ensured if 5% neutral formalin is brushed into the inside of the carapace and tail. If the shell cracks during the drying process, fill the cracks using some form of silicone putty or water based crack filler. When dry and set, specimens are retouched with oil based paints, applied lightly. To fix specimens into cases, bolts may be used either by cutting a key hole shape into the ventral surface of the shell, inserting bolt heads, then glueing them in, or by drilling holes into the shell before the carapace is glued back into position. Specimens prepared in either of the aforementioned ways may have to be painted to show correct colour patterns. It is important to keep a second specimen of the same species in a freezer to be used as a colour reference.

C. Specimens such as prawns may be prepared in an acrylic wet box. After a fresh specimen is killed it should be injected with 10% neutral formalin and stored in 5% neutral formalin until it is needed for display. The procedure for final preservation and storage is the same as that described in Volume 1, Chapter 7.

D. Freeze drying is a very good method for preparing crustaceans for display if the specimens are fresh. The procedure detailed in Volume 2, Chapter 10 should be followed. Various crustaceans are shown in Fig. 7.5.

E. If small specimens such as prawns, copepods and isopods are to be handled and looked at closely, then they can be embedded in clear plastic. The Antarctic krill *(Euphasia superba),* as with other specimens with a thin exoskeleton, takes on a slightly translucent form which may be distracting. The injection of catalysed resin into body cavities may cause this problem. Consult Volume 2, Chapter 12 for information on embedding and follow the instructions for biological material.

Colour preservation

Toyama and Miyoshi (1963) used a prawn *(Pandalus borealis)* preserved in 10% formalin to test butylated hydroxytoluene (BHT) as a solution for preserving colour. Their results show that BHT provided good colour retention after a short period, but the long term effects are not yet known.

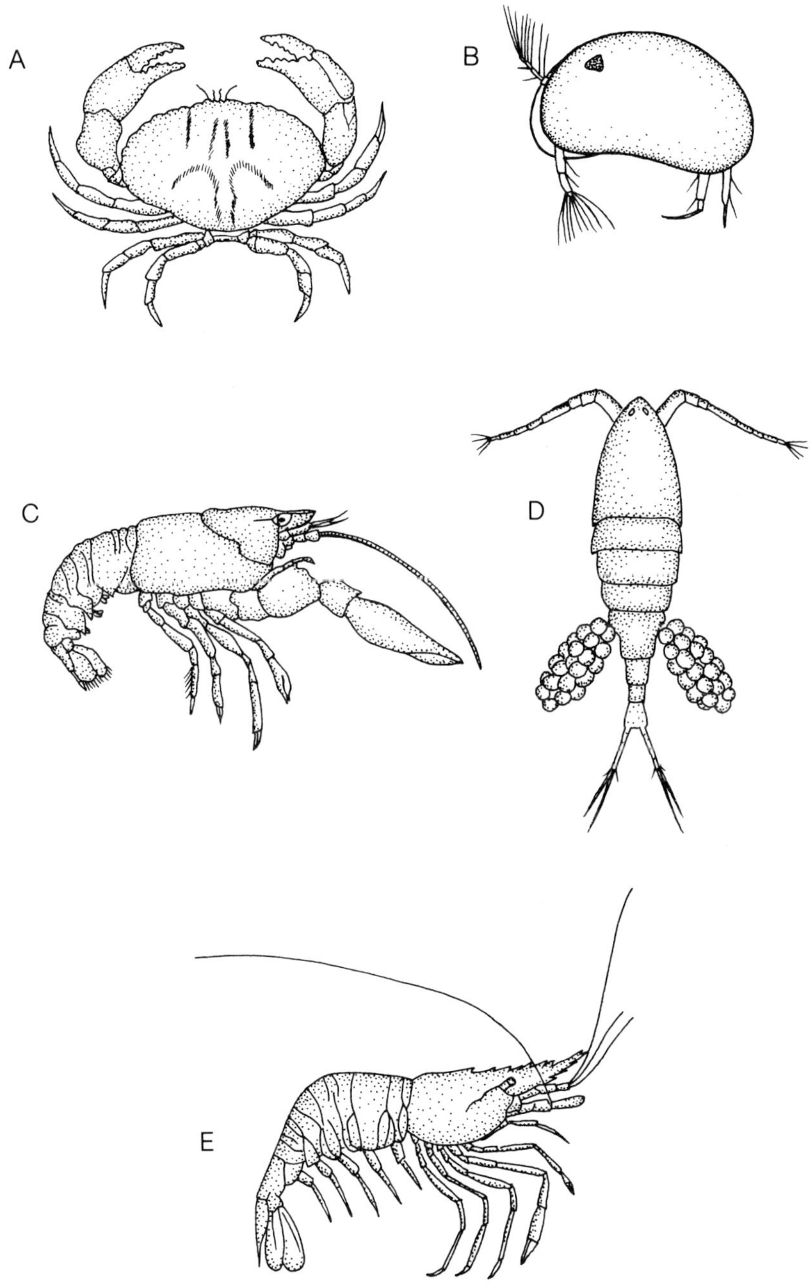

Fig. 7.5. Several types of crustaceans: A. Crab. B. Ostracod. C. Lobster. D. Copepod. E. Prawn or shrimp.

CHORDATES

The animals of this phylum are almost always vertebrates, but two subphyla are sometimes regarded as invertebrate. All are marine. At some point in their life cycles members of the two subphyla possess a notochord, pharyngeal gill slits and a hollow dorsal nerve chord which are three characteristics of chordates.

UROCHORDATES

Members of this subphylum are placed into three classes: Ascidiacea, Larvacea and Thaliacea. Ascidians, or sea squirts as they are common known, are sessile creatures during their adult stage of their life cycle and they are found as solitary or colonial forms attached to rocks and seaweeds. They are best collected during low tide when they become exposed. If they are stepped on or squeezed, they shoot out a jet of water; hence the name sea squirt. They inhabit the intertidal zones as well as the great depths of the oceans. They vary in colour from grey through to green and their external appearance can be smooth, leathery or encrusted with sand and shell fragments.

The Larvacea and Thaliacea are basically transparent planktonic animals. The Larvacea are tadpole like animals found throughout the world's oceans and the Thaliacea are principally tropical and subtropical.

Collecting

Ascidians can be removed from the substrate using a strong blunt knife. Part of the substratum should be removed with the specimen, especially when collecting colonial forms. If difficulty occurs whilst trying to remove the rocky substatum, a chisel may be used to make the task easier.

Larvacea and Thaliacea are collected using a plankton tow net towed behind a boat through the upper surface of the sea. Towing times depend to a great extent on the amount of plankton floating in the collecting area, but a general rule is to tow the net for 30 minutes to one hour, after which it should be hauled aboard and examined for specimens.

Killing and preserving

The urochordates may be killed and fixed by plunging specimens into 5% buffered neutral formalin for 48 hours. Specimens are then transferred into fresh 5% buffered formalin or 70%–90% alcohol. Time and care will result in relaxed and well preserved specimens if they are relaxed with an anaesthetic before being killed. Anaesthetizing can be by various methods; experimentation may be required to produce optimum results. The simplest method is to place specimens into a bowl containing sea water and allow them to relax and expand. This is partly due to the dissolved oxygen in the sea water being used up. When the specimens are fully relaxed, enough neutral formaldehyde should be added to the sea water to make a 5% solution of formalin which should kill and fix the specimens. Other methods used to anaesthetize specimens all use chemicals. These are added drop by drop (for liquids) or a few crystals at a time to the sea water containing the specimens until relaxation occurs. Suitable chemicals include a 1%–2% solution of 2-phenoxyethanol, magnesium sulphate and menthol.

CEPHALOCHORDATES

Animals belonging to this subphylum are known as lanceolates, but are more generally known as Amphioxus. They are smallish, laterally compressed animals occurring in marine waters around the world. Some species swim freely with fish like body movements, whilst others are found with their posterior partially buried in the sand or mud; they feed by filtering the sea water through their mouths to collect food.

Collecting

Adults can be collected by dredging the sea floor, whilst the larvae are collected just after sunset using a tow net to skim the upper layers of the sea.

Killing and preserving

Specimens are placed directly into 5% neutral formalin for initial killing and fixation. They are then transferred to fresh 5% neutral formalin after a few days, or can be washed in tap water for 30 minutes and transferred to 70%–90% alcohol for permanent storage.

REFERENCE

Toyama, K. and G. Miyoshi: Prevention from colour fading of aquatic animals under preservation. *Journal of the Tokyo Fisheries* 50/1 1963: 43–48.

SOURCE MATERIAL

Bianco, S. L.: The methods employed at the Naples Zoological Station for the preservation of marine animals. *Bulletin of the U.S. National Museum* 39/1899.

Knudsen, J. W.: *Collecting and preserving plants and animals.* Harper and Row, New York 1972.

Lincoln, R. J. and J. G. Sheals: *Invertebrate Animals. Collecting and preservation.* Cambridge University Press, London 1979.

MacGillivray, J.: Hints on the preservation of specimens of natural history. Pamphlets Vol CXVII. Australian Museum, Sydney n.d. (but before 1916).

8

Arthropods: Insects and others

THE NUMBER of insect species known to science, and newly discovered species, described everyday, make insects the largest and most diverse group within the animal kingdom. They are seen by humans more often than any other group of animals. They have used every niche within the world's environments to their best advantage, conquering the land, water, and air, wherever plants grow.

Most insects appear more active in spring and summer than in winter and autumn. In the tropics, where there is a wet and dry season, they are more easily found when wet and warm weather coincide. Not all insects adhere to these situations. Their behaviour depends on whether they are active during the day, night, dusk, or morning, or if they are terrestrial or aquatic. Collecting methods and the necessary equipment vary a great deal especially when pursuing a particular group of insects. Specialized equipment is often necessary, may be made up by individual researchers to cater for their needs.

If collectors are not aiming at a specific order of insects, they will be able to seek insects at any time of the year employing simple collecting techniques and equipment.

COLLECTING TERRESTRIAL INSECTS

These are found in just about every place that you care to look. For example, there are those which can be caught flying through the air, those which can be picked off flowers, those which can be knocked off foliage into trays, those attracted to lights at night, those which can be picked out of leaf litter, those found under rocks, logs and bark, those located in animal faeces and carcasses, and those which are parasites on other animals.

SINGLED COLLECTING

The simplest way to collect insects is to find them individually and pick them up with the hand or a pair of forceps. Specimens can be found on vegetation, often on the underside of leaves, on flowers, and on treetrunks. A very common and most prolific method of collecting is to search under rocks and logs. Many species of beetles, earwigs and cockroaches can be collected this way. The upturned rocks and logs must be replaced after inspection, otherwise the delicate environment under them is destroyed (and the landscape marred).

In temperate zones, spring and early summer are excellent for singled collecting. Many species of insects can be found on forest paths. Some will be ground dwellers, and some will be seeking a temporary resting place on the ground or warmth on a sun-warmed patch of the road. Large predatory beetles, such as *Carabus* and *Calosoma* can be detected if the collector walks quietly in the forest, listening for the faint noise these insects make while hunting for their prey in the dry leaf litter.

The dung of larger mammals is food for many insect species, especially beetles. Interesting collections can be made from dung piles; the collector should use a pair of forceps.

KILLING

Most collectors kill their catch immediately after collection. The method of killing should be carefully chosen, since some which are practical for one particular group of insects can render others useless for further study. As an example, butterflies and moths must be killed quickly in a dry killing jar in order to prevent, or at least minimize, injury to the delicate specimens. More sturdy insects, such as bettles and bugs, can be dropped in a jar containing alcohol or paper crumpled with a few drops of ethyl acetate. Various killing methods suitable for individual orders are described later.

Cyanide killing jar

The most efficient killing agent for insects is potassium cyanide. It is also a deadly poison for warm blooded creatures, including humans! The utmost care must be taken to avoid skin contact, especially contact with the mucous membrane, since the smallest amount of this substance can kill a person. Prolonged inhalation of the fumes can also cause severe illness or death. However, if the jar is made properly, it can be used safely. A very good method of making up a cyanide jar was demonstrated to the authors by Dr Eduard Diehl, the noted German lepidopterist from Sumatra, Indonesia. In his method, the jar is made of a length (approximately 180 mm) of acrylic (perspex or plexiglass) pipe of approximately 60 mm diameter. A circular piece of acrylic is cemented on one end to form a flat bottom. The mouth of the jar is fitted with a cork which should be coated with paraffin wax in order to render it airtight. (Screw top bottles and jars are not suitable for collecting purposes as they cannot be opened and closed quickly enough. The screw top also squashes specimens trying to escape from the jar while it is being screwed closed.)

A 20 mm thick layer of dry sawdust should be poured in the jar. A small well can be made in the middle of this and a lump of potassium cyanide placed there. Dry, powdered plaster of Paris should be poured in next until a 20 mm thick layer is built up. The plaster must be levelled, and water can be sprinkled on top until it penetrates the top 3 mm–5 mm to form a hard crust. Small, circular pieces of paper towelling or blotting paper should now be placed on top of the plaster to absorb any excess moisture. Once the plaster has set, the jar is ready to be used. It should kill insects very rapidly and can remain potent for years. It should be clearly marked "Poison: Potassium Cyanide" and locked away safely when not in use (Fig. 8.1).

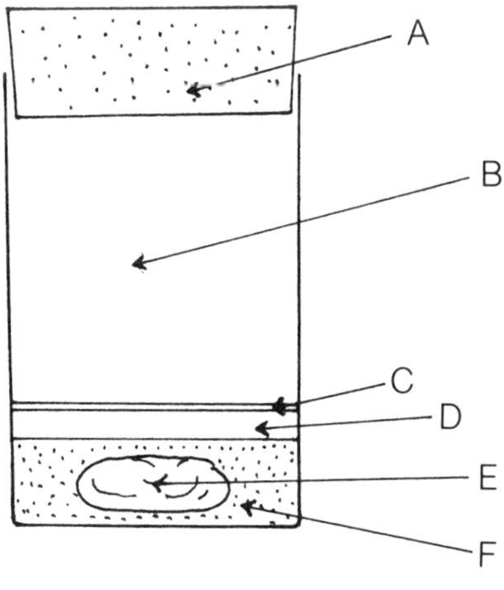

Fig. 8.1. Cyanide killing jar. A. Cork. B. Killing area. C. Paper or cardboard. D. Plaster of Paris. E. Lump of cyanide. F. Sawdust. Drawing G. Hangay.

Butterflies and moths are killed with this jar, but no more than two to three specimens should be kept in the jar at the same time. As soon as the specimens have quietened down they can be transferred to another similar jar where they should remain for a few hours. If too many specimens are crammed into one jar while they are still fluttering around they will damage each other.

Other insects can also be killed with the cyanide jar, but some crumpled paper (preferably paper towelling) should be placed in it first to separate the specimens and thus prevent injury. As the paper becomes dirty, it should be changed and the wall of the jar cleaned with a damp cloth.

Some insects with metallic colours (for example, a few of the Chrysomelidae species) suffer colour changes in the cyanide jar. Such changes must be observed and noted and when these species are collected again they should be killed with other methods.

Vapour killing jars

These are less dangerous for humans than the cyanide jars, and therefore more suitable for student use, especially for junior students and interested children.

The simplest jar consists of a wide mouthed glass container with a tight fitting cork. The cork should be boiled in melted paraffin wax in order to provide an airtight seal. The jar should be half filled with loosely crumpled paper to which a few drops of ethyl acetate have been added (Fig. 8.2). The jar is now ready to use. The collected insects are simply placed into it and the ethyl acetate kills them in a short time. After repeated opening and closing the chemical will evaporate. The jar must then be recharged. It is practical to leave the insects in the jar for approximately 12–24 hours, depending on the size and the nature of the species. Some small, delicate insects expire almost immediately in the ethyl acetate vapour, while others, such as large beetles, although immobilized while in the jar, could recover even after they have been pinned and cause havoc in an insect box filled with other pinned insects.

An ethyl acetate jar filled with crumpled paper is not suitable for killing butterflies and moths. These delicate insects would be damaged by the damp paper so a different design must be employed.

There are many different types of killing jars constructed in such a manner that the killing agent (for example, ethyl acetate) cannot come in contact with the insect in liquid form, but only as vapour.

In an emergency, any glass jar will suffice if it is fitted with an airtight cork or even a screw top. Some cotton wool should be pressed into a tight ball and a few drops of ethyl acetate applied to it. It should be placed in the jar and a larger wad of dry cotton wool pressed on it, thus keeping it on the bottom of the vessel. A circular piece of cardboard should then be cut and pressed

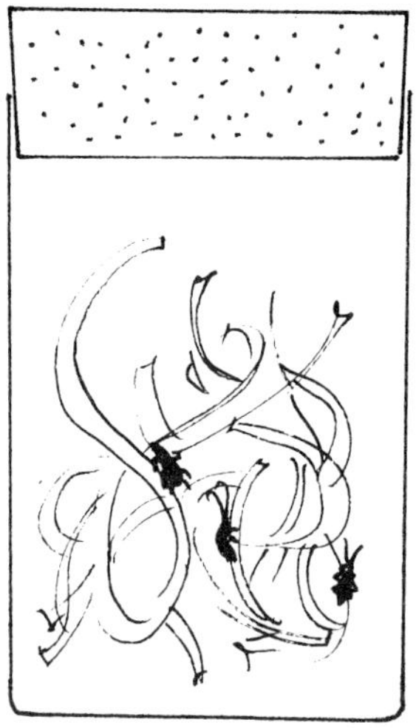

Fig. 8.2. Killing jar with shredded or crumpled paper. Instead of paper, wood wool can be used. Drawing by G. Hangay.

on top of the wool to compress it and keep it in place. The jar must be recharged from time to time. This is best done with a glass syringe or eye dropper. The chemical can be squirted on the wool by lifting the cardboard a little on one side.

A much better killing jar can be made the following way: Select a wide mouthed jar of approximately 0.3 litre capacity, fit it with a wax impregnated cork, and sprinkle sawdust in it approximately 15 mm–20 mm thick. Cut a glass tube of small diameter (2 mm–3 mm) 25 mm–30 mm long and set this upright in the sawdust. Mix a small amount of plaster with water and pour it on top of the sawdust, making sure that the glass tube is not covered completely (Fig. 8.3). Allow the plaster to set and dry. After the water has evaporated the jar can be charged with ethyl acetate. This is carried out by injecting the chemical into the glass tube, with either an hypodermic syringe or an eye dropper. The glass tube can be stoppered with a small cork or plug of paper to prevent the smaller insects getting into it. The charged bottle should be potent enough for a few hours of constant use.

Alcohol jars

Another simple way of killing collected insects in the field is to drop them into 70% alcohol (ethanol). The alcohol is carried in vials and jars. Here, again, tight fitting corks are essential. Vials with plastic stoppers are also suitable, but only new stoppers, which fit tightly into the vials, should be used.

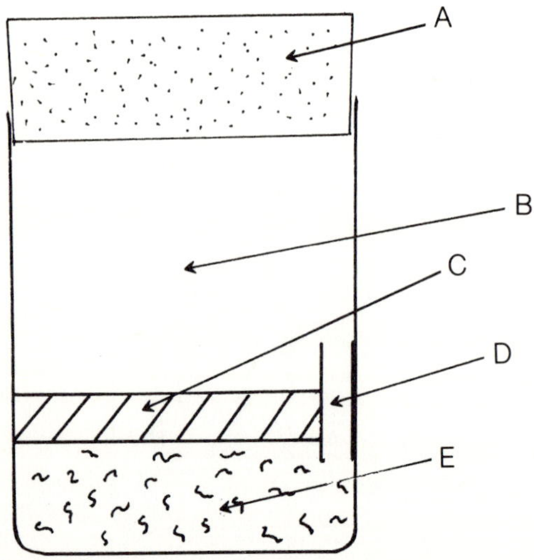

Fig. 8.3. Killing jar with liquid killing agent. A. Cork. B. Killing area. C. Plaster of Paris. D. Glass tube. E. Sawdust.

KILLING AGENTS

Potassium cyanide is the best killing agent. It is available from chemical suppliers in granular form or in small, cake like lumps. The latter form is more suitable. Sodium cyanide is equally effective. Calcium cyanide, which is also often used by entomologists, gives out toxic fumes at a greater rate, especially in the presence of moisture. This renders the killing jar very potent, and also more dangerous. Rapid sublimation means the jar must be recharged sooner than those charged with potassium or sodium cyanide.

Ethyl acetate is believed to be a very good killing agent for general insect work. Most insects can be killed with this substance. The specimens die relaxed, which makes setting easier. However, some insects discolour in ethyl acetate vapour (some species of Orthoptera). The vapours of ethyl acetate are not extremely dangerous for humans. However, repeated inhalation can cause drowsiness, nausea, or unconsciousness. It is flammable.

The vapours of benzol also kill insects. Benzol is especially recommended when collecting Orthoptera. Some Acrididae tend to turn reddish-pink if killed with benzol fumes.

Carbon tetrachloride is also commonly used to kill insects. The chemical is used like ethyl acetate. Although its vapours are not as toxic to insects as those of the former, it is popular amongst entomologists because it kills specimens with fair to good relaxation.

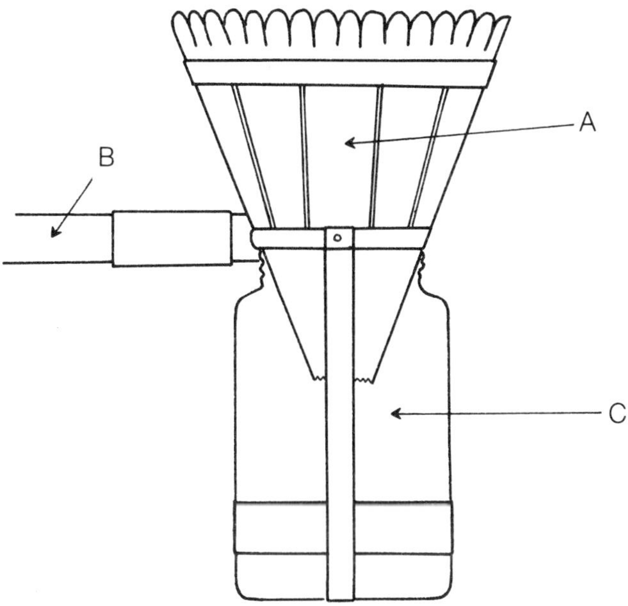

Fig. 8.4. Capturing and killing jar. The jar is fitted with a long handle (B) to enable the collector to reach moths settled high on tree trunks. The funnel like rubber section is placed over the moth which flies into the jar (C).

Chloroform is excellent for killing butterflies and moths. The substance is highly volatile and rapidly immobilizes the insects, thus preventing a lot of damaging flapping in the jar. A disadvantage of chloroform is that the specimens become rigid, which makes immediate setting very difficult, if not impossible. However, since most Lepidoptera specimens must be relaxed before setting anyway (because they dry out quicker than many other insects), this does not cause unsolvable problems. Chloroform is especially useful when collecting specimens of moths (for example, Catocala) during the day. These large moths often settle on tree trunks, walls of buildings, and other high places where they are out of reach. A small killing jar fastened on a long pole can be used to catch them. The open jar is very carefully manipulated over the resting moth. The chloroform fumes quickly anaesthetize the specimen; the collector secures a perfect specimen which otherwise would wildly thrash about in the jar. The jar can also be fitted with a rubber funnel with toothed edge (Fig. 8.4). This will aid the capture of specimens settled on rough bark.

Ammonia can also be employed to kill small insects such as Microlepidoptera. The ammonia is used in the same way as any other liquid. In its liquid form, the chemical should not come in contact with the specimens. Only the vapour is needed. Here again, the main advantage of the chemical is that it kills the specimen in a relaxed state. The jar should be lined with blotting paper to avoid condensation on the wall. This would ruin the specimens.

This is especially important in the case of Microlepidoptera, which are usually immediately pinned and the wings positioned in a spread position. All these substances are dangerous to humans. Collectors must avoid skin contact and inhalation of vapours.

KILLING AND PRESERVING SOLUTIONS

Alcohol

Institutes are usually able to obtain ethyl alcohol at reasonable prices. Industrial methylated spirit or methyl alcohol is also suitable for killing and preserving insects, but the (usually coloured) domestic variety is not because its colour can stain specimens and a cloudy, white flocculent precipitate forms if it is mixed with water. The alcohol should be diluted with water to about 70%–80%. Insects can be dropped into it live and stored in the same liquid. Soft bodied specimens should first be placed in 30% alcohol for a few hours, then transferred into a 50% solution for the same amount of time. After that, they may be placed in a 70% solution for a few days. They should then be transferred into a fresh 70% alcohol solution. If this procedure is not followed and the animals are placed into 70% alcohol immediately, shrinkage and hardening may occur. The strength of the final storing liquid should not be higher than 80%, and not lower than 50%.

Lacto alcohol

Some Hemiptera (aphids, and scale insects) are killed and stored in a formula consisting of: 80 mL lactic acid 87.5%; 74 mL 95% alcohol; and 46 mL water. Although the lacto alcohol has some macerating effect, specimens can be kept in it indefinitely without harm.

Carnoy's fluid

Carnoy's fluid consists of: 60 mL 95% alcohol; 30 mL chloroform; and 10 mL glacial acetic acid. The larvae of various orders (Neuroptera, Mecoptera, Lepidoptera, Hymenoptera, and Coleoptera) are usually killed and fixed in this formula. It penetrates rapidly and does not harden the specimens. Larvae and pupae of Lepidoptera can be injected with the liquid prior to further preparation.

Oudeman's fluid

This formula consists of: 88 mL 70% alcohol; 4 mL glycerine; and 8 mL glacial acetic acid. Nymphal stages of Odonata are usually dropped in the fluid and killed that way. The fluid is also suitable for storage.

Peterson's KAA solution

This solution is made from: 10 mL kerosene; 100 mL 95% alcohol; and 20 mL glacial acetic acid. This is a very good killing and temporary preserving solution for larvae, especially Diptera. The mixture produces some inflammation of the tissues and this causes the larvae to extend fully. The acetic acid keeps maggots white. Other larvae, especially those with a green colour, tend to fade. Some become translucent if kept too long in KAA. Specimens may need 30 minutes or a few hours in the fluid depending on their size and the toughness of their exoskeleton. After this, the specimen should be removed, blotted, and placed in a 95% solution of alcohol for storage.

Huffaker's XA mixture

This is a killing agent for Lepidoptera larvae consisting of: 100 mL xylene (xylol); and 100 mL alcohol. Specimens die in a distended state; after that they should be placed in a 95% solution of alcohol. Larvae tend to become translucent if kept in this mixture for a prolonged period of time.

Dornisthorpe's fluid

The original description of the formula called for absolute alcohol. Wagstaffe and Fidler (1955) modified the formula and their version is: 5 mL glacial acetic acid; 37 mL 95% alcohol; 58 mL distilled water; and 0.12 g corrosive sublimate. It is a killing and preserving mixture into which ants are dropped in order to kill them instantly and keep them in a soft and relaxed state. It is presumed that the mixture could be applied to many other insects with similarly good results.

Kahle's fluid

This solution (100 mL 95% alcohol; 7 mL glacial acetic acid; 40 mL formalin) achieves rapid penetration due to the alcohol content. The formalin hardens the tissues of the specimen killed and fixed in this liquid.

Pampl's fluid

This fluid (4 mL glacial acetic acid; 30 mL distilled water; 6mL formalin; 15 mL 95% alcohol) is used to kill and fix larvae, especially Trichoptera. It is a slow penetrating liquid which sometimes allows some internal decay. It also hardens the specimens over the long term.

Van Emden's fluid

This is basically a modified Pampl's fluid, designed for the killing and fixing of beetle larvae. It consists of: 2 mL glacial acetic acid; 30 mL distilled water; 5 mL formalin; and 15 mL 95% alcohol.

Bles' fluid

Bles' fluid is another variation on Pampl's fluid designed to kill and fix beetle larvae. The formula (3 mL glacial acetic acid; 7 mL formalin; 90 mL 70% alcohol) assists the hardening of the specimens, which is desirable for dry preparation.

KILLING WITH INJECTION

Some large insects are killed by injecting ethyl acetate, carbon tetrachloride, chloroform, or ammonia. This method immobilizes these larger specimens very quickly. A large beetle, if placed in a killing jar which already contains dead specimens, may take a long time to die. In the process, it may wreak havoc in the container. Large moths can thrash around so vigorously that they ruin their wings. The best place to inject an insect is at the thorax. If hard exoskeleton prevents this, a larger dose should be injected into the abdomen. Large moths or butterflies usually die after administration of a minute amount of any of the above mentioned substances. These insects are also very vulnerable to the smallest amount of nicotine. Wiegel (1958) described the preparation and usage of "killing pins": tobacco is soaked in a small amount of water, then the liquid is separated and reduced (by heat) until it becomes syrup like. Insect pins are dipped in this syrup and allowed to dry. The pins are stuck in the thorax of Lepidoptera specimens and kill them.

KILLING WITH HEAT

In this method, the specimen is placed in a dry test tube and, by pushing a plug of cotton wool or tissue paper down the tube, confined in a small space at the bottom. The tube is dipped in a vessel containing hot (near boiling) water. The insect expires within 20 seconds. It can then be prepared or stored.

This method is useful in laboratory conditions. More robust specimens, such as beetles, or grasshoppers, can be dropped into near boiling water for a few seconds and killed that way. However, the method is inconvenient under field conditions, and unsuitable for an unsorted, mixed collection of insects.

TRANSPORTING LIVE INSECTS

It is often necessary to keep the captured insect alive for study purposes, or to delay killing and preparation until it can be carried out in the laboratory.

Most insects can be kept for a short time in a screw topped jar, if the lid is perforated. If the insects are very small, a larger hole should be cut and a piece of fine mesh is glued over it from the inside.

Some insects are very sensitive to changes in their environment, especially to changes in humidity, temperature and light. For these, specially designed transport boxes and cages are made, sometimes incorporating a food plant, water, or even humidity and temperature controlling devices. Peterson (1959) describes various methods and equipment designed for transporting and keeping live insects.

COLLECTING EQUIPMENT

Collecting with bare hands or a pair of forceps may be effective in many cases, but the serious collector should have some basic equipment at his or her disposal. Much of this is available from entomological suppliers, or can be made easily.

Aerial or butterfly net

This instrument consists of three major parts: the frame or hoop; the handle; and the net bag.

The hoop is about 45 cm diameter and fixed to a long handle made of wood, cane or aluminium. The length of the handle depends on the ability of the user to swing the net around with ease, and on the insects being sought. Handles are normally 0.9 m–12 m in length with a diameter of about 25 mm. Anything longer or thicker makes the task of sweeping the net through the air very cumbersome. The hoop, made of a light metal (5 mm diameter), suspends a cotton, nylon, bretonne, or marquisette bag, twice the length of the hoop diameter and with a rounded end, with rings (Fig. 8.5A). Some handles are telescopic or have screw fittings so that extra lengths can be added to enable the collector to reach insects at a greater distance (Fig. 8.5B).

Some entomologists have a cloth bag which hangs outside the net attachment to the hoop (Fig. 8.6A). This is the same length as the net and has open

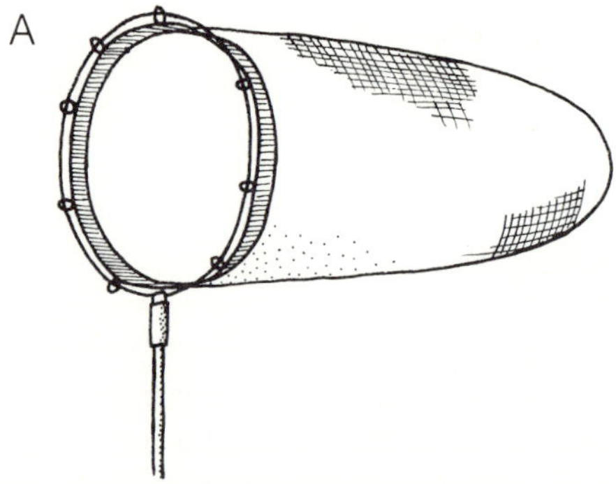

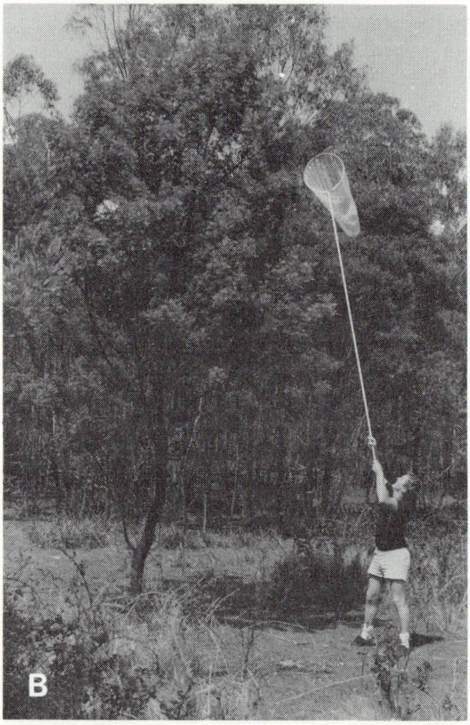

Fig. 8.5. A. An areal or butterfly net. This illustration shows a model which is fitted on the metal loop by rings. Shaded area indicates reinforcing at the edge. **B.** A butterfly net with extended handle is being used to collect insects from the high branches of a tree. Photo G. Hangay.

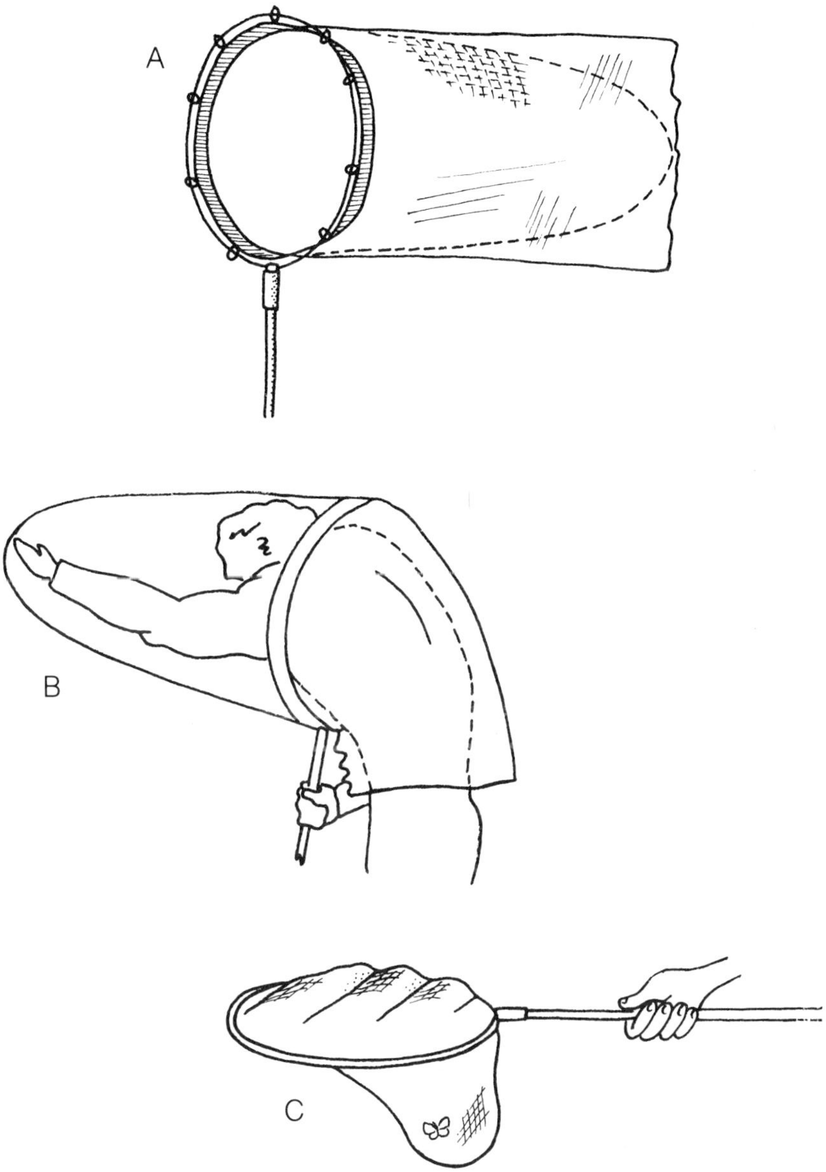

Fig. 8.6. A. Butterfly net with an outer cover made of strong cloth. **B.** The method of capturing small or other, agile insects from the net. Note the outer cloth cover which is draped over the collector in order to prevent the escape of the specimens. **C.** The method of holding the net with a captured insect in it.

ends. It can be used as a sweep net and also, once small insects have been caught, can be pulled up over the collector's head, thus preventing the escape of the netted insects (Fig. 8.6B). These, usually small, can be extracted with an aspirator or sucking tube.

Use of the butterfly net requires more experience than at first realized. When the net has swooped upon the target the handle should be twisted so that the net will fall over the rim of the hoop, preventing escape (Fig. 8.6C).

Sweep net

The frame of this net should be somewhat sturdier than that of a butterfly net. The net bag should be made of strong, but not too thick, canvas (the shape and dimensions can be the same as those of the butterfly net). If the bag is sewn on the loop directly (which is desirable), it is best to sew a strip of strong calico all around the loop to prevent quick fraying of the material. The net is used as follows: the collector slowly walks through the field while continuously sweeping side to side with the net. The rim of the net disturbs the insects (usually small species) as it brushes through the vegetation and as they fall or take to the wing they are swept into the bag of the net. After 10 to 15 sweeping strokes, the bag should be examined and the specimens extracted. This can be done in a number of ways. Small, agile specimens are difficult to capture. Therefore, it is sometimes necessary to place the entire net in a large plastic bag and kill all the captive insects with vapours of ethyl acetate. After a few minutes, the entire contents of the net can be poured out onto a sheet of white cloth or in a white enamelled dish and the material sorted with ease. This however is not often necessary and the insects can be extracted by an aspirator or sucking tube. There are various designs for sweep nets, but the most commonly used is the simplest. Fig. 8.7 shows some of these various designs, and a method of collection.

Mobilized nets

It is sometimes necessary to collect a large number of insects in a reasonably short time. To carry out this, nets may be mounted on a vehicle (preferably a four wheel drive) and then moved through the field. A large aerial net can be mounted on top of the vehicle. This will capture flying insects. The net bag should be at least 1.5 m long. If it is shorter, many insects can escape. Sweep nets can be mounted on both sides of the vehicle (mounted on the front bumper bar). The bags of these must be made of strong calico to withstand vigorous bashing by the vegetation. The bags should be long (1.5 m–2.0 m) and their ends supported in outrigger fashion (Fig. 8.8) in order to prevent snagging. An alternative to this is to fit the bags with large screw top plastic

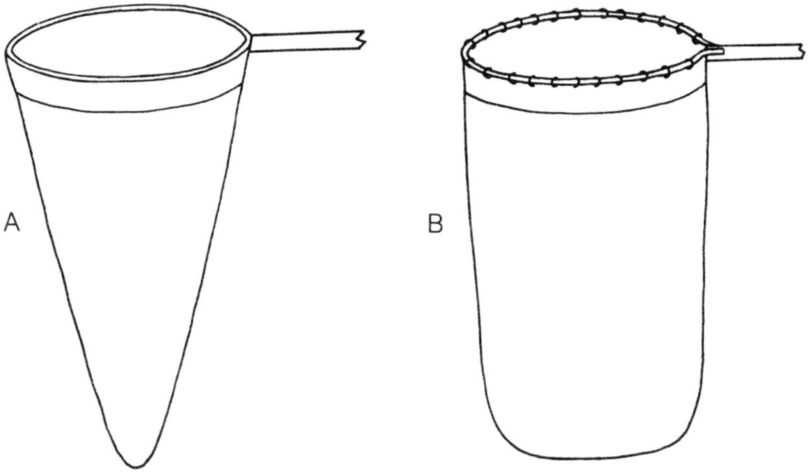

Fig. 8.7. A. Sweep net design. The pointed shape of the bag assists the concentration of the specimens at the bottom of the net. **B.** Sweep net design. Specimens are more widely distributed in this bag.

jars which collect most of the captured insects. If the bags are long enough, these jars can even be kept inside the vehicle. Then the collectors may travel a little faster so that the insects will be carried by air into the jars. This technique is especially useful for quantitative collecting of swarming insects and for general sweep netting on grassy plains.

Beat net or tray

A beat net or tray can be constructed very easily from thin canvas or similar cloth cut into a diamond or oval shape approximately 750 mm × 600 mm. Four double hems should be sewn at the furthest points on each axis so that two rods can be inserted crossways to form a shape similar to a kite (Fig. 8.9A). The beat net is then held under an overhanging branch or between shrub growth and the foliage is beaten with a stick, causing any insects to fall and land on the canvas where they can be picked off. Some collectors use a square beat net; some use a white umbrella with a modification to the handle so that it can be held upside down whilst beating the bushes (Fig. 8.9B).

Specimens can also be collected using a sheet of white cloth. The cloth is spread under a tree or shrub and the insects fall on it as the foliage is jarred (Fig. 8.10A). They show up well against the white cloth and can be picked

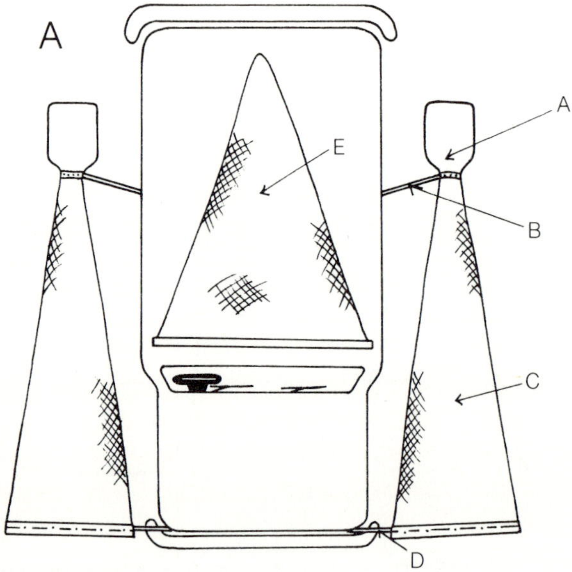

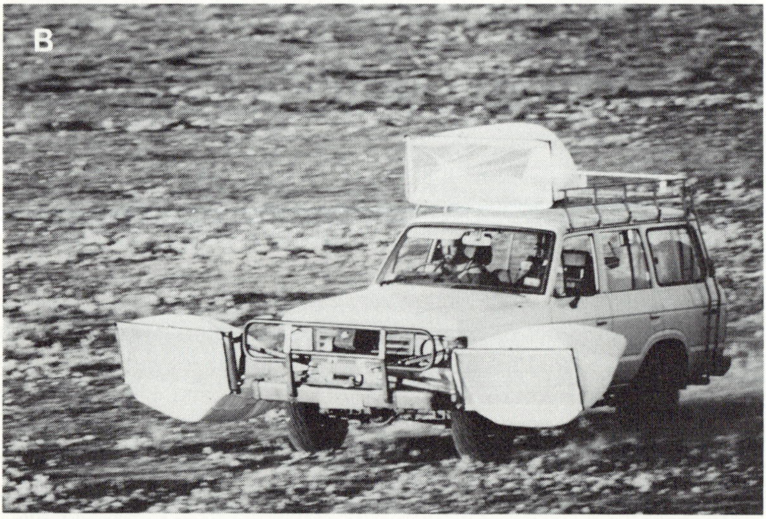

Fig. 8.8. A. Vehicle with fitted nets. A. Collecting jars. B. "Outrigger" for sweep net bag and jar. C. Sweep net. D. Fitting to hold net. E. Large areal net on top of vehicle. **B.** Vehicle fitted with sweep and areal nets in the field. Photo J. Fields. Courtesy of The Australian Museum.

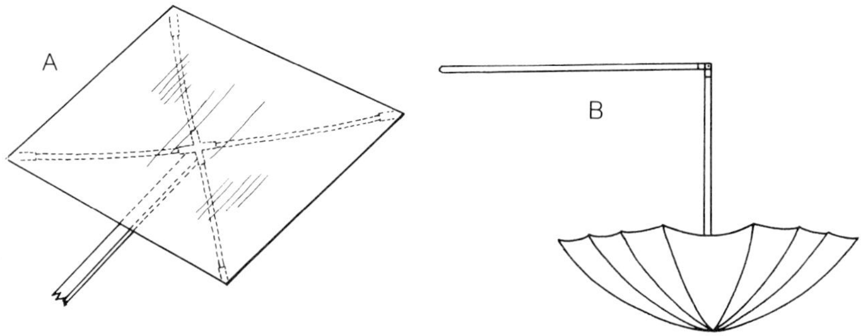

Fig. 8.9. A. Beating tray. **B.** Converted umbrella for catching falling insects.

up either by hand or, if they are small and numerous, with an aspirator. The cloth is easy to carry in a haversack, and so may at times be more practical than the net or umbrella.

The technique can be combined with poisoning. A tree or shrub should be selected (preferably one in flower and so attracting great numbers of insects) and the cloth is spread underneath it (Fig. 8.10A). A spraycan of household insecticide (containing tetramethrin and bioresmethrin) should be taped to a long pole. The jet of spray is started by taping the release button down. The can should then be raised near the flowers or parts of the foliage where the insects are congregating in the largest numbers (Fig. 8.10B). As much of the area as possible should be sprayed. Under optimal conditions (good weather, no wind, and well selected terrain) great numbers of insects can be collected in this way.

Aspirators

For small or fragile specimens, an aspirator or sucking tube is a valuable piece of equipment for any insect collector. There are many very good and well made aspirators on the market, but one can easily be made. The instrument consists of a glass or plastic transparent jar 7.5 cm × 2.5 cm with a rubber bung or cork in the top. Two holes must be drilled through the stopper to take two slightly bent glass tubes (Fig. 8.11A,B). One tube is short and has nylon gauze tied over the end inside the jar. This tube leads to a long rubber or plastic tube for insertion into the operator's mouth. The other glass tube is left open at both ends. The open ended tube is placed close to the specimen which can then be sucked into the jar. The nylon gauze prevents captured insects being sucked into the operator's mouth (Fig. 8.11C).

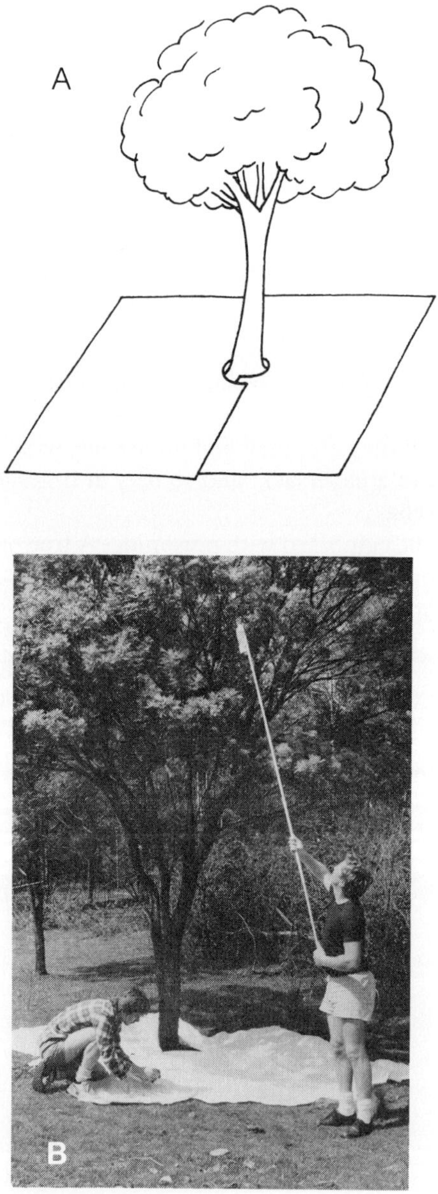

Fig. 8.10. A. The method of covering ground under a tree prior spraying. **B.** Spraying a flowering tree in order to kill insects. Photo G. Hangay.

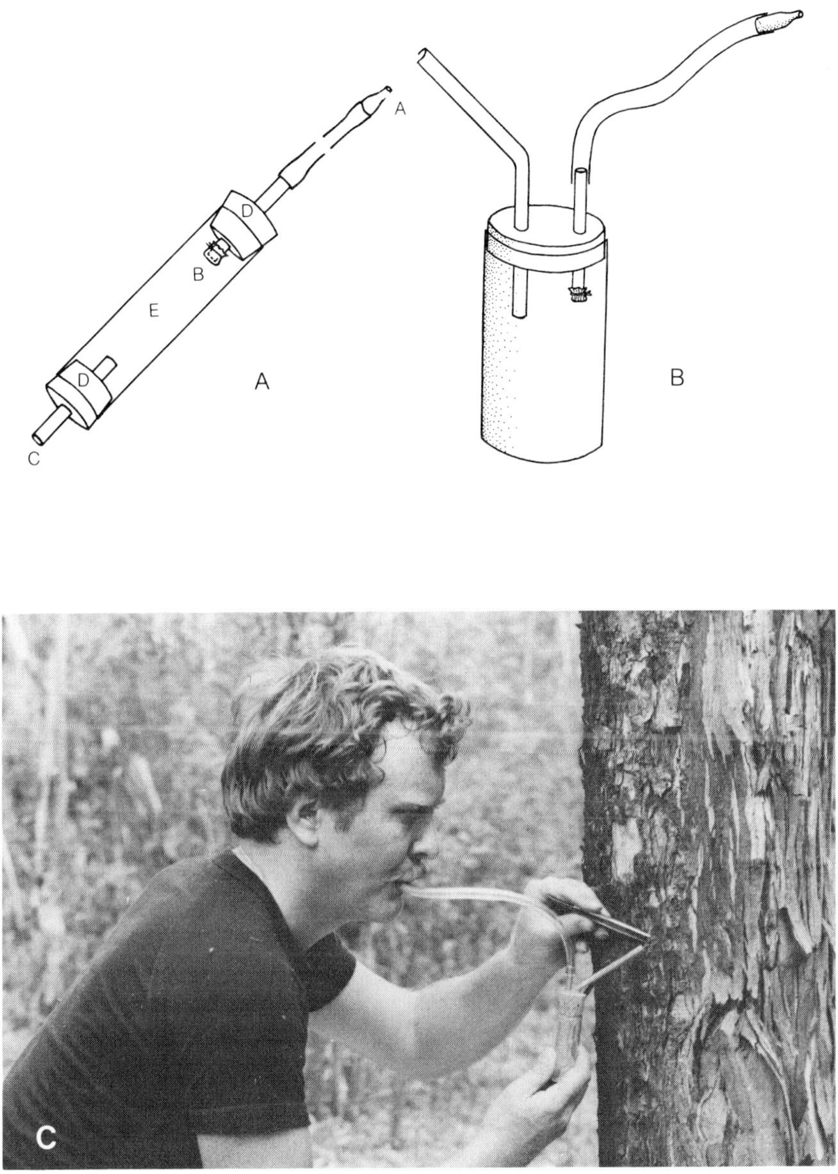

Fig. 8.11. A. An aspirator or sucking tube. A. Mouth piece. B. Fine mesh to act as filter. C. Point of entry for specimens. D. Cork or rubber stoppers. E. Glass or acrylic tube. **B.** Alternative design for an aspirator. **C.** The aspirator is a most practical piece of equipment for collecting small insects. Photo G. Hangay.

The aspirator can be used to pick small insects individually from the beat tray or for removing specimens from butterfly and sweep nets. Other specimens can be taken from flowers, bark, or anywhere insects live. The aspirator can also be used to pick up small invertebrates for other collections. When the aspirator has picked up a sufficient quantity of insects it should be emptied into a killing bottle.

Reitter sieve

Many insects and other arthropods live in areas of permanent humidity — under leaf litter at the base of trees, under rocks and stones, under fallen trees or branches, and under loose bark still adhering to trees. In order to collect specimens from these areas, large amounts of organic matter must be sorted. The Reitter sieve enables large amounts of organic matter to be collected and sorted quickly. Fig. 8.12A shows the Reitter sieve made up into two separate divisions. The top section is separated from the bottom by a mesh 10 mm + 10 mm which is connected to a handle. The entrance to the top section is held open with a thick gauge wire ring, also connected to a handle. This wire is flattened or bent slightly concave opposite the handle for easy positioning against tree trunks (Fig. 8.12B). The bottom section is tapered and tied so that sieved material can be transferred to Moczarsky–Winkler selectors, or Berlese funnels, for the separation of specimens. The sieve is operated by filling the top section with organic material and moving the two handles quickly backwards and forwards until enough small material has been separated from the larger twigs, leaves, or bark. The shaking should ensure specimens clinging to the debris fall through the mesh.

Leaf litter accumulates under trees, and provides a cool, dark, and damp environment for many insects and other invertebrates. There are several methods of separating the fauna from the leaf litter. The simplest method is to throw a handful of litter onto a white cloth laid out on the ground and to pick out specimens using fingers or forceps. For collectors who require information on population sizes within specific areas, the litter must be placed into a white dish and sorted carefully. All surfaces of leaves and twigs must be inspected before being removed from the dish. As specimens are collected, they can be placed into a killing bottle or reagent.

Another quick method is to place the litter into a Reitter sieve with a mesh size of 25 mm × 25 mm. When shaken, small animals and some leaf litter fall through, whilst larger vegetable matter is retained.

Berlese funnel

The Italian entomologist, A. Berlese, invented this instrument, which aids the sorting of animals from leaf litter and soil. The original design has been

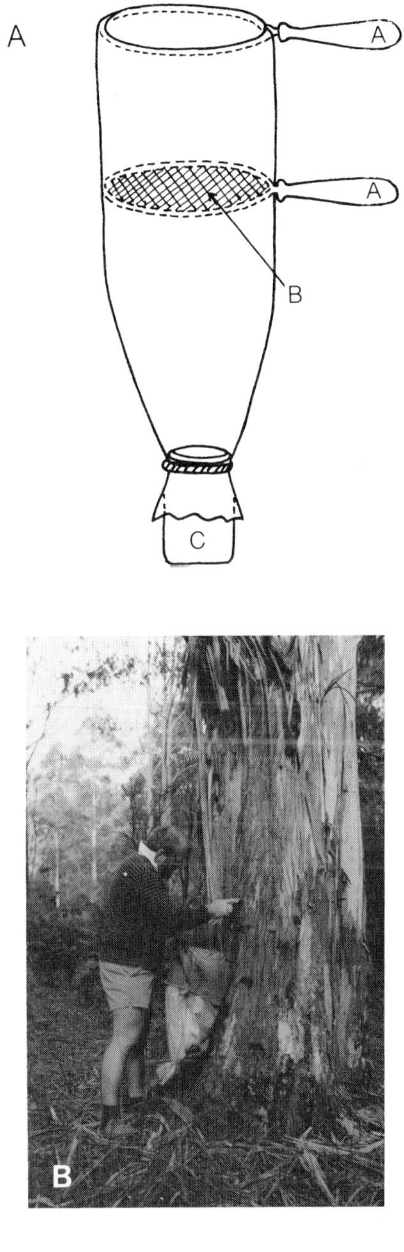

Fig. 8.12. A. A Reitter sieve or sampler. A. Handles. B. Wire mesh. C. Jar. The jar tied to the bottom of the bag is not absolutely essential. It is only used when small amounts of material are collected. B. Reitter sieve (without jar) being used to process flaking pieces of bark and debris underneath. Photo M. Dingley.

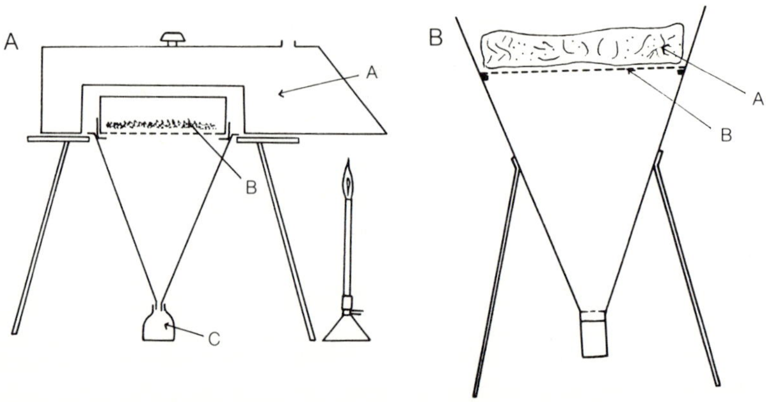

Fig. 8.13. A. Berlese funnel. A. Metal container with water which is heated continuously. B. Leaf litter on wire mesh. C. Jar to collect specimens. **B.** Modified Berlese funnel to process soil samples. No artificial heat source is used and the top of the funnel is left open. Slow drying of the sample forces the insects and arachnids to move towards the lower areas until they fall into the specimen jar at the bottom. A. Soil sample. B. Wire mesh.

modified by various workers. The most widely used variety consists of a funnel with a smooth inner surface, a sieve near its opening, a supporting frame, a heat source, and a collecting or preserving jar (Fig. 8.13). Leaf litter is spread across the sieve and the heat element is switched on. As the litter begins to dry out from the top, the animals move downwards to stay in the damp conditions. As the litter completely dries the animals fall through the sieve to collect in the jar. The procedure may take several days. If a collecting jar is used instead of a preserving jar, the specimens must be removed at least once a day for sorting. A Tullgren funnel is similar to the Berlese funnel, but it uses a light source to produce light and heat in order to sample the specimens (Fig. 8.14).

Regardless of which funnel is used for the purpose, they are more effective if the leaf litter is presorted through a Reitter sieve before being placed in the funnel. The Reitter sieve separates most of the small debris and animals from the larger pieces of the litter. Therefore, it reduces the volume of the material to be sorted and concentrates animal content.

However, when complete soil samples must be examined, this presorting is not desirable. A disadvantage of the Berlese or Tullgren funnels is that some of the insects, especially beetles which are very slow moving, cannot dig their way down through the leaf litter in time and die of dehydration or exposure

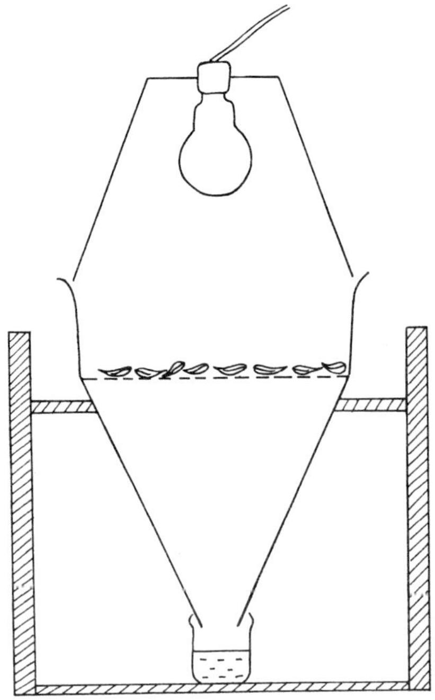

Fig. 8.14. Tullgren funnel. Based on the Berlese funnel, a 40 W or 60 W lightbulb is used to dry the sample until the specimens fall into the jar at the bottom of the funnel.

to heat. Others, which are not repelled by light, simply fly away through the gap between the funnel and the top part of the instrument.

Moczarsky-Winkler selector

The Moczarsky–Winkler selector is made from a square or rectangle of canvas. Sorting is automatic, but it takes several days to sort all specimens. The size of the selector can vary to suit individual needs, but a good size is 300 mm × 300 mm × 700 mm. A 6 mm × 6 mm mesh is held inside the bag (Fig. 8.15). The end of the bag can be tapered or tied around the top of a jar. This jar can be used to sort live insects at a later time or can contain 5% formalin. The bag and jar should be suspended above the ground and, as the moisture slowly departs from the litter (from top to bottom), the insects move downwards until they fall into the jar. The speed at which this works depends on the surrounding temperature; the warmer the air, the quicker the litter will

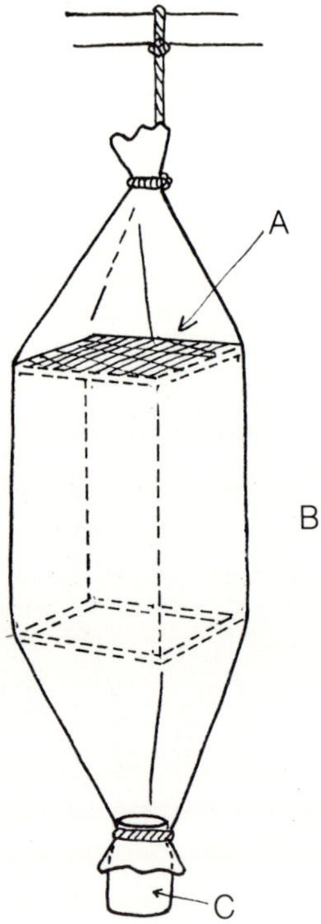

Fig. 8.15. A basic design for a selector, mostly used by coleopterists. A. Wire mesh to hold sample. B. Internal frame. C. Specimen jar.

dry out. If drying is too fast it can cause the premature death of specimens, which will remain in the unsorted material; if drying is too slow, fungus will develop, and this can also destroy the specimen.

Mesh bags can also be employed to hold the material. The bags should be 150 mm × 250 mm diameter and made of a relatively stiff material. The walls of the bags can be stitched together at two or three spots with one stitch at each place. This is to prevent the bags bulging when they are filled and allow the accumulation of material which can hold moisture for a long time. Usually, six bags can be hung in each selector, vertically (Fig. 8.16). To facilitate drying, the bags should be emptied in a container every day, or every second day, the material mixed, and the bags refilled. Any specimen which becomes visible during the emptying and refilling process should be captured. Premature death due to adverse conditions can contribute to loss of specimens, but the Moczarsky–Winkler selector is one of the most useful instruments for the entomologist, especially for the collection of beetles.

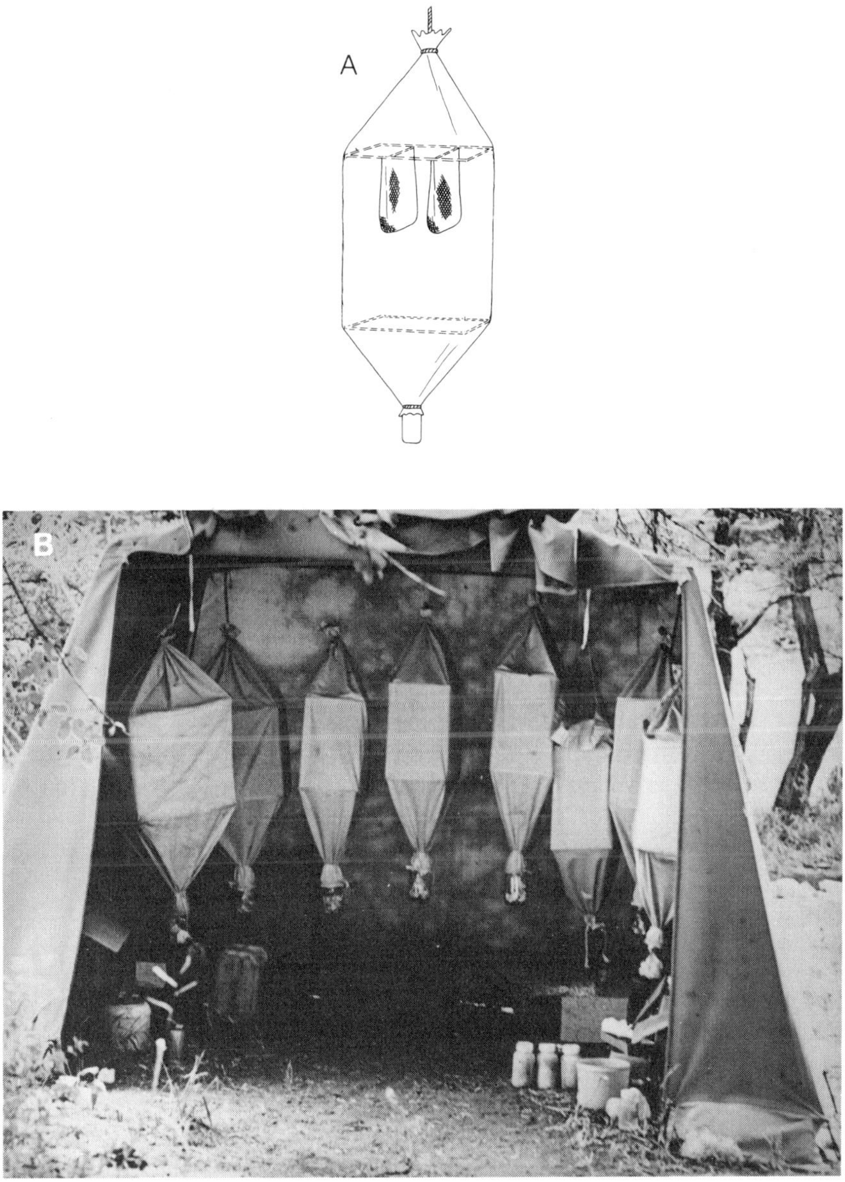

Fig. 8.16. A. Moczarsky-Winkler selector. The most frequently used selector, ideal for fieldwork. The sample is placed in net bags and suspended within the canvas bag. Most are designed to hold six bags. **B.** Laboratory tent with Moczarsky-Winkler selectors. Photo Dr S. Endrödy-Younga. Courtesy of the Transvaal Museum, Pretoria, Republic of South Africa.

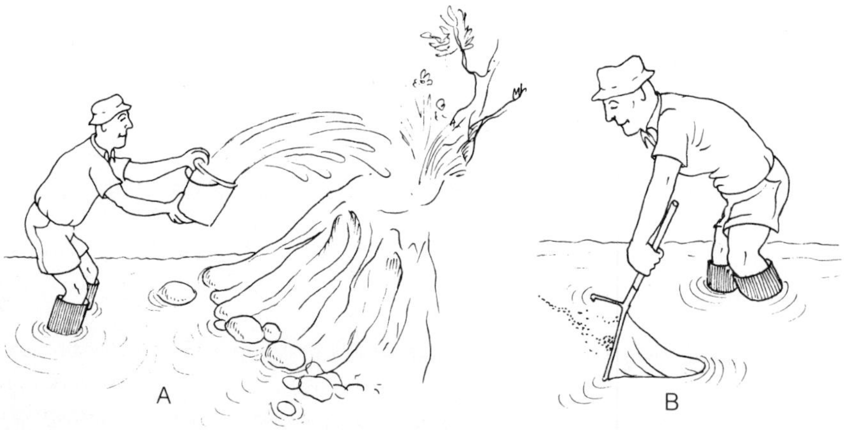

Fig. 8.17. A. "Shore washing". **B.** The insects, which were washed into the water are collected by a hand net. Drawing by G. Hangay.

Shore washing

This technique enables the collector to capture terrestrial insects and other terrestrial invertebrates that live around ponds and lakes. The method is as follows: find a lake or pond which has a bank sloping towards the water. Stand in the water and fill a bucket with water and throw the contents onto the bank. The water will drain back into the pond (Fig. 8.17). This should be repeated several times. The water will carry the specimens from the land and onto the surface of the pond, where they float helplessly. A skimmer can be used to collect all of the floating matter which should be transferred to a container for sorting at a later time.

Sorting can be done in various ways, but the most practical method is to place the material in an emergence box (Fig. 8.18) and, once the specimens have stopped emerging from it, transfer them into a Moczarsky–Winkler selector. The method can be applied to streams as well, although the assistance of another person is desirable. The process is started upstream, where one person throws buckets of water on the shore. The other person stands a few metres downstream and collects the floating debris with a net. If no assistance is available, a seine net can be spanned across the stream to collect the debris. However, some of the insects clinging to the debris are very agile and take to the wing readily. As more of the flotsam accumulates at the seine net, these can escape.

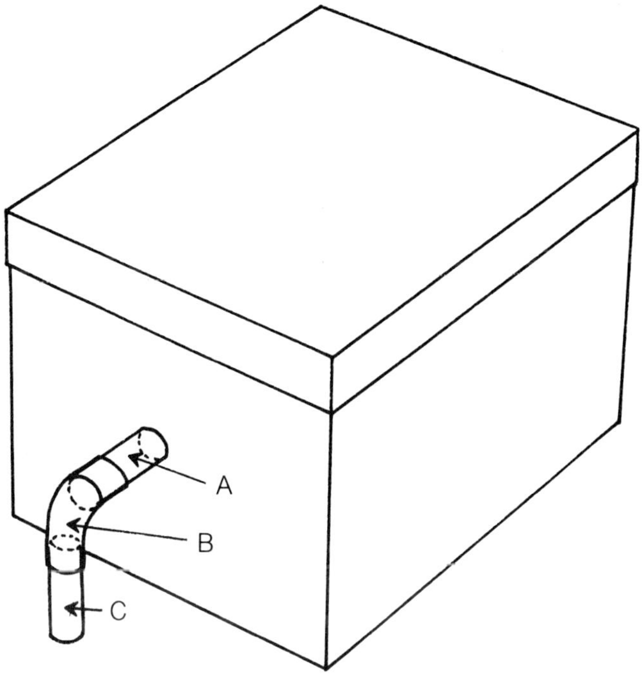

Fig. 8.18. Emergence box. A. Glass tube. B. Flexible, transparent plastic tube. C. Glass vial.

Emergence box

To extract the insects from the debris collected with the method described above, it is best to place the debris in a light tight box. The box should have a hole drilled into one side towards the top (Fig. 8.18), and this can be fitted with a short length of clear, flexible plastic hose. A glass vial should be attached to the end of the hose. As the freshly collected debris is placed into the box, many of the insects emerge from it and head for the light entering the box through the hole. They travel into the glass vial. Once 30 to 40 insects have accumulated in the vial, it can be changed, and the captured specimens can be preserved or kept live for further studies. Instead of the vial, a larger jar can be fitted to the hose. Some crumpled, wet absorbent paper should be placed in it to separate the specimens from each other. A larger jar can be left for longer. After a while (depending on the nature of the fauna, temperature, and relative humidity), insects stop emerging from the box. This does not mean that there are no more insects in the material. Many of them prefer to hide in the wet debris, away from the light. These can be sorted by placing the debris in one of the funnels or selectors described above. The sorting of wet debris can take a long time. Material should be skuffled occasionally to assist drying.

COLLECTING AQUATIC INSECTS

There are many habitats in which to search for aquatic insects. The simplest method of collection is to pull up aquatic plants and lay them on a white sheet or plastic tray. They can be inspected for larval or nymphal forms which will move around the vegetation and drop off onto the white background. Insects that live in sand can be found by dropping the sand into a bucket containing water. The sand will sink quickly, leaving the insects floating on the surface.

Dredging

There are many dredge nets available on the market, so it is best to buy or make one to suit individual needs. These nets are used to dredge the bottom of lakes, ponds, and streams and should be made of fine material and preferably covered by a canvas bag to prevent snagging and tearing. A bar on the base of the open end disturbs rocks and other bottom material when it is pulled along the bottom of the river or pond. Specimens are forced upwards into the path of the net. After dredging, the contents of the net are emptied into a white tray and sorted through for larval and nymphal forms.

Dip net

This net is useful for aquatic insects as well as many other kinds of aquatic animals. Basically, it has a depth equal to the diameter of the ring and is made of marquisette or bobbinet. The top of the net should be reinforced with canvas. A canvas bag should hang loosely around the net to prevent tearing (Fig. 8.19). It is swept through vegetation and finally upwards through the water. If the mesh size is too small it is difficult to sweep through the water, but if the mesh size is too large the specimens will escape. Other useful tools for this purpose are flour sifters and tea strainers. They can be very useful for collecting insects that hide under submerged rocks and logs.

Seine net

This is a rectangular piece of netting attached to two vertical poles. The poles can be pushed into the bed of a stream so that the net spreads out across the stream flow. Any small specimens carried by the current are caught in the net. If rocks, logs, vegetation, or other debris upstream are disturbed, any animals that make their home in such places will be carried down to the net (Fig. 8.20).

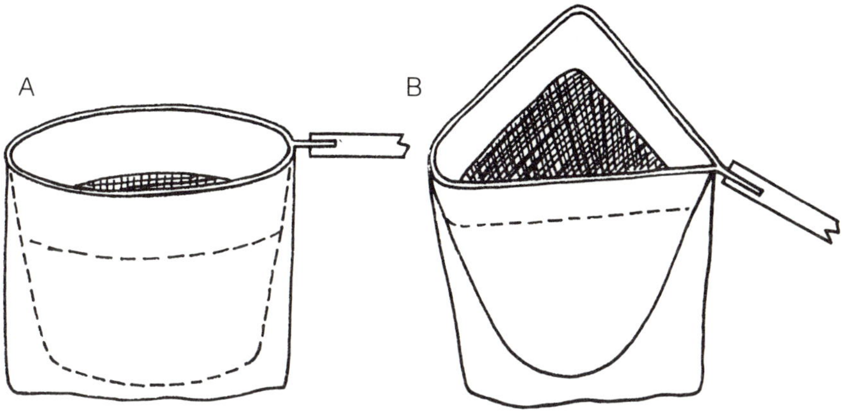

Fig. 8.19. A. Net for collecting aquatic insects. Circular design. Note the outer cloth covering, which protects the net. **B.** Net designed to collect aquatic insects which live on the surface of the water.

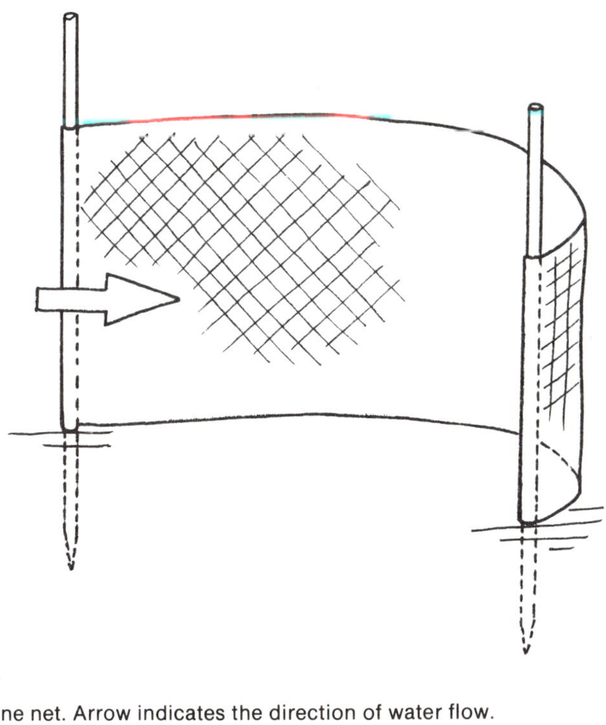

Fig. 8.20. Seine net. Arrow indicates the direction of water flow.

COLLECTING MOSQUITOES

Mosquitoes must be sought with some assiduity. However, the ideal source of adults is laboratory hatched larvae which have been field collected, either as larvae or egg rafts. Many species of mosquito are extremely habitat specific. Some breed only in water trapped in the base of bromeliaed leaves, others in water in tree holes, and a few are even specific to the temporary pools of rainwater on dead leaves on the forest floor.

If egg rafts or larvae are taken back to the laboratory, they can be bred quite successfully. In this way accurate locality and habitat data can be obtained, and the individual successive casts from each larval (and the pupal) moult can be kept (either under cover slips on glass slides, or preserved and tied to their resulting image).

COLLECTING PARASITIC INSECTS

Parasites of vertebrate animals are best collected in collaboration with collectors of these animals. A description of a trap which secures parasites of trapped mammals is given in Volume 1, Chapter 6. Freshly killed mammals and birds, if placed in a cloth bag, usually yield insect parasites. Most of these leave the host animal soon after its death, and can be collected from the bag. Some, however, do not do so and these must be collected by carefully combing through fur or plumage and picking out the individual specimens with a pair of fine forceps or the wetted tip of a camel hair brush.

Nests and the litter in burrows can also yield parasitic insects. Material suspected of hosting parasites should first be placed in an emergence box, then in Moczarsky–Winkler selectors. The method of collecting arthropods which live as ectoparasites on vertebrates is described later in this chapter.

Parasites of invertebrates are more difficult to collect. Many adult forms can be collected through the previously mentioned methods, but, if the worker wishes to study and collect information about the parasite's life history, the host animal must be collected and kept alive as long as necessary for the development of the parasite. The most common parasites are parasitic wasps. Some of these can complete their development in the egg of another insect species, but more often in the larva or the pupa. If the collection and study of these insects is embarked upon, the collector must be familiar with the methods of collecting and rearing the host animals. Information on the subject can be gained from the multitude of appropriate publications.

ATTRACTING AND CAPTURING INSECTS

There have been many techniques and pieces of equipment invented by specialized collectors and students of entomology. Descriptions of these can be found in the appropriate literature. Here, only those which are the most frequently used and can be considered as well proven methods for general collection work are mentioned. These standard techniques are usually modified by individual collectors.

Light traps

These traps work on the principle that many insects are attracted to light at night. Once insects have been lured to the source of the light, they can be captured manually or by mechanical means.

Mercury vapour lights are most effective for this type of work. There are numerous types on the market. Mercury vapour lights usually operate from mains electricity or a portable generator. Most designs necessitate the use of a choke. The user must make sure whether or not a choke is needed before switching the light on, otherwise the bulb may blow up. Blended or mixed lights are also available. These do not need a choke and can be connected directly to the power supply. Mixed lights are almost as effective as mercury vapour lights. Black light fluorescent tubes are also suitable for insect collecting. Their disadvantage is that it is almost impossible for the collector to distinguish the specimens as they come for the light, since this light is not received well by the human eye.

The simplest trap consists of a light source and a white sheet. The light should be suspended from a suitable prop, such as a low tree branch. A white cloth, two metres by two metres, should be hung, vertically or at a slight angle, using two long poles and guy ropes so that part of it lies on the ground. As the insects come up to the light they can be picked up by the collector (Fig. 8.21A). There are many designs for light traps which capture and retain specimens, without the need for personal attendance during the night, and some are shown in Fig. 8.21.

One basic design consists of a large (450 mm × 450 mm × 600 mm) wooden or metal box with a circular hole on the top. A funnel is fitted in this hole and the light is fitted in the funnel. The slope of the funnel should be 60° and the bottom opening 60 mm diameter (Fig. 8.22). The insects approach the light, usually circle it, and eventually fall into the funnel and, consequently, into the box. Depending on the nature of the work, the specimens can either be killed by toxic fumes on entry or retained alive until they can be examined and then released.

If toxic fumes are used, a layer of plaster of Paris should be incorporated at the bottom of the chest — not unlike in a killing jar — to be saturated by ethyl acetate or carbon tetrachloride (the latter is especially useful since its fumes are heavier than air, and therefore stay in the chest). If all the interior surfaces of the box are lined with plaster the fumes will be more efficient in killing insects. The greatest disadvantage of these unsupervised traps is that

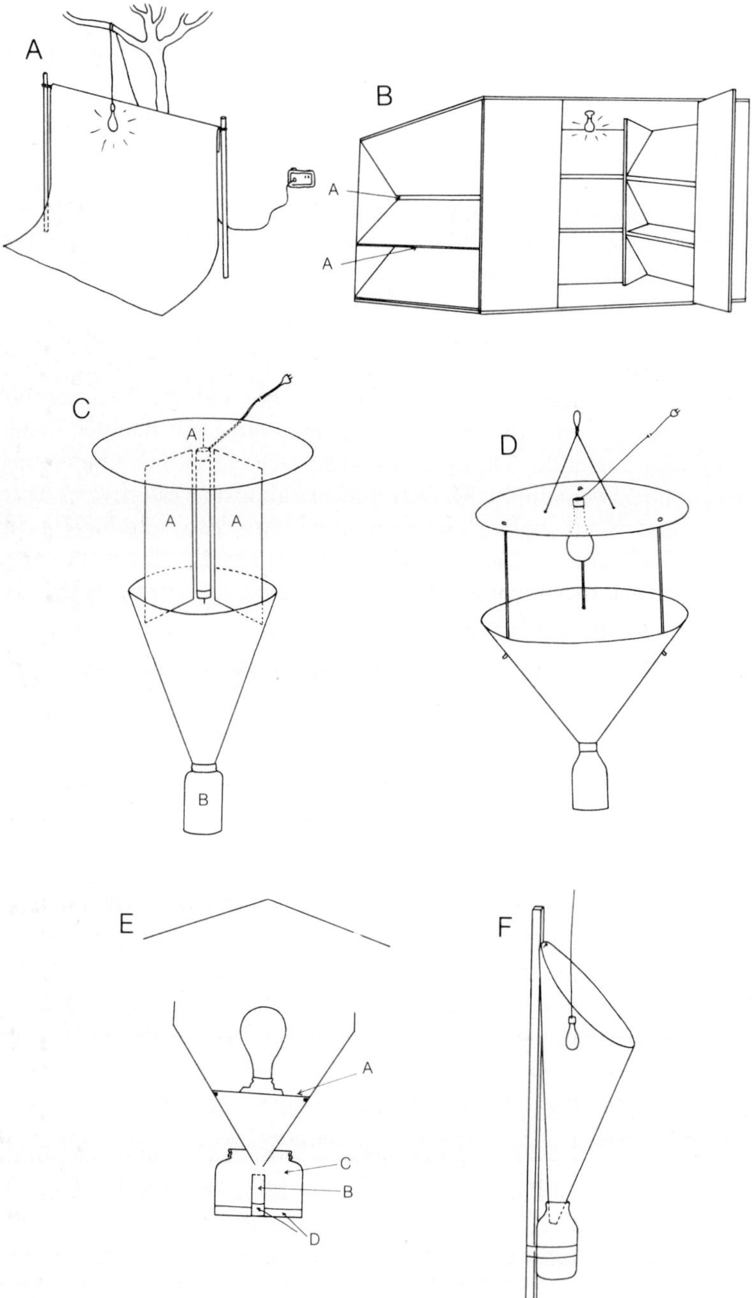
A
B
A
A
C
A
A
A
B
D
E
A
C
B
D
F

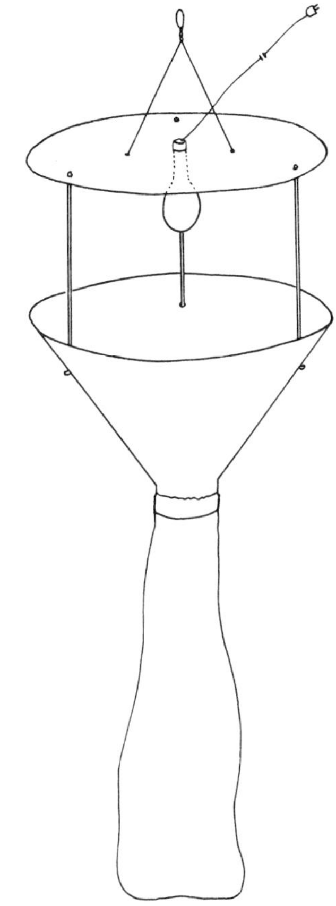

Fig. 8.22. Light trap with large canvas bag. Moths may settle inside the bag which contains no killing agent. After examination, the catch can be released.

Fig. 8.21. A. A simple method of setting up the light, and sheet for attracting insects at night. **B.** Shed-like light trap. The light is visible through the glass or acrylic wall panels and the arriving insects enter the shed through the gaps (A) between the panels. **C.** Light trap. Arriving insects collide with baffles (A) and fall into the killing jar below (B). **D.** Light trap. Insects settle on inside wall of funnel, become anaesthetized by fumes of killing agent from below and fall into the jar. **E.** Light trap. Attracted insects fall through mesh (A). Small beetles, bugs and alike fall into small vial (B) moths, which may fly more readily find their way into larger jar (C), thus become separated from the other more robust insects which may damage them. (D) Layer of plaster of Paris, saturated with killing agent. **F.** Simple light trap. Jar contains killing agent.

if beetles (which can take a relatively long time to die) fall in it they can ruin the vulnerable Lepidoptera specimens. In some areas large Orthoptera can also cause severe damage to other more fragile insects.

Various baffles and shelves can minimize this danger; however, no trap has yet been designed that overcomes this problem. The authors are presently experimenting with wire mesh placed horizontally across the box (mesh size 25 mm × 25 mm) on top of which is placed a loose layer of wood wool. As the moths fall into the trap, they are overcome by the fumes and come to rest on top of the wood wool. If beetles or other similarly damaging insects fall into the trap they try to find their way to the bottom, going into the wood wool, and through the screen, to congregate at the bottom of the box. It is hoped that in most cases they get through the wood wool (where all the dead moths are lying) without inflicting serious damage. Most of the damage occurs when beetles repeatedly walk over Lepidoptera specimens.

COLLECTING WITH LIGHT

Every collector who is experienced in field work usually adheres to his or her methods and often considers variations of the same method ineffective. As an example, the method described below, which was originally developed by Dr A. Vojnits, Curator of Lepidoptera in the Hungarian Natural History Museum, is practised by the authors with minor, but important, modifications. These modifications assist the collector to obtain maximum numbers of specimens and keep them in perfect condition until further treatment. The routine is as follows:

For optimal results a moonless, humid night should be chosen for collecting. A period just before rain is ideal, or, in arid areas, just after a good soak. Cold windy nights during a long, dry period with a clear, moonlit sky are the least promising. Undisturbed, natural areas with varied flora, preferably near a body of water (fresh, not marine) are good for collecting, while heavily disturbed agricultural or urban areas are not. The lamp and white sheet should be set up in a suitable spot. The rest of the quipment can be laid out on a portable table nearby, with folding chairs provided for the collectors. Other equipment should be stored so that everything can be found quickly in case of emergency (breakdown of light or sudden rain). Electric torches should be placed to be located in the dark if necessary. It is most important that there are a sufficient number and type of killing jars available. For average collecting under good conditions, one can anticipate collecting some hundreds of specimens. In the tropics, the number could be several thousand in one night. In order to process large numbers without damaging the specimens, one collector needs at least four small (0.3 L) cyanide jars and two large (1 L) cyanide jars. Two or three additional jars can be prepared with ethyl acetate and a few glass vials (2.5 cm diameter) containing 70%–80% ethyl alcohol and with plastic stoppers should be ready.

The light should be switched on at dusk (not only after dark) to attract the early flying insects. As a rule, the quick moving individuals are picked up first (that is, Carabid beetles, horse-crickets, and so on, which usually do not sit still but tend to run around aimlessly and then disappear). Great care must be taken to use appropriate killing jars: alcohol vials should be used for small insects; ethyl acetate jars should be used for the medium size and large insects of durable nature; and the cyanide jars should be used for Lepidoptera. The latter group requires the most attention, since incorrect handling can render the entire catch useless for further study. The proper method is to capture two to three specimens in one of the small jars, put the jar on the table, take the next jar, repeat the capturing, and keep doing so until all four jars are filled with two or three specimens. The contents of the first jar can then be emptied into one of the large jars and the small jar used again to capture two or three specimens. Repeat the process with the other small jars through the entire collecting session. This way the specimens do not damage themselves and each other in the jar. Once they are immobilized they can be stored on top of each other in the large jar.

Casual observers and assistants must be prevented from moving the large jar to inspect the catch. This can ruin the specimens; as they rub against each other the scales are rubbed off their wings. A Lepidoptera specimen should never be handled by hand, except when injecting large individuals (see below). If assistance is available, roles should be defined; for example, one person should capture Lepidoptera, and the other everything else. In this way the proper sequence for using the small jars will not be disturbed and specimens will not be damaged.

Large specimens can be injected. The best method for this is to allow the insect to settle on the sheet. Then, in the case of a large moth, carefully clasp it between index finger and thumb, just at the thorax (Fig. 8.23). With some experience one can learn how much pressure to apply to hold the specimen securely, without causing too much damage to it. If the syringe is held in the other hand at the same time, capture and injecting can be done in one swift action without damaging the specimen. Overinjecting will cause a specimen to drip, which is damaging for other specimens stored nearby.

Large beetles, which can cause much damage to more fragile specimens in a killing jar, should also be injected. They will not die as suddenly as moths, but at least will be immobilized. This is especially important with the largest Dynastids and Cerambycids in tropical areas. On some nights, especially before a rain storm, thousands, even tens of thousands, of small insects (mainly Diptera, Coleoptera, and Hemiptera) could come to the light. It is impossible to collect selectively on such occasions.

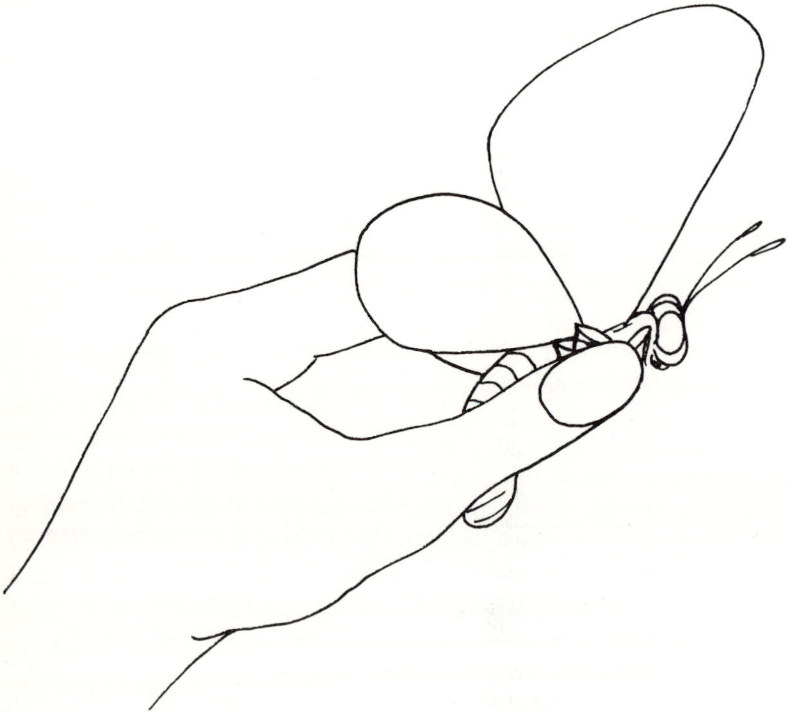

Fig. 8.23. The method of squeezing the thorax of a large Lepidoptera.

If the material is to be captured, as it always should be in entomologically unexplored areas, the best and quickest method is to capture most of the Lepidoptera from the sheet with the above method, then spray the rest with a household insecticide. The two bottom corners of the sheet should be lifted and the dead insects shaken into the middle. If an assistant is available, he or she should hold the sheet; a teaspoon can now be pressed into the deepest part of the cloth and vibrated. Insects will quickly fill the spoon and can then be transferred into a vial of alcohol. This process should be repeated until most of the insects have been collected from the sheet.

Alternatively, a collecting jar can be fitted to the sheet. To do this, suspend the sheet as it would be during night collecting. Mark the centre of the part of the sheet that lies on the ground and sew the lid of a jar with the centre underneath the sheet at the marked position. A hole will have to be cut into the sheet to allow the lid to be sewn. When the sheet is set up for collecting, the jar can be placed into a hole in the ground and screwed onto the lid. After insects have collected and been sprayed with insecticide the sheet should be picked up and the insects funnelled through the hole and into the jar.

If a mass collection such as this is not required, insects can be collected singly by scooping those wanted from the sheet with an open vial. An aspirator can also be used, or, for the smallest specimens, a wetted fine brush.

After the collecting has been completed, the Lepidoptera material must be packed for the journey back to the laboratory. The specimens cannot be transported in the large jars since they would rub against one another. Other insects collected during the session can be left in the collecting (killing) jars and in vials until preparation takes place in the laboratory.

Not all nocturnal insects are attracted to light. However, many of these can still be caught with the aid of light. Night flying moths can be detected with a hand held spotlight and captured with an aerial net. Ground dwelling insects, especially the larger species, can also be spotted at night. For this, a strong portable light is needed. A kerosene pressure lamp seems to be the most suitable. Sand dunes and other environments where the ground cover is sparse are good collecting areas. On sandy ground, it is often possible to follow the tracks of an insect.

BAITING INSECTS

In order to attract insects, a variety of substances can be used as bait. Many species of moths, which are otherwise difficult to encounter, can be lured with a technique commonly called "sugaring". The bait is prepared from: 200 mL beer, 2 tablespoons sugar or honey, 1 tablespoon molasses; and a few pieces of fruit. This should be mixed and allowed to mature in a closed container for two days. Pieces of sponge 20 mm + 40 mm should then be tied to 400 mm long pieces of string. The sponges must be saturated with the bait and tied to the branches of shrubs and trees in convenient spots, preferably along a clear path which can be followed even in the dark (Fig. 8.24). At dusk and after dark the moths are attracted by the aroma of the bait. They hover around the sponges or settle on them. The collector can carefully capture the moths using a wide mouthed killing jar. A net can also be used. Peterson (1959) describes one with a fitted light (Fig. 8.25), which enables the collector to inspect the catch immediately. It is practical to place a white cloth underneath the baits. This assists the collector to spot specimens which may fall from the bait. Some species become intoxicated quickly from the bait and sit quietly on it or simply fall to the ground. Some others seem to hold their liquor better and can be startled by clumsy handling or light.

The same "sugaring" liquid can also be used, especially if it is made a little thicker by adding more honey or molasses, on tree trunks. The liquid is simply painted on the bark in the hope that it will attract insects. In tropical

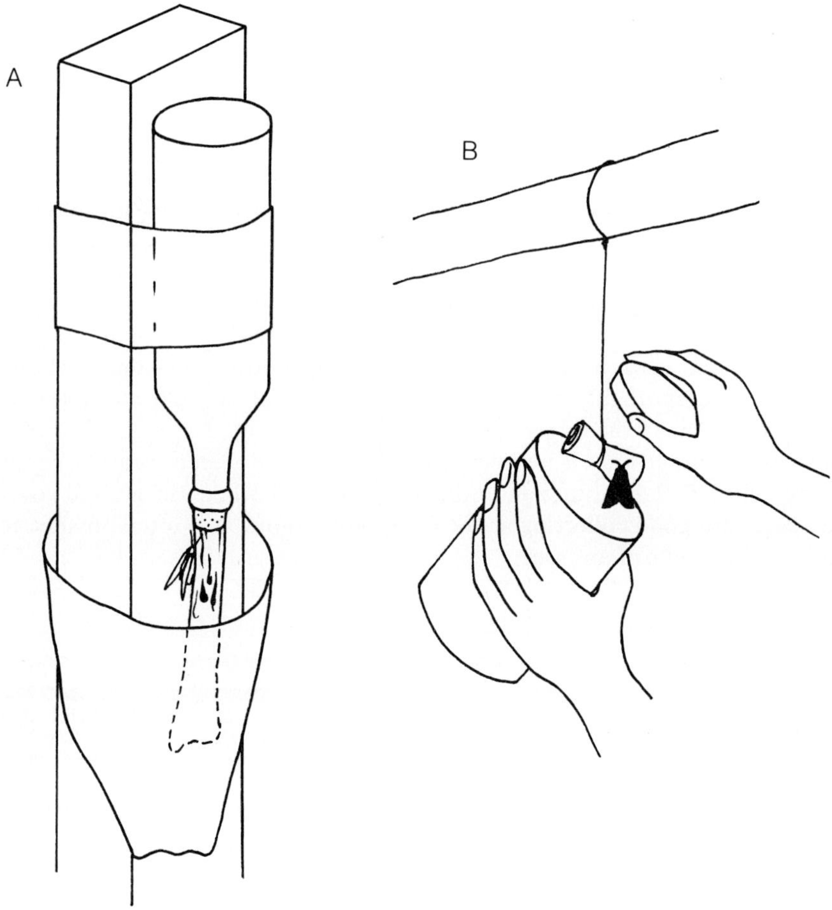

Fig. 8.24. A. The method of capturing a moth from bait. **B.** Baited trap. Bottle contains sugaring mixture, which slowly soaks through strip of cloth fastened to the bottle with drilled cork. As moths become intoxicated from the bait, they fall into the bag below. Trap must be emptied frequently.

and temperate areas, this can give excellent results; spectacular specimens of Lucanids and Dynastids can be collected this way, as well as many other species of insects.

Strong cheese, urine, and human faeces can also attract insects, especially butterflies. A common, although somewhat unsavoury, practice of butterfly collectors in tropical areas is to designate particular spots (usually bare patches of ground) for urination. The cumulative effect of the urine brings in a good variety and number of butterflies.

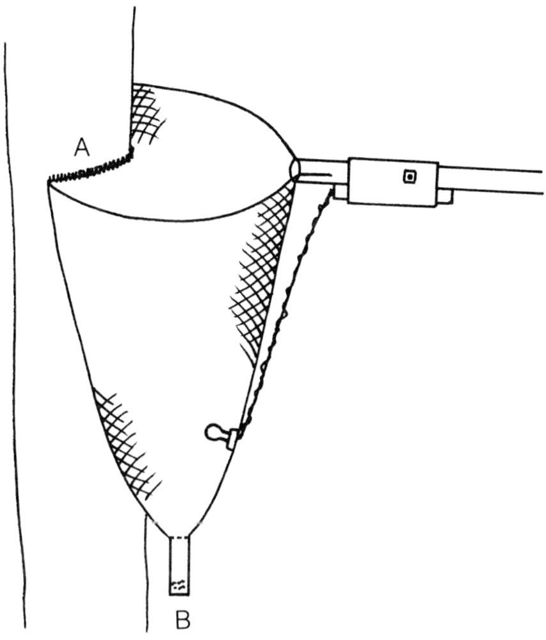

Fig. 8.25. Net with battery operated light. This net is designed for night collecting and to capture insects from tree trunks. The flexible section (A) of the frame allows the collector to press the net against the trunk of a tree. The captured insect falls into the small jar (B) fitted to the bottom of the net.

CARCASSES

Decaying carcasses are favourite collecting spots for entomologists. The fauna of a small carcass can be very much different from that of a large one. The age of the carcass is also an important factor. There are different insects on a relatively fresh cadaver, compared with those living on or under an old dehydrated one. In order to capture all insects associated with the carcass, a fine wire mesh tray should be dug 300 mm into the ground. The carcass can then be placed on top of this and allowed to decay. The colonizing insects can be picked off as they arrive. Finally, as the carcass is dehydrating, the last inhabitants dig themselves into the ground underneath. These can be captured by lifting the mesh out and sorting the soil from the insects. If no wire mesh is available, or the carcass was found in the field, it always pays to turn the carcass over, inspect its underside and dig the ground underneath.

PIT TRAPS

Many ground dwelling insects can be collected with the aid of these traps which consist of glass or plastic jars or buckets of 2 L capacity (plastic ice cream containers are ideal) sunk into the ground to leave their mouths 30 mm above ground. A small slope of soil should be built up to the rims. This is to prevent flooding of the traps by surface water. As they walk about the ground searching for food, a mating partner and so on, insects fall into the containers and are unable to escape as they cannot climb the smooth sides. If the traps are covered with a lid or flat stone (lifted on one side), dehydration of the captured specimens is unlikely and the danger that the containers will fill up with rainwater is minimal. If live specimens are required, no preserving liquid should be used in the traps. In this case, the traps must be inspected every day. For quantitative survey or when the traps cannot be inspected every day, a great variety of preserving liquids can be used. Preserving liquid should be poured into each container to a depth of 30 mm–50 mm. The cheapest preservative is 2% formalin. A disadvantage of this substance is that it hardens proteinaceous tissues permanently and the specimens become fixed in unnatural positions. Insects collected this way cannot be set nicely and could be cumbersome to handle. They also can lessen the aesthetic values of a collection. Ethylene glycol, according to Endrödy-Younga (1979), is the most suitable preserving liquid. It is odourless, very slow evaporating, and provides perfect preservation in the field to a tested period of up to three months, provided that the liquid can cover the trapped specimen and is not diluted.

The potency of the pit traps can be greatly increased if they are used with bait. A very effective bait consists of rotting flesh, human faeces, beer, and sugar, mixed in equal proportions. A fermenting mixture of chopped fruit, beer, sugar, and rum is also effective. Strong cheese, fish, and the dung of various large mammals can also be used. The bait should be placed in a tall, narrow container in the middle of the trap (Fig. 8.26). The insects are lured into the trap by the odour, but, as the bait container is placed in the middle, they usually fall into the preservative. Although baited traps can be used without a preservative, flying specimens often escape from them.

The employment of a small drift fence can increase the number of trapped specimens. The fence should be made of metal or stiff plastic flyscreen, 300 mm wide, and as long as the trap line. The fence has to be erected with pieces of wire (3 mm diameter, 500 mm length — Fig. 8.27) for support, and spanned across each trap. The bottom of the fence must be buried in the ground. Insects on the ground reach the fence. They try to walk around the obstacle, and fall into one of the traps.

The erection of a drift fence can be hard work, especially in a hot climate, but the results are usually rewarding.

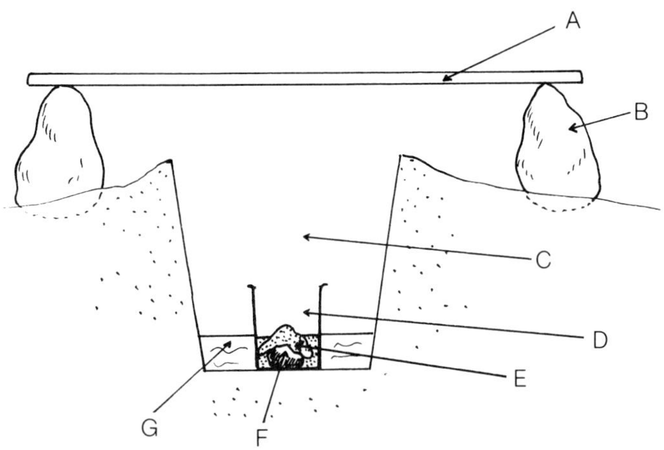

Fig. 8.26. Baited pit trap with preservative. A. Cover to keep rain out. B. Small stone to support lid. C. Container sunk into the ground. D. Bait container. E. Bait. F. Small stone to weight down bait container. G. Preservative.

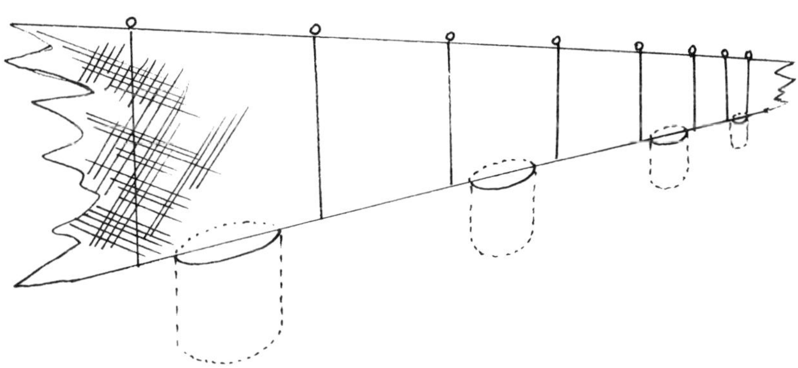

Fig. 8.27. The method of arranging a drift fence and series of pit traps.

DUNG TRAPS

The use of dung to trap insects may not appeal to many, but to catch insects such as flies it is a must. A trap is constructed of mosquito net so that a tray containing the dung is held under a funnel shaped section which is ultimately the passage the insects follow to become trapped in a larger chamber. The trap should be hung from a branch of a tree and inspected every day. (Fig. 8.28).

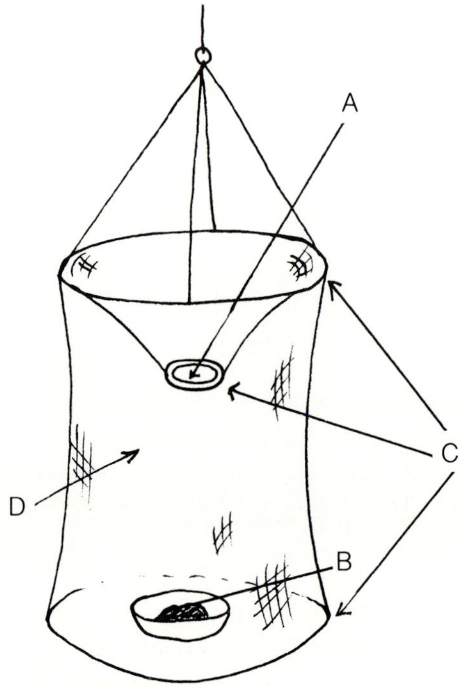

Fig. 8.28. Dung trap. Traps of this kind are usually baited with decaying meat, human faeces or strong cheese and are especially suitable for trapping flies. A. Entrance to the trap. B. Bait. C. Wire loops. D. Gauze.

Human faeces are very attractive for many flies. Instead, a piece of rotting flesh may also be used, although in many areas this will attract mostly blowflies.

ATTRACTING WITH FEMALES

This is a very old, but very effective, technique for the capture of many species of insects. Once a female of the species has been captured, she must be placed alive in a wire mesh cage and put out to attract males. The arriving males can be collected with an aerial net. This technique is mostly practised by butterfly collectors.

TRAPPING FLYING INSECTS

To collect insects which are active during the day and travel on the wing, like many of the Diptera and Hymenoptera, the Malaise trap should be used. The

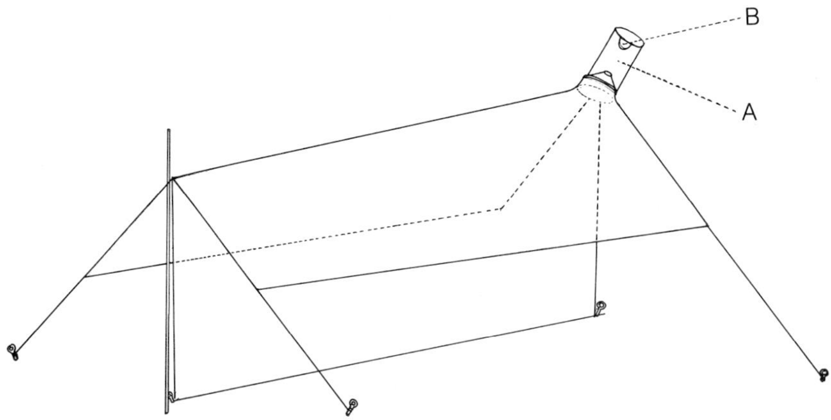

Fig. 8.29. Malaise trap. The main principle of this trap is that the entering insects usually endeavour to reach the highest point within the enclosure. A. Specimen chamber. B. Killing agent (usually ethyl acetate).

trap resembles a tent made of fine netting (Fig. 8.29) and should be set up in a spot frequented by flying insects. With experience, the collector can recognize the flight paths of insects, which often are near a body of fresh water. If the trap is set up in a "good" spot, the results can be astonishing: thousands of insects can be caught in a day with little effort. Wads of cotton wool or tissue paper soaked in ethyl acetate kill the specimens which enter the collecting chambers. In warm weather, the ethyl acetate has to be replenished from time to time. Most specimens caught in a Malaise trap are small and dry out rapidly. The collecting chambers should be emptied as soon as a number of insects has accumulated.

PRESERVATION AND STORAGE IN THE FIELD

PINNING

If relatively few insects were collected on a field trip, most of the specimens can be pinned into a storage box immediately. The box should be strongly constructed of timber, with white paper covered cork or rigid foam lining. Some boxes have the lid lined with the same material to double the pinning space. The internal depth must be enough to accommodate the largest size pin upright when the lid is closed (Fig. 8.30A).

A cube of naphthalene or a lump of paradichlorobenzene should be fastened in one corner under a small piece of fine wire mesh (Fig. 8.30B). Alternatively, small sachets may be made from strong material such as cotton; these can

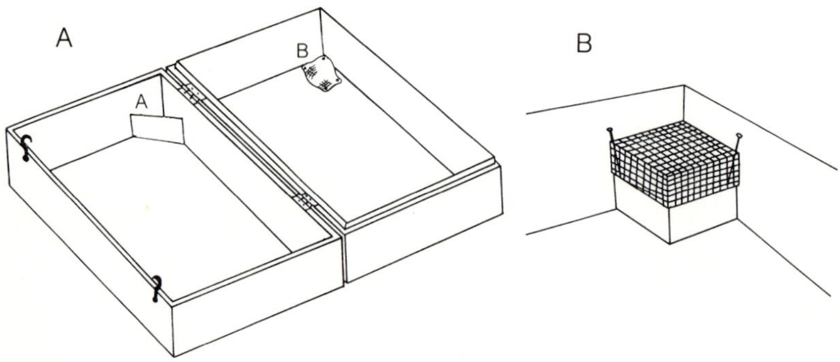

Fig. 8.30. A. Storage box, with alternative naphthalene containers. A. Corner partitioned off to hold mothballs or cake of naphthalene. B. small gauze bag to hold naphthalene flakes, fastened by pins. **B.** Small naphthalene container with removable wire mesh lid, positioned in a storage box.

be filled with naphthalene and pinned into one corner of the box to keep out insect pests. These boxes may be bought commercially, but can just as easily be made. A common size is 350 mm × 250 mm × 85 mm.

The method of pinning and the selection of pin sizes are explained later. The disadvantage of this practice is that pinned specimens do not travel well; they may spin around on the pins (this can be prevented by pinning them with more pins) and suffer damage with accidental bumping or jarring. The pinning box is also an awkward piece of luggage.

The Microlepidoptera is perhaps the only group of insects to benefit from immediate pinning. These minute, extremely sensitive insects can be pinned more easily immediately after their death than later on in the laboratory. These insects dry out very rapidly and, if their wings are not spread and dried in position, it is very difficult to achieve. The minute wings can only be spread well if the insect is pinned. The best method is to press thumb and index finger together, place the moth in the groove formed this way, and pin the insect through the thorax with a minuten pin. The wings separate and spread if gently blowed upon. The moth is then pinned into the box. Since these small pins are very small and thin, it is often difficult to pierce the paper lining of the box; therefore, it is practical to use boxes without paper lining. Once the moth has dried with its wings spread it can be relaxed later on and set properly on a miniature setting board. However, it is often very difficult to carry out such tedious work in the field and transport safely the storage boxes filled with fragile specimens. The collector often must make some compromises and package the small moths in the same way as other, larger Heterocera.

LAYERING

The usual method of treating larger specimens is to lay them neatly on thin, even layers of cotton wool in a tight closing box. Once the layer of wool has been covered with moths (which should not touch each other) another layer of wool is placed on top of it and more moths laid down. This process can be repeated until the box is full. Paradichlorobenzene and thymol crystals keep fungus growth at bay. Fungus can ruin the entire catch within a few hours. To prevent this many collectors use wooden or cardboard boxes, but the authors have experienced very little trouble of this nature while using metal containers. Our specimens were packed in these containers between layers of cotton wool and, despite the high humidity and hot, tropical conditions (Sumatra, Borneo), the specimens kept well for weeks.

Another method of preventing the growth of mould is the British Museum mixture, which is made up as follows: Flake naphthalene is added slowly to a quantity of chloroform until the latter is saturated. An equal bulk of creosote is then added, and the mixture is kept well stoppered. The mixture is then painted on the inside of storage boxes and allowed to dry. The result is a thin film of creosote and naphthalene which prevents the growth of mould and attack by other insects.

Other orders of insects than Lepidoptera can also be stored with the layered method. Some workers prefer to use tissue paper instead of cotton wool.

If specimens (other than Lepidoptera) become mouldy, they may be painted with a solution of 1% thymol in water or in alcohol. Mouldy Lepidoptera specimens are practically impossible to restore.

ENVELOPES

Butterflies can be packaged and stored the same way, but many collectors prefer to store these and other winged insects, such as dragonflies, damselflies and larger wasps, in triangular paper envelopes. The wings of the insects are folded together over their backs, and the insect is placed in the envelope. The envelopes are made of greaseproof paper or Cellophane. Once the insect is placed inside, the corners are folded over (Fig. 8.31).

PAPER CONTAINERS

Larger insects, mainly Blattodea, Coleoptera, Hemiptera, and Hymenoptera, can be packed into empty cigarette or pill boxes. The insects should be tightly packed and the remaining space filled with tissue paper.

Medium to large specimens of Odonata, Grylloblattodea, Orthoptera, Phasmida, and Mantodea need special treatment, immediately after killing.

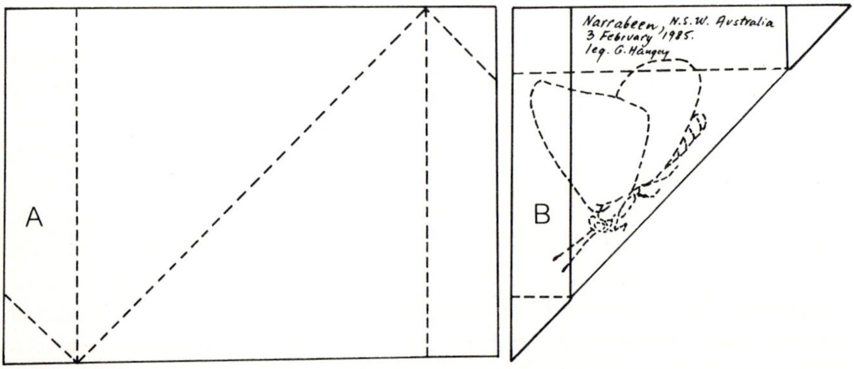

Fig. 8.31. A. The method of folding a Lepidoptera storage envelope. Dotted lines indicate the folds. **B.** The folded envelope with specimen inside. Data can be written on the outside of envelope.

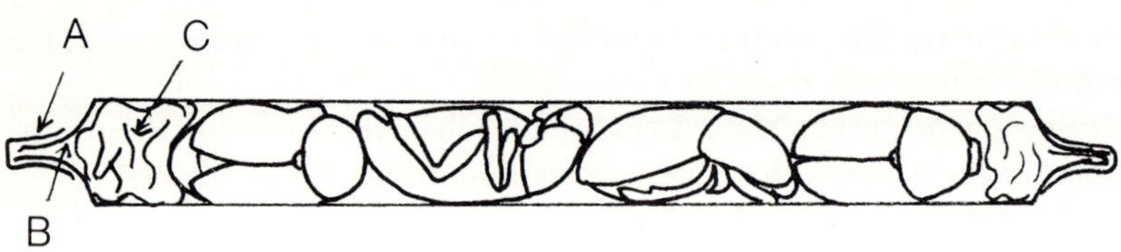

Fig. 8.32. Beetles in a cardboard tube, temporary storage. A. Masking tape seal. B. End of tube pinched in. C. Excesss space is taken up by crumpled paper tissue.

The abdomen should be cut along the ventral surface with a pair of fine pointed scissors (some workers prefer to make the cut on one side of the abdomen), and the insect eviscerated with a pair of fine pointed forceps. The delicate inner lining of the abdomen must not be disturbed and the genitalia must be left intact (including the external accessory genitalia). A small stick, somewhat shorter than the abdomen, should then be wrapped with cotton wool to the shape and dimensions of the abdomen and coated with a borax–glycerine paste. It can then be placed in the abdomen, with one of its points embedded into the thorax. The cut edges of the abdomen can then be pressed together and the specimen placed in a cardboard tube. Several specimens can be stored in the same tube, but it is essential that the tube diameter is not too big (Fig. 8.32). The tube should be plugged with tissue paper.

Once packed, the specimens are dried over a heater. In the field air drying can be difficult, mainly because it can take a considerable time during which the specimens could be damaged by ants, wind, or other factors. An improvised heater can be constructed from a kerosene lamp and a metal tray, or a camp gas stove and a cooking pot.

PRESERVATION AND STORAGE MEDIA

Silica gel

Insect groups which require dry mounting are usually killed in ethyl acetate vapour and stored in dry conditions. After selection, small specimens (chiefly Coleoptera) of the size group to be card mounted are placed in Petri dishes and left to dry until almost rigid. Each batch of specimens is stored in glass vials of adequate size, labelled, and kept at the bottom of the vial by means of a cotton wool plug. The filled vials are kept in an airtight container with silica gel until the specimens are completely dry.

The above technique is practised by Dr Sebastian Endrödy-Younga in the Transvaal Museum, Pretoria, Republic of South Africa, and presumably could be applied to orders other than Coleoptera. However, many small insects (for example, Diptera, and Hymenoptera) are extremely fragile and could not survive this method of storage. If kept dry, these should be packed into small glass vials with each specimen separated by small plugs of cotton wool or tissue paper. The vials are stored in airtight containers with silica gel.

Alcohol

The simplest method of storing insects is in alcohol. Solutions of ethyl alcohol (70%–80%) are suitable for killing and storage; therefore, the specimens may be dropped into vials containing alcohol immediately after capture, and stored in the same vial until sorting and preparation can take place in the laboratory. The filled vials are kept in larger jars, filled with alcohol. Alcohol specimens tend to fade quickly, especially if subjected to light.

Ethylene glycol

Specimens collected from pit traps containing ethylene glycol are usually stored in the same liquid until further preparation.

Smart's embalming fluid

Smart (1949) recommends a solution of 60 mL toluene (or xylene), 25 mL 3% butyl alcohol, 15 mL ethyl alcohol, 5 g phenol, 20 g paradichlorobenzene, and

10 drops of Canada balsam in xylene which can be injected into the bodies of large insects with a hypodermic syringe. The needle is passed through the posterior segments to the thorax and up into the head. It is slowly withdrawn and the fluid injected into the body. The fluid prevents rapid decay of the internal organs and assists the drying of the specimen. This fluid also retains colours in some specimens. Therefore, it may be suitable to preserve Orthoptera specimens which are too small for easy eviscerating and stuffing in the field.

Oudeman's fluid

Specimens killed in this fluid can be stored in the same.

Formalin

This is a very powerful preservative and usually available in 40% strength (formaldehyde in water). Since it is slightly acid, it may damage specimens containing calcareous matter. This can be avoided if 200 g of hexamine is added to each litre of full strength (40%) formalin. The hexamine neutralizes the formalin, and a constant pH of about 7.8 can be maintained. Insects, especially larvae, can be stored in 3% formalin (3 parts by volume of 40% formalin and 97 parts by volume of water) practically indefinitely, but specimens kept in formalin tend to harden and cannot be manipulated (set) properly. If killed in Pampl's fluid, various forms of Trichoptera are usually stored in 3% formalin.

RECORDING DATA

Natural history specimens have no scientific value without appropriate data. Insect specimens are no exception. The data which should accompany each batch of specimens collected from one location and habitat includes: locality (name of the actual place or the distance from the nearest major landmark, the province or state and the country, geographic coordinates, and altitude); date of collecting (day, month, year); name of the collector (initials and full name); and collection number. The number should refer to an entry in the collection log or diary in which more details are recorded: habitat, collecting method, meteorological notes and any other data which is considered relevant and important by the collector.

The importance of recording data with the utmost accuracy cannot be overemphasized. Incorrect data are even more damaging to scientific work than no data at all, and a few wrongly labelled specimens can undermine the credibility of an entire collection and the entomologist connected with it.

Data labels written out in the field should be enclosed with the specimens in their vials and plastic bags, if these are stored in liquids. The information must be written in India ink on small, oblong cards. It is wise to test the ink before usage, since some India inks are not waterproof. The writing must be completely dry before the label is placed in the liquid. The label should be turned so that the writing is visible from the outside, thus enabling the worker to read it without removing it from the container.

If live specimens are kept in containers during transportation the labels can be damaged. Data for live specimens can be written on the outside of their containers.

Dry preserved specimens should also have their labels enclosed in their containers. India ink is necessary for labels for these specimens also as they are relaxed in a high humidity container before mounting.

LABORATORY PRESERVATION AND MOUNTING

Insects preserved by drying are always mounted dry, which means that the prepared specimen can be stored in a dry environment. This method of preservation is only effective for the external skeletal cuticle; the internal organs are allowed to decay or shrink to an unrecognizable formation within the exoskeleton. Most internal organs of a dry preserved specimen are not suitable for further studies. Wet preserved specimens are wholly preserved, their soft tissues saved from decay by the preserving fluids. Specimens of some orders, although preserved in liquids for convenience or practicality in the field, are often prepared as dry mounts for a collection. The soft internal organs of these cannot be studied after drying. Some other groups, mainly small, soft bodied insects, cannot be prepared as dry mounts; these must be stored in fluids. Others are mounted on slides after wet preservation.

RELAXING

Most insect specimens have to be set in order to comply with the uniform appearance of a collection. Fresh specimens, killed with one of the agents which do not render them rigid, can be set practically right away. Those which were allowed to dry out or were placed in alcohol or some other fluid usually need to be relaxed before setting.

Dry specimens

Larger specimens which have been temporarily stored in paper tubes and envelopes are relaxed in airtight containers with saturated air humidity. The container (preferably a glass or plastic jar) should be big enough to accommodate at least several large insects and their paper surrounds. On the bottom of the jar, 30 mm of clean river sand should be placed. This should be saturated with water and a few drops of carbolic acid. The carbolic acid controls bacteria and fungus in the humid environment. A piece of fine plastic flyscreen should be placed on the sand, supported on three or four small wooden blocks so that it will not come in direct contact with the wet sand. The specimens can be laid on this screen and the jar sealed. Relaxation usually takes place within a few days. If the specimens can be extracted from their paper surrounds they may relax sooner. Most of the time it is wiser to leave them in the containers and allow a longer time for the process than to hasten it and risk injury to the specimens.

Larger specimens which have been stored in envelopes can be very carefully taken out and laid on a piece of blotting paper in the relaxing jar. Lepidoptera stored in layers must not be disturbed before complete relaxation. The entire box of layered insects must be placed in a relaxing jar.

Endrödy-Younga (1979) steams Coleoptera specimens to relax them. The beetles are kept in closed paper containers (cigarette boxes) during the steaming. In this way, the vapour condensation on the specimens is minimized and delicate pubescence kept intact. Endrödy-Younga (1979) also describes the relaxation procedure for small, silica gel dried specimens which are to be cardmounted. These are transferred into 10% acetic acid and left to relax for a period of three to four weeks. Acetic acid prevents any decomposing or moulding and does not affect the condition of material, even over a much longer period of time. These specimens will also be cleaner than those relaxed by other methods. Specimens should be rinsed with ample distilled water before being blotted and mounted. Acetic acid cannot be used for specimens to be pinned because it corrodes the part of the pin passing through the body of the specimen.

Wagstaffe and Fidler (1955) recommend wood naphtha as a potential relaxant. Coleoptera specimens can be immersed in it or subjected to its vapour in a confined space. Wood naphtha can be used alone or mixed with equal parts of alcohol and a few drops of carbolic acid.

Coleoptera specimens can also be relaxed if they are boiled in water. Prolonged boiling, which may be necessary to relax some specimens, may cause loss of natural gloss. Boiling also can prove to be drastic for delicate or old specimens.

Specimens which have already been mounted and dried may need resetting. These can be relaxed in the relaxing jar, but strips of cork should be provided to accommodate the pins on which the insects are mounted. The best method is to fasten a piece of cork sheet or rigid foam onto the underside of the jar's lid, then pin the specimens into that.

The relaxing jar must not be subjected to direct heat or sunlight when containing specimens. It should be kept at a steady room temperature.

Wet specimens

Insects stored in alcohol may become too rigid for convenient mounting. They can be softened if soaked in water with a little detergent added, or in a 2% aqueous solution of carbolic acid.

Alcohol preserved specimens may be transferred into 10% acetic acid until relaxed. Once the specimen become pliable, the acid should be rinsed out with distilled water, and the insect blotted dry, then mounted. Here again, acid treated specimens cannot be pinned because the pin corrodes.

Insects collected in pit traps with ethylene–glycol must be cleaned. The specimens should be thoroughly washed on a dense cloth sieve under running water and placed in a white enamelled or plastic dish. Specimens, except Coleoptera, should be sorted from the debris and soil particles and transferred into 75% alcohol. After that, they can be treated as any other alcohol preserved material.

The beetles are selected into two groups by size: the group consisting of small specimens which are to be mounted on cards are transferred to 10% acetic acid; the larger ones, which are to be pinned, are transferred to 75% alcohol. The ethylene–glycol must be thoroughly removed from the specimens before they are mounted. The temporary storage liquids (acetic acid or alcohol) must be changed four or five times before this is achieved. Only after this can the specimens be mounted. Specimens containing ethylene–glycol remain relaxed or dry over a very long period of time.

SORTING ALCOHOL STORED MATERIAL

Endrödy-Younga (1979) contributed the following description of the most practical technique:

"Alcohol-stored samples often contain quantities of soil particles and debris that makes straightforward selection difficult or impossible. Such samples are the Moczarsky–Winkler and Berlese selector samples, water samples, and often ground trap samples.

"For the enrichment of such samples a floating process is applied as follows: A saturated salt solution (NaCl) is prepared. The raw samples are drained and transferred into 96% ethanol. After 24 hours the samples are transferred into tall cylinder glasses (for example, a measuring cylinder) in a layer not thicker than the diameter of the cylinder. The alcohol is drained from the samples, and the cylinder, held in a tilted position, filled up with the saturated salt solution. Soil particles stay in the bottom of the cylinder and all organic matter floats. The method utilizes the density difference between an absolute alcohol and a saline solution and the period of flotation of the particles will depend on the rate of exchange of the two substances. Open entities such as decaying plant particles or arthropod shells will start to sediment immediately whereas closed entities such as arthropods, seed, and so on would float longer. Without stirring it, the solution can be left for a few minutes to allow further sedimentation of debris. However, as soon as the slow sedimentation of the first specimen is observed the process should be interrupted. The floating layer can be poured with a swift movement onto a dense cloth-sieve, rinsed thoroughly under running water, blotted, and transferred into 75% ethanol. The group selection from the reduced quantity of organic debris is done under a microscope. The material, selected into orders or families as is practical, is labelled and stored in vials".

PINNING

Insect specimens which are placed in a dry collection are either pinned through their bodies or mounted on small cards with glue or some other mounting media and pinned through these cards.

Through two centuries of entomology, a great array of different types of insect or entomological pins has been developed. The best pins are the widely distributed British stainless steel pins, with heads formed from the same material as the pin. These come in various lengths and thicknesses, but unfortunately their numbering is not uniform and could lead to confusion. The most popular thicknesses (approximately) are 0.34 mm, 0.38 mm, 0.43 mm, 0.46 mm, 0.50 mm, 0.59 mm, 0.66 mm, 0.70 mm, and 0.78 mm. The best length to use is 34 mm, since these pins will fit in most insect boxes. There are shorter and longer pins available, but unless some special reason warrants their use, it is best to stick to one length. Short pins cannot accommodate larger insects or more than one data label and the longer ones are often too big for standard insect storage boxes. As a rule of thumb, the larger the specimen, the larger the pin that should be used to pin it. Specimens under 8 mm–10 mm body length usually are not pinned with regular insect pins but with headless minuten pins. These are made in three lengths: 10 mm, 15 mm, and 20 mm. They range from about 0.14 mm to 0.30 mm in thickness. Here again, it is practical to use one length for all the specimens.

Thickness is selected according to the specimen's size. Large insects are pinned through the thorax or the right wing (Fig. 8.33). The pin should run through the body vertically. This will result in a perfectly horizontal specimen. Care must be taken that the pin does not skewer any of the legs. The back of the insect should be exactly 10 mm below the head of the pin. This

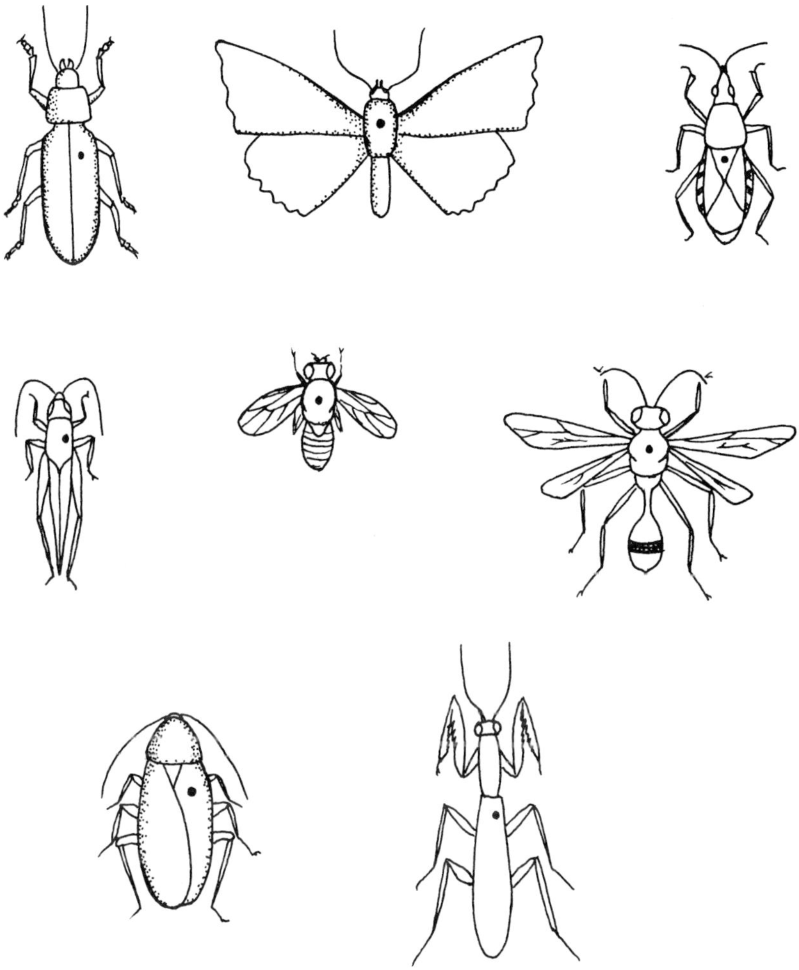

Fig. 8.33. The pinning of specimens belonging to various orders. Pinning methods may vary, according to special needs of the entomologist (dots mark pinning sites).

can be achieved with perfect uniformity in all specimens if a pinning block is used. (Fig. 8.34). The same block can be employed to pin the data labels under the specimen. Uniformity is a great asset in a collection, not only from an aesthetic point of view, but also because it assists workability.

Small insects are pinned on minuten pins. The small blocks that support the minuten pins are of rigid foam called "polyporus". This material is available from entomological suppliers. In the past, small pieces of thick

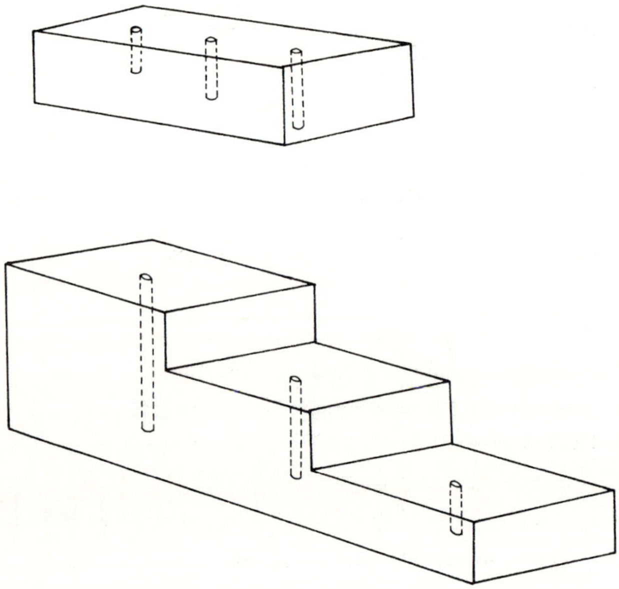

Fig. 8.34. Pinning blocks.

blotting paper, pith, cork and other similar materials were used and examples of these can be seen in many collections.

The blocks should be pinned on regular insect pins and the height adjusted so that the specimens on the minuten pins are exactly 10mm below the head of the insect pin (Fig. 5.35A). This corresponds with all the other specimens, regardless of size, in the collection.

CARDING

Small species of some orders are hardly ever pinned. It is very unusual to find minuten pin mounted beetles or bugs, for instance, although they could be prepared that way. In a number of museums and collections, most small insects are not mounted on minuten pins (although Microlepidoptera are always pin mounted), but on small cards. The shape of the card is usually oblong, with rounded corners.

Good quality white card (Bristol board) is used and three fine lines mark where the insect pin, on which the card is mounted, should be inserted. Usually, cards come in three sizes and can be obtained from entomological suppliers. Punches can also be bought to allow collectors to make their own

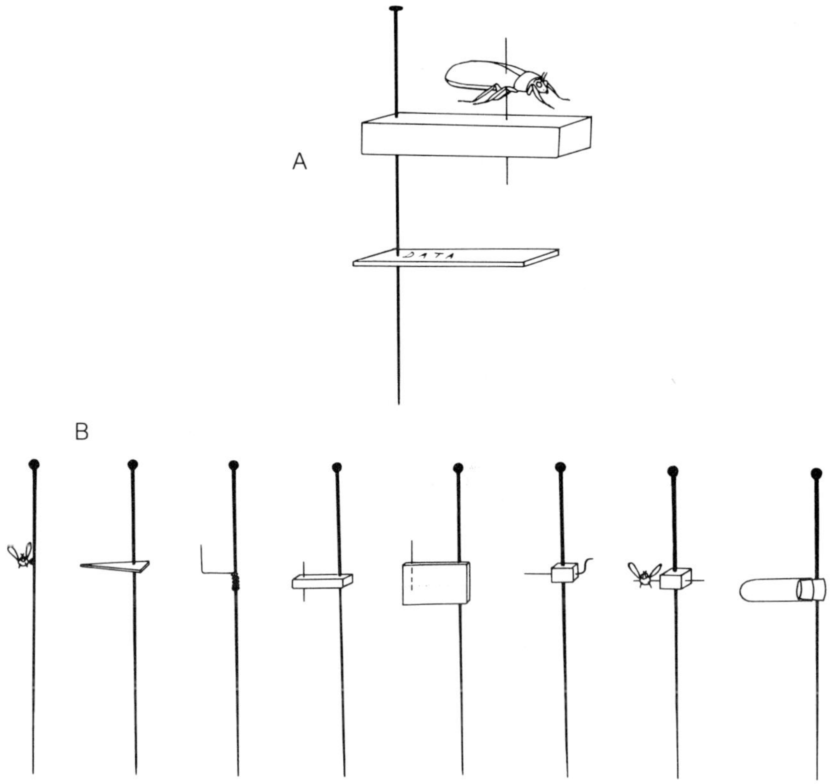

Fig. 8.35. Double mounting. **A.** Typical and most common double mounting method, using minuten pin. **B.** Various examples of double mounting.

cards. However, the printed lines are difficult to duplicate and, if specimens are mounted on cards without these, they may lack the uniformity of a regular collection. This may sound like a trifling matter for some, but it can seriously handicap a collector when it comes to exchanging specimens with another collector. Many collectors (professionals or amateurs alike) place great importance on aesthetics.

If a larger number of beetles or bugs is to be card mounted, the method is as follows:

The specimens are sorted in size groups in order to correspond with the three standard size of cards. One size group is worked on at a time. Ten to 12 cards are laid out in a row on a piece of cork or rigid foam sheet. An equal number of insects is taken out of the container, blotted gently if stored wet, and laid on a piece of cheesecloth or gauze. The first insect is turned on its back with a fine setting spatula or Leonhard forceps (Fig. 8.36), and the insect is gently pressed down. The spatula or the Leonhard forceps must be so flexible that, no matter how hard the insect is pressed, it will not be squashed. A fine setting needle may also do the job. This can be made of a fine insect pin, set in a small wooden handle (Fig. 8.37). While the insect is held fast, its legs, antennae, and mouth parts are spread with a fine, camel hair artist's brush. If the specimen is sufficiently relaxed, these parts will stay in the required position. If they spring back, the specimen must be further relaxed. After all the specimens have been treated, small droplets of water soluble glue should be applied to the cards. This can be done with a setting needle or a piece of fine wire. The insects should then be picked up one by one with a wet artist's brush and gently pressed on the glue. The droplet must be sufficient to bond the insect to the card, but not so much that it will be visible once the specimen is placed on top of it.

The insects should be placed on the cards quickly. The cards can then be pinned to the cork in a neat row with medium thick insect pins. If necessary, the legs can be readjusted using a fine camel hair brush. All parts of the insect must be on the card; under no circumstances may the insect overhang the edges. If possible, the legs should be spread (Fig. 8.38A). In the case of long legged specimens, or those with long antennae, they should be slightly folded (Fig. 8.38B). However, they must not be obscured by being pushed under the body. Insects which must be examined on their ventral surfaces for identification may be glued on their backs. It is an accepted procedure that card mounted material be soaked off the cards before examination (identification); therefore, it is important that only water soluble glue is used for card mounting insects. With the above method, an experienced insect preparator can mount several hundred specimens in a day.

An alternative method is to use triangular cards, or cardpoints, instead of oblong cards. The insects are glued to the very point of these cards. The advantage of this method is that only a minimal area of the ventral surface of the insect is obscured by the card and it can be examined without having to be soaked from the paper. The disadvantage of it is that the delicate extremities are not protected by the card and injury can be expected.

Triangular cards are usually used in Anglo-saxon countries, while the oblong cards are more common in Europe.

Acetate sheet cards and points are also used, especially by beetle collectors. Clear, cellulose based cement, diluted with ethyl acetate, is employed to glue the insects to the acetate. The same solvent is used to soak the specimens off if necessary.

Regardless of which method is followed, the carded or point mounted specimens should always be pinned through the card or acetate and positioned exactly 10mm below the head of the pin.

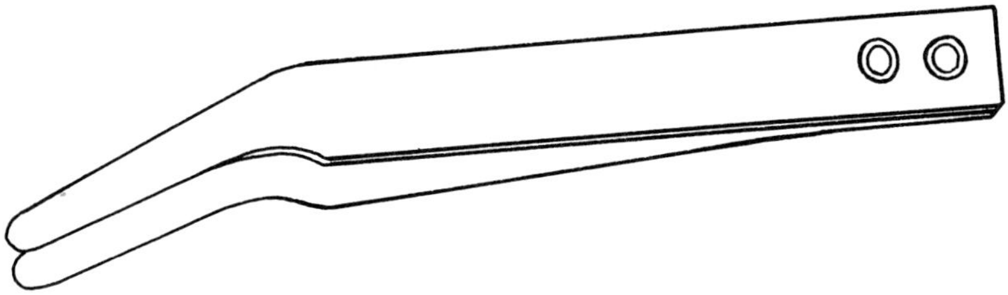

Fig. 8.36. Leonhard or "feather light" forceps.

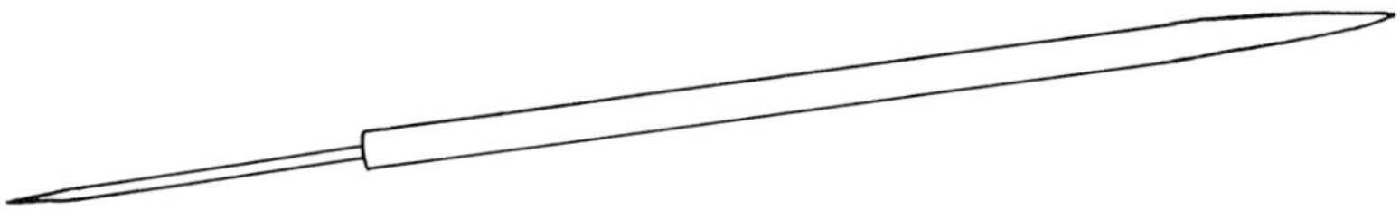

Fig. 8.37. Setting (or dissecting) needle.

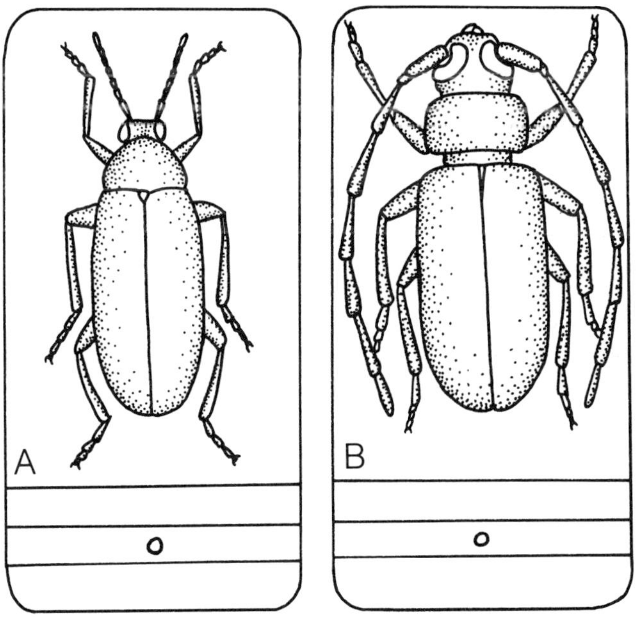

Fig. 8.38. A. The method of glueing a typical beetle specimen on card. **B.** Specimens with extremely long antennae are set this way.

In some collections, the larger species are also carded. According to Wagstaffe and Fidler (1955) all species of Orthoptera, Dermaptera, Ephemeroptera, Plecoptera, Heteroptera, Neuroptera, and Coleoptera should be carded, using good quality card or celluloid. The advantage of this method is that the specimens are better protected from accidental breakage than those which are pinned through their bodies. Carded insects are prepared with their appendages spread; therefore, these are more visible. The disadvantage of the technique is that specimens can take up more than twice the space of those pinned through their bodies and mounted with folded legs and antennae. The cards of larger specimens must be held in position by two or more pins, thus making their removal or transfer awkward. The method is rarely practised; it is peculiar only to a few institutes or collectors. However, if the material is used in exchange, the recipient is faced with the task of remounting the specimens in order to fit them into a regular collection.

SETTING

If a relaxed insect is positioned in a particular way and held there until dry it will maintain that position permanently.

Larger specimens which are pinned must have their wings, legs and other appendages set in such a way that they can be easily inspected. A dry insect is extremely fragile; therefore, it is also important that the appendages are not spread out unnecessarily, inviting injury. Saving space is also very important in an insect collection. Specimens set with neatly folded legs take up much less space than those with spread limbs. Uniformity is important; the general appearance of a collection can be ruined by untidily set specimens.

Pinned specimens of the orders Grylloblattodea, Orthoptera, Phasmida, Dermaptera, Blattodea, Mantodea, Hemiptera, and Coleoptera are usually set with their wings (if they have any) folded. The insects should be pinned (some should be eviscerated and stuffed as described earlier) on a block of rigid foam. Rigid polyurethane blocks, 30mm × 200mm × 300mm, are ideal for this work. The pin is pushed in until the ventral surface of the insect touches the foam. The appendages are arranged as illustrated (Fig. 8.39), and fastened by pins. These are not pushed into the appendages, but next to them or crossed over them, thus fixing the part in the required position.

Sometimes it is desirable to set a specimen with its wings spread, or, more often, with wings at one side (usually the left side) spread. This is to show some important or interesting wing formation, colour, or pattern. Spread winged specimens are set on butterfly boards, as described below.

Members of the orders Ephemeroptera, Odonata, Plecoptera, Embioptera, Hemiptera, Neuroptera, Mecoptera, Lepidoptera, Trichoptera, Diptera and Hymenoptera are usually prepared with all their wings in a spread

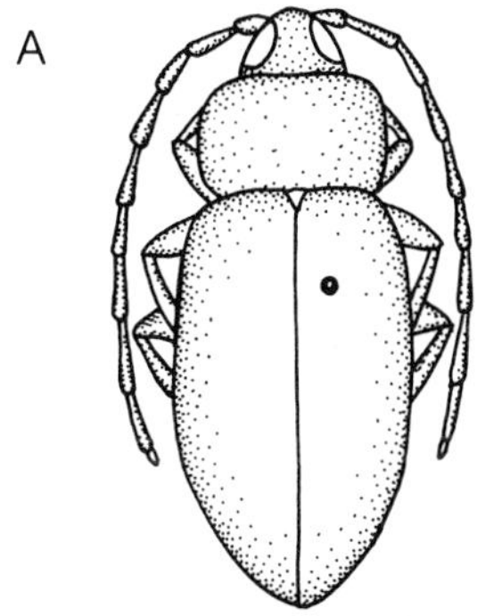

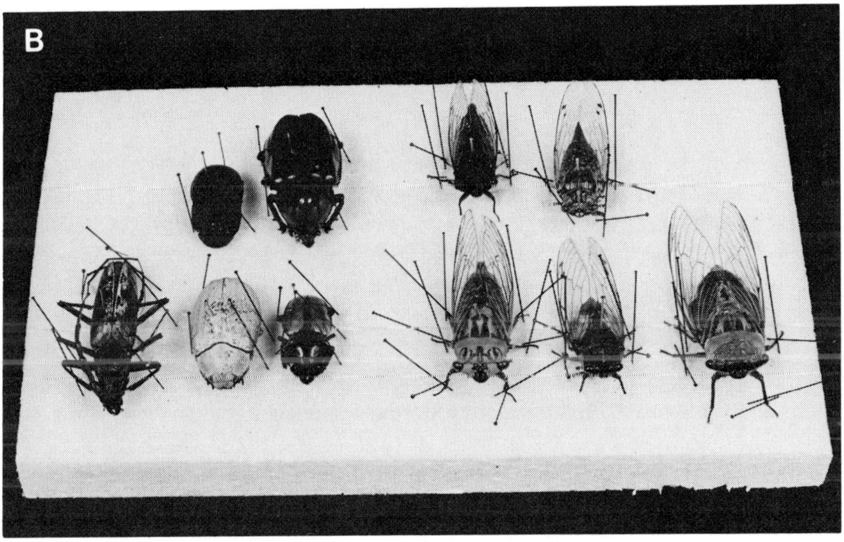

Fig. 8.39. A. Legs and antennae are set this way in order to prevent damage to the specimen and also to save space. **B.** Specimens on the left are set for display purposes, while those on the right are set for a regular collection.

position. To assure perfect symmetry and uniformity, setting boards of various size are used. There are different designs of boards available from entomological suppliers. Some types are easy to make, but others are more complicated. (Fig. 8.40). English boards are suitable for setting specimens with their wings perfectly horizontal. The disadvantage of this system is that some specimens (especially the larger ones) may eventually sag due to humidity, and the wings may drop at various degrees, depending on the nature of the specimen. The European boards have wing supporting slats

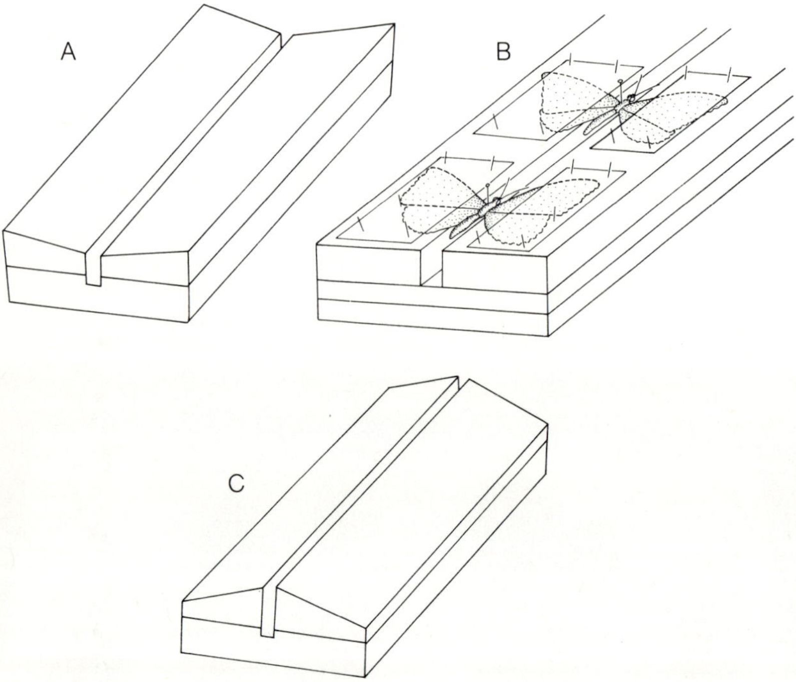

Fig. 8.40. A. Setting board with horizontal surfaces. **B.** Setting board to set butterflies or moths on their backs. **C.** Setting board with sloping surfaces.

sloping slightly away from the central groove, thus setting the specimens' wings on a slight angle. This is accepted setting form in non-Anglo-saxon countries. If sagging occurs at all, it is of a much lesser degree than that of horizontally set wings. It is also more pleasing aesthetically, since live insects, especially butterflies, tend to spread their wings with a slightly upward angle.

Setting boards with sliding slats (Fig. 8.41) are handy, since the width of the central groove can be altered at will.

It is necessary to keep a good variety and quantity of setting boards handy. Insects should be set on the board with the following technique (the description is based on the method employed for setting a butterfly or moth and can be applied to any other winged insects).

The butterfly is pinned through the thorax vertically and the body is positioned exactly 10 mm under the head of the pin. The specimen is pinned into the central groove and the pin is pushed in until the wings lie comfortably on the slate. It is important that the body is not pushed in too deep, nor left

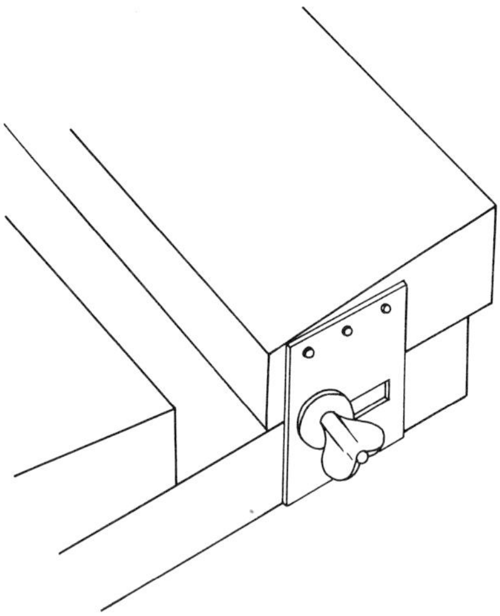

Fig. 8.41. Setting board with adjustable side.

too high (Fig. 8.42A). The pin must be vertical. A 5 mm strip of transparent paper is pinned on the board above the butterfly and one forewing is manipulated to the required position. Using a blunt needle, the wing is pushed at one of the larger veins at its base. Often, the wing tends to pull or hinge back to its former position. This is prevented by placing the paper strip on it, pulling it tight (gently), and pinning it just under the butterfly. The gentle pressure of the strip is usually enough to keep the wing in place. Now the hind wing is pulled up the same way. If necessary, more pins should be used to keep the wings in position. Care must be taken not to pin through the wings. The procedure is repeated on the other side. The forelegs and antennae are carefully arranged and fastened in position with pins. The abdomen may be supported by crossed pins. Small squares and triangles of paper can also be used to hold appendages in position (Fig. 8.43). The setting board is filled with butterflies this way. Once it is full another two strips of transparent paper are pinned on in order to cover all the wings. If this is not done, the wing tips may curl up during drying.

There are other techniques for setting butterflies or other winged insects, using small cards, setting bristles and fine thread to hold the wings in position. These techniques are demonstrated in Fig. 8.44.

A well set specimen should always follow the rule of the cross (Fig. 8.45).

The specimens must be thoroughly dry before they can be taken off the setting boards. Drying may take a few days for small insects, or several weeks

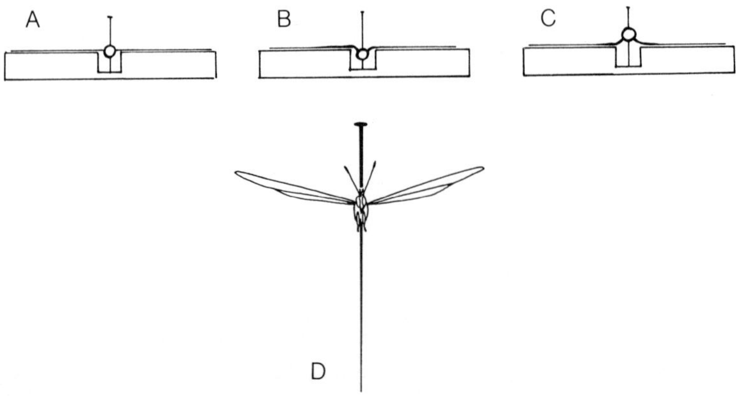

Fig. 8.42. The method of adjusting the height on the pin of an insect in relation to the setting board, set with its wings open. **A.** Correct height. **B.** Insect positioned too low. **C.** Insect positioned too high. **D.** Correctly set Lepidoptera specimen, set on a board with sloping surfaces.

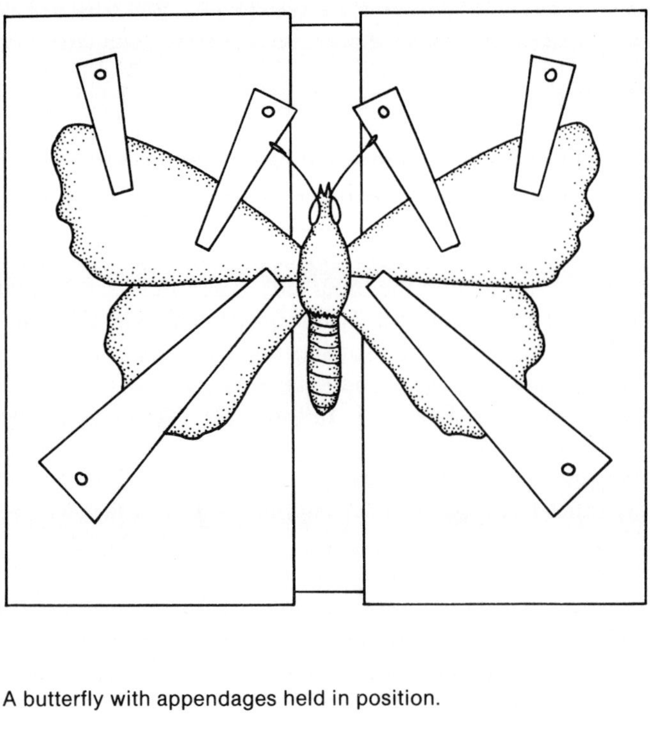

Fig. 8.43. A butterfly with appendages held in position.

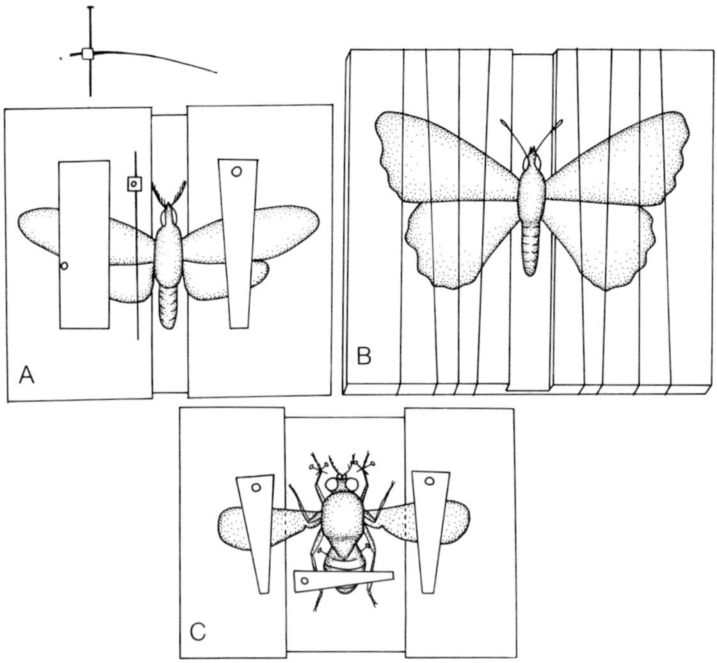

Fig. 8.44. A. Microlepidoptera set on a miniature setting board with the aid of small pieces of paper and a setting bristle. Detail of setting bristle above. **B.** Lepidoptera specimen set on board, wings held in position with carefully wrapped thread. **C.** Diptera specimen set with the aid of small pins and pieces of paper.

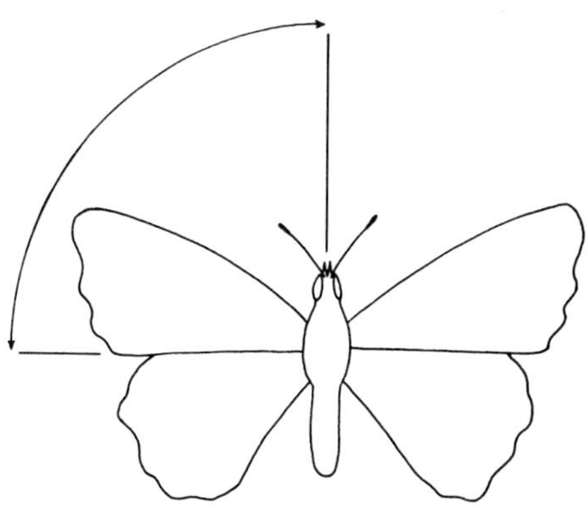

Fig. 8.45. Correctly set Lepidoptera.

for large ones or those in a humid environment. The drying specimens often omit an unpleasant odour which attracts various pests. These can range from small mites to mammals. The danger is especially great in the tropics where ants, large cockroaches and rodent are common and very active. Naphthalene, paradichlorobenzene and other insecticides usually deter or kill insect pests. The best defence against rodent pests is to place the specimens in a strong metal trunk with a metal insect screen on it. The screen provides enough ventilation for drying. Under very humid conditions, thymol or paradichlorobenzene must be placed near the specimens to prevent mould.

Once the specimens have dried the pins, cards, and paper strips can be removed and each individual insect is labelled.

PREPARING INSECT EGGS

Most eggs can be killed and preserved in 75% to 95% isopropyl alcohol. Peterson (1959) has killed eggs in a kerosene, acetic acid, and alcohol mixture. The kerosene prevents the eggs collapsing, but if the formula contains more than 1% the eggs may burst.

Eggs with firm shells can be dry preserved. In order to prevent the embryos from hatching, the eggs should be subjected to heat for a few minutes, or boiled in water for about a minute. Dried eggs are usually carded, then placed in the collection. White or light coloured eggs are sometimes glued on black cards.

PREPARING PUPAE

Pupae may be killed if immersed into boiling water for a few seconds. Those which are enclosed in cocoons can be carefully extracted through a neat cut (using a surgical scalpel) of the cocoon and either boiled briefly, or injected with Carnoy's fluid or isopropyl alcohol. Degreasing and dehydrating with ether may also give good results, especially when treating the pupae of Coleoptera (further details below in larval section). Pupae of many species of Lepidoptera may be preserved dry. Dried specimens are pinned and placed in the collection.

PREPARING SOFT BODIED LARVAE

The best method of preparation of soft bodied insects, either larval or other forms, is freeze drying. Most of the literature dealing with soft bodied insect preparation was written before the freeze drying of whole specimens was perfected. The freeze dryer apparatus is no longer considered a rare instrument in museums or institutes dealing with animal preparation.

Freeze drying techniques are described in Volume 2, Chapter 10. If there is no freeze dryer available, many larvae can be prepared successfully with the old techniques. A few examples are described.

Inflating

Most medium to large caterpillars, beetle larvae, and larvae of other orders can be prepared with this method, although it is somewhat tedious and requires a great deal of dexterity.

The larva should be starved for 24 hours, then killed in ethyl acetate. As soon as the larva appears to be dead, it should be placed on a piece of blotting paper and gentle pressure applied by a round implement, such as a glass tube or pencil, near the head. The pressure should be directed towards the anus and as the viscera start to appear, they are punctured with a fine pointed scalpel. This enables the contents of the intestines to ooze out freely. Pressure is now applied by rolling the tube or pencil along the body until it is completely flattened and all contents removed. A small, tapered glass tube can then be fitted into the anus while the skin is held tight by a small clamping device (Fig. 8.46). The tube should be lightly oiled or coated with Vaseline before insertion to prevent the drying skin adhering to it. The skin can now be carefully inflated by air being blown into it. The inflated larva can then be held in a small metal oven (Fig. 8.47A) or in a lamp cylinder (Fig. 8.47B) and gently heated with a small Bunsen burner. Sudden, strong heat must be avoided as this may singe the specimen and destroy its colours. Drying may take up to an hour, or even longer, in the case of large specimens, and perhaps ten minutes or so for small ones. The specimen is dried once the head cannot be moved and it can then be removed from the glass tube. This can be replaced

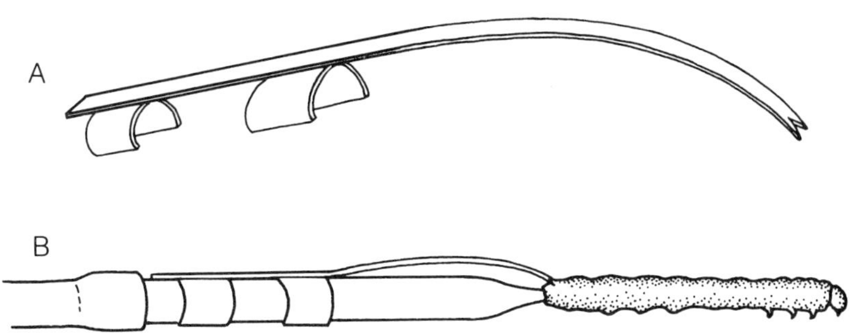

Fig. 8.46. A. Clamp made to hold the skin of a larva while drying. **B.** The clamp, fitted on a glass tube, is holding the inflated skin.

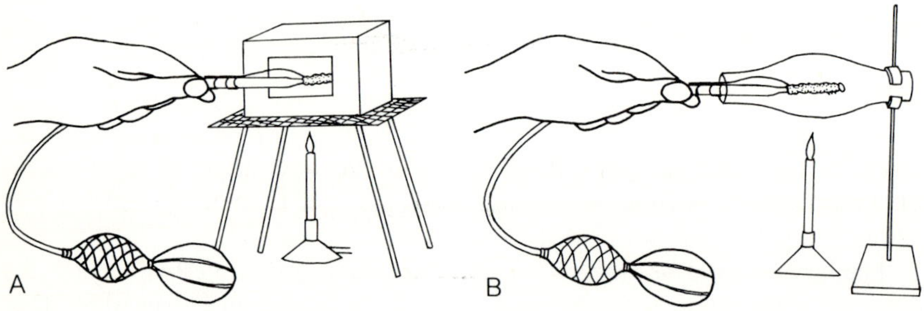

Fig. 8.47. A. The inflated skin is dried in a small oven. **B.** Instead of an oven, a lamp chimney can also be used.

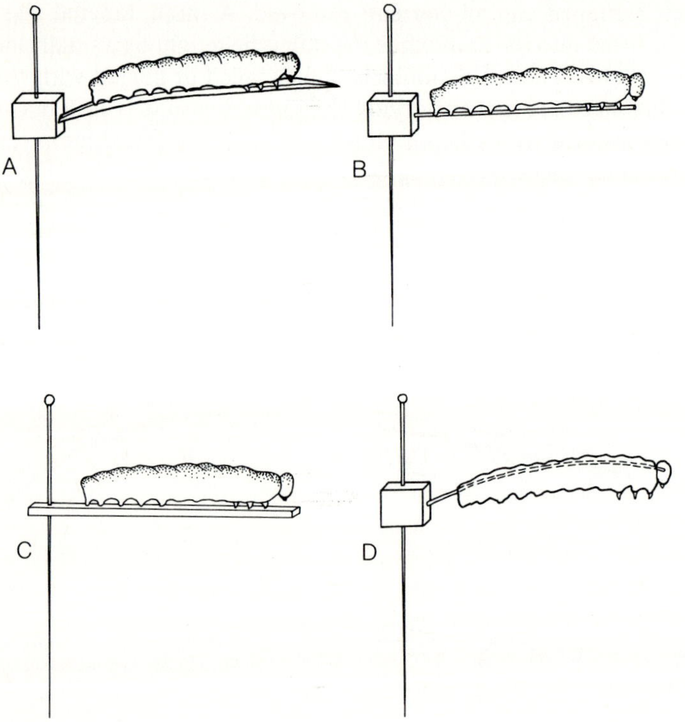

Fig. 8.48. A. The dried skin of a caterpillar mounted on a stalk of grass. **B.** Caterpillar skin on straight wire. **C.** Caterpillar skin on polyporus block. **D.** Caterpillar skin on straw.

with a stalk of straw or dried grass and the skin glued to it by a drop of glue at the anus. The specimen can be pinned through the straw and, after the data labels are added, it is ready to be placed in the collection.

It may be impossible to pin a larva mounted on dry grass. In that case, a small polyporus or elder pith block should be cut and the grass pinned into that. The block can then be pinned with an insect pin (Fig. 8.48A). Inflated larvae can also be mounted on fine wire (Fig. 8.48B), or on polyporus blocks cut a little longer than the specimen (Fig. 8.48C).

Wax filling

The empty skin of larvae can be filled with wax instead of air. The skin should be prepared as described above, then two parts paraffin wax should be mixed with one part beeswax (any low melting point petroleum wax can be substituted) and heated in a double boiler until liquid. A disposable syringe with a heavy gauge needle should be preheated in an oven or in hot water, and filled with the liquid wax. The needle is fitted into the larva's anus while the skin is held in position with the aid of a pair of fine point jeweller's forceps. Great care must be taken to hold the skin only by its extreme edges. The wax is slowly injected until the skin is completely filled. The filled skin and the needle of the syringe are then dipped in cold water to solidify the wax. The needle is removed and the finished specimen can be pinned and placed in the collection.

Degreasing and dehydrating

Peterson (1959) recommends ether as a degreasing agent. The specimen should be placed in it and allowed to cure. When the ether becomes yellowish from the dissolved fat, the specimen should be transferred to a fresh ether bath. This process should be repeated until the ether remains clear. The fat free larva can then be removed from the ether and placed on a piece of absorbent paper under a lamp. The proximity of the hot light bulb will determine the rate of evaporation of the ether. It should be rapid enough to inflate the specimen. If the bulb is placed too near the larva, the larva may even burst. Once all the ether has evaporated, the specimen is ready to be pinned and placed in the collection.

It is also possible to degrease and dehydrate larvae by placing the specimens in acetone. The liquid should be changed until it remains clear. Even large caterpillars degrease and dehydrate nicely. Afterwards, the larvae are air dried and pinned. Although some larvae tend to shrink to some extent, this method is a very good one for large hairy specimens where some minimal shrinkage cannot be easily detected.

Critical point drying

Alcohol stored material can be prepared by critical point drying. Dr Don Horning of the MacLeay Museum, Sydney, Australia, uses the technique to prepare mostly small Hymenoptera specimens. He has passed on the following information: the specimen previously stored in alcohol is dehydrated in a succession of acetone baths, and it is placed into the chamber of the critical point drier apparatus. The chamber is filled with acetone and closed. Liquid carbon dioxide is pumped into the chamber, and this replaces the acetone (which is pushed out through an outlet). The chamber is heated for approximately 15 minutes and, when the pressure within rises to 7900 kPa, which is the critical point, the liquid becomes gas and the drying of the specimen begins. The pressure is raised to 9500 kPa; this permits the specimen to dry completely.

When the insect is removed from the chamber, it resembles a live specimen; there is no shrinkage and all the hair and bristles adopt their natural position. Many specimens can be treated in the apparatus at the same time. After treatment, some may be mounted on points or cards, or stored *en masse* in small trays until proper setting or mounting becomes necessary. Large specimens can also be treated in the apparatus, providing they fit into the chamber.

LABELLING

The recording of data in the field has been described already. After the insect has been prepared these data must be attached to it. Usually, the information is transferred to two 8 mm × 12 mm white cards, which are pinned under the insect. The first label carries the data on locality (country, place, altitude, and geographic coordinates), while the second the date, collection number (which should correspond with the log book containing more detailed information), collecting method and the name of the collector. Often, more labels are attached to the specimen giving more detailed information, such as reference to host animals (in case of parasitic insects), food, plants and so on. These labels can be handwritten in India ink with a fine pen, offset printed, or typed then photographically reduced. Labels can be partly printed, leaving blank spaces to be filled in by hand. Such labels are practical for a collector who frequently works in one area. The locality, the year and the name of the collector can be printed on the labels and the day, month, and method of collecting can be written in.

THE PREPARATION OF GENITALIA

The examination of genital organs can help differentiate certain insect species. Often, species can only be recognized after thorough examination of their genitals. These are often not visible on the exterior of the insect; therefore, they must be dissected and prepared specially.

In many cases, a gentle squeeze or push is sufficient to extend and make visible the genitalia of a relaxed, freshly killed insect. If the specimen is then carded or set, the genitalia dry and remain visible.

However, it is often more difficult. The genitalia of beetles must be extracted most times. The larger species need a lateral incision in the sixth and seventh abdominal segments. The cut is carefully pryed open with dissecting needles and hooks (Fig. 8.49) and the aedeagus pushed out with a blunt dissecting needle. It can be removed without wrecking the specimen and the incision usually closes itself as the exoskeleton springs back. The genitalia are mounted on a card with water soluble glue and pinned under the specimen on the same pin.

Small beetles are handled in a somewhat different manner. The sixth abdominal tergite is loosened with a hooked dissecting needle, while the body of the insect is held down with a straight needle or a fine flexible spatula (Fig. 8.50A) and the genitalia are extracted through the opening. An alternative is to dislodge the segments on the side and hinge them open (Fig. 8.50B), thus gaining access to the genitalia.

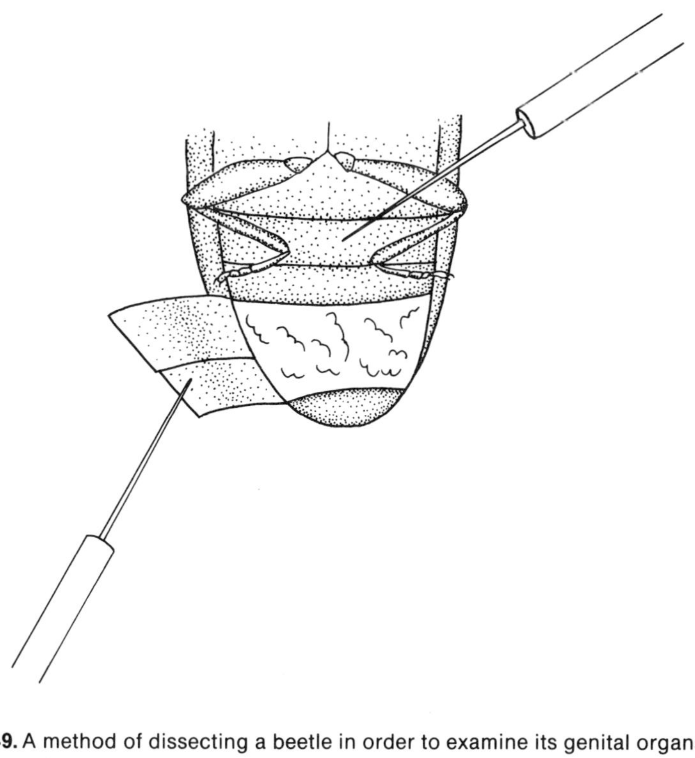

Fig. 8.49. A method of dissecting a beetle in order to examine its genital organs.

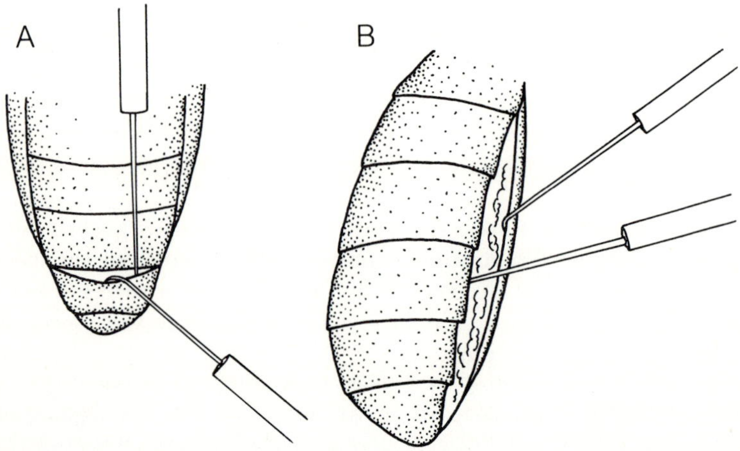

Fig. 8.50. A. Another method of dissecting a beetle. **B.** Access to the genitalia can also be gained this way.

The removal of genitalia from Lepidoptera specimens often causes the partial destruction of the specimen. The extraction is usually executed by cutting through the abdomen between the fifth and sixth abdominal segments and macerating the section in 10% caustic soda in water. Once the section is soft and well macerated, the genitalia are extracted with the aid of a fine brush and dissecting needle.

Sometimes, specimens have been successfully treated with wood naphtha. A drop of the substance can be applied to the genital cavity of the Lepidoptera specimen. Naphtha usually relaxes the dried tissues immediately, and the genitalia may be extracted. This method, when it works, does not destroy the specimen.

The extracted genitalia can be mounted in their natural state on cards or they can be macerated, stained, and mounted to suit the requirements of the entomologists working with the material. Usually, the prepared specimens are set in Euparal and Canada balsam and placed between slips of cellulose, or prepared as microscopic slides.

MICROSCOPIC PREPARATIONS

Specimens belonging to the orders Thysanura, Diplura, Protura, Collembola, Zoraptera, Psocoptera (small species), Mallophaga, Siphunculata,

Hemiptera (some), Thysanoptera, Diptera (some), Siphonaptera, and Sterpsiptera are usually prepared as microscopic preparations. Although preparation of a microscopic slide of an insect (or parts of it) can vary a great deal, the description below will provide the reader with basic information about the subject. Individual groups may require specialized treatment and more detailed descriptions of these can be found in the relevant literature.

A wet preserved specimen should be transferred to a test tube containing a 10% aqueous solution of caustic soda. In 12 hours the specimen will have thoroughly macerated (all soft parts disintegrate and only the chitinous skeleton remains). It should then be rinsed in distilled water with a few drops of glacial acetic acid. Afterwards, it must be transferred to glacial acetic acid and gently agitated with a fine, flexible tool, such as a bristle or fine shim spatula until all the macerated soft parts have been released and washed out. Since colourless chitin is almost invisible, specimens with very little or no pigmentation must be stained. The stain (there are many different stains which can be used) is dissolved in the acid and the specimen allowed to absorb it. The specimen is transferred from the staining bath to clove oil or xylol for clearing. This last step can be omitted if the specimen is going to be mounted in Euparal.

The next step is to transfer the specimen to a glass microscopic slide. A variety of mounting media can be used; the best is Euparal. This is a mixture of camsal, sandarac, eucalyptol and paraldehyde; it is available from entomological suppliers. It is a clear, syrup like substance which can be diluted with Euparal essence. Euparal has clearing qualities, therefore, prior clearing of the specimen is not necessary.

To mount the specimen in a slide, a small drop of Euparal should be placed on the glass, and the specimen is immersed in it. Some groups may present a ventral view, and others a lateral view. The cover slip is very carefully positioned over the slide and gently lowered on it, settling on the specimen by its own weight. The mountant eventually will harden, but before that happens the slide must be stored horizontally.

Canada balsam is another mounting medium, perhaps the oldest in entomological use. It has a better refractive index (1.536) than Euparal (1.485). It is a resinous substance dissolved in xylol, and is not miscible with water. Specimens must be absolutely water free. This can be achieved if the specimen, after maceration, is put through a succession of alcohol baths of increasing strengths (30%, 50%, 70%, 90% and finally 100%). Then the specimen should be transferred to xylol, and from this it can be placed in Canada balsam, following the same procedure as described for Euparal mounting.

Gum chloral was invented by Berlese last century. It is known as a good clearing mountant which penetrates well, but is somewhat unsuitable for permanent mounts, since it has a tendency to crystalize when set. It also needs heat drying for several days, and the slides should be ringed.

Ringing is a procedure when the slide and the edge of the adherent cover slip are coated with a sealing agent, thus preventing air and humidity from affecting the mountant and the specimen. There are various sealing agents available from entomological or biological suppliers. One of the most popular brands is Glyceel, consisting of linseed oils, alcohol, nitrocellulose, butyl acetate, and toluol.

Specimens mounted on slides are usually very thin, therefore, when the cover slip is placed on top it will lie more or less flat and level. However, the specimen can be a little more bulky. Since squashing or flattening it should be avoided, it is necessary to prop up the cover a little. This is usually done with four slivers of glass, which form a frame around the insect, or a paper frame. Small metal rings (cells) of various sizes and thicknesses can also perform the function. They may be purchased from supply houses.

A cell can also be built of a ringing agent. The substance is painted in a circular pattern on the slide and built up to the required height by applying several layers. The mountant is placed within the ring, the specimen embedded in it, and the cover slip is positioned on top of it. Cells formed like this need to be ringed again, once the cover slip is in position. Data labels are glued to the slides. These must be written in India ink and contain the same kind of information described earlier.

STORAGE OF DRIED SPECIMENS

The most widely accepted method of permanently storing dry, mounted specimens is to pin them in glass topped, tightly closed drawers. There are many designs for storage boxes and drawers, and the cabinets which accommodate them. Insects can be stored in a temporary storage box for as long as necessary, but they are not really suitable for permanent storage.

A popular design for a drawer to house insects on a permanent basis is as follows: The size of the drawer is 600 mm × 400 mm × 50 mm. The top, which consists of a tight fitting frame with glass, should not be hinged. It should be able to be lifted off. The tray is double walled with a gap of approximately 7 mm between the two walls. This gap is filled with melted naphthalene which serves as an impenetrable barrier for any invading insect pests. The bottom of the tray is lined with a 7 mm thick polyethylene foam sheet. This foam closes after a pin is removed from it, thus avoiding any trace of pin holes. Natural sheet cork, which is also excellent, needs a fine, white paper covering. The paper is pasted on, wallpaper fashion, with gum arabic. If the interior of the drawer is papered, the paper should run up the walls as well. Composite cork sheets are not suitable since the adhesive used in manufacturing the sheets can corrode some types of insect pins.

The drawer should be made of hardwood (soft wood may extrude resin which can damage delicate specimens). It should be fitted with one or two knobs and a label holder at the front. It is practical to have uniform boxes for the entire collection. This will allow the drawers and tops to be swapped around if necessary. The drawers should be placed in cabinets with tightly fitting doors. If the doors are glazed, the labels on the drawers can be read without opening the cabinet. This makes work easier for the entomologist working with the collection.

ARRANGING THE SPECIMENS

An insect collection is not only a working tool in the hands of an entomologist, but also a permanent source of enjoyment. Many amateur collectors establish collections of considerable size and quality purely for recreational reasons and without having any great scientific ambitions. Therefore, the important scientific value of a well curated collection should not cause aesthetic values to be overlooked.

A typical 600 mm × 400 mm cabinet drawer is divided into six 100 mm columns (Fig. 8.51) either by fine lines drawn on the lining paper or by lengths of black cotton stretched between pins. An internationally accepted classification system should be adopted for arrangement of the insects.

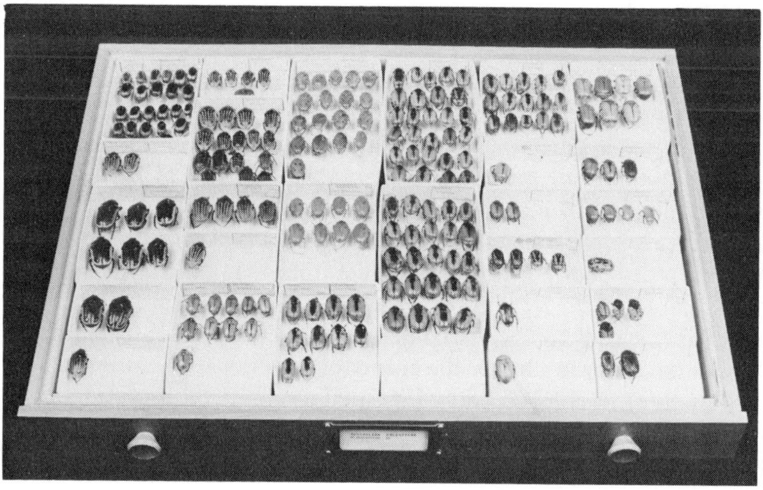

Fig. 8.51. Cabinet drawer with unit trays. Photo A. Pitman. Courtesy of The Australian Museum.

A label bearing the family name should be pinned at the top left hand side of the drawer. This label with the name of the genera should be pinned underneath. This is followed by neat rows of specimens of the first species of that genus. The specimens can be lined up by size, by colour, or by sex. Another label should be pinned after specimens of the same species; this label should bear the name of the species and of the person who originally described it.

In countries where the insect fauna is well explored (for example, in Europe) it is common practice to divide the drawers with pencil lines into as many sections as the species which live in the area. This way, the collection does not need to be rearranged as more and more species are collected. These are simply pinned in the vacant places until the collection is complete.

The contents of each drawer should be written on a label placed in the label holder on the front. This need not list all the species within the drawer, but merely the name of the family and, perhaps, the names of the genera.

Rearranging the contents of a drawer can be time consuming and, though great care is taken, sometimes detrimental to the specimens. Dried insects are extremely vulnerable and delicate antennae or tarsi can break easily during such a shift. To avoid this, many collectors omit the foam lining from the drawer. Instead of a sheet of foam, foam lined, standard size, open topped cardboard trays are fitted in the drawer, each intended to hold a number of specimens from one species. If rearranging becomes necessary the trays can be swapped around without the specimens being touched.

MAINTAINING A DRY COLLECTION

The colours of dried insect specimens fade rapidly if subjected to natural sunlight or artificial light with high ultraviolet levels. Ideally, the specimens should be stored in the dark. Most collections are only subjected to light when examined. Exhibition specimens suffer more because of high light levels. Insects in museum displays are often faded beyond recognition.

Mould is also very damaging to prepared insects. The collection should be housed in a dry room at an even temperature. Mould can be controlled if thymol, paradichlor benzine or chlorocresol crystals are placed in each drawer. The crystals can be placed in a small open topped cardboard tray and pinned in the lower left hand side corner of the drawer. A saturated solution of naphthalene in chloroform, with equal amounts of creosote added, can also be used as a mould preventative. The liquid is poured in a small nonspill glass container fitted on a pin (Fig. 8.52), and pinned in the corner of the drawer. Carbolic acid is also a good mould deterrent. The crystals are melted, then a wad of cotton wool is dipped in and pinned in one corner of the drawer. A disadvantage is that the carbolic acid may stain the paper or foam lining of the drawer.

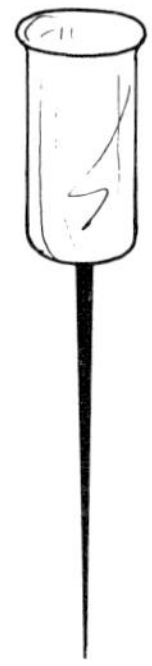

Fig. 8.52. Small glass vial mounted on a pin.

Mites and insects are the chief enemy of a dry collection. Mites can penetrate seemingly well sealed drawers and, because of their minute size, can start their damaging work undetected. The damage soon becomes noticeable. Brown dust is deposited under the pinned specimens; later, the more fragile insects in the collection disintegrate. Insects are equally damaging; moths, beetles and some other pest species of other insect orders can ruin an entire collection in a relatively short time.

The most popular deterrent is naphthalene. Naphthalene is not a lethal poison for many insects, but it is an effective deterrent. Other insecticides can also be used in the same way, but their disadvantage is that they dissipate after a few months and have to be removed and exchanged with a fresh batch. The more effective ones are usually dangerous for humans too. Naphthalene also evaporates and must be replenished from time to time, but not more often than once every 12 to 18 months. Well sealed drawers prevent the access of larger pests.

If a collection becomes infested it must be fumigated. In the past, carbon disulphide was frequently used. It is a most effective fumigant, but can be very dangerous. It is highly volatile, flammable, and, in a certain concentration, explosive. It is presently used for general fumigation work. It is a dangerous substance and, usually, only authorized persons are permitted to use it.

Private collectors, or institutes lacking appropriate staff and equipment, should employ the services of pest control firms. These firms can perform services on location or the infested drawers can be transported to their premises where they can be treated in a fumigation chamber.

DEGREASING AND CLEANING INSECT SPECIMENS

Specimens containing fat may discolour (mostly darken) and the oily seepage may stain cards, labels or even the lining of the insect drawer they are stored in. To prevent this undesirable effect, the fat must be removed from the specimen. A great deal of fatty tissue can be removed from the insect while it is still fresh by evisceration (see above) and consequent "stuffing". Orthoptera and related orders are usually treated this way, but some groups of other orders may also be handled the same way. The females of oil beetles (Meloe), which have huge, soft abdomens, must be eviscerated and stuffed. Their viscera is very fatty and could excrete large amounts of oil if left intact. Moths and some butterflies also have large abdomens containing large amounts of fat. One group in particular (Cossidae) is very fatty; their abdomens must be cut open, the contents removed, and cotton wool dipped in boric acid placed inside.

Hard bodied beetles usually are not opened up and eviscerated. The body fat is dissolved by immersing the specimen in a degreasing solvent such as chloroform, xylol, benzol, benzene, carbon tetrachloride, trichloroethane, diethyl carbonate, or ethyl ether. Mounted specimens can be immersed together with their pins and labels and left until all fat dissolves. Gentle agitation and several changes of the solvent speed up the process. Solvents which are not flammable, such as carbon tetrachloride, may be heated to 50° C for more effective degreasing. Fumes of all these solvents are harmful to humans and the work should be carried out in a fume cabinet.

McKillop and Preston (1981) describe the vapour degreasing of entomological specimens. The insects are placed in a vapour degreaser, and trichloroethane is boiled and vaporized. The vapours condense on the surfaces of the specimens, fat is dissolved, and the contaminated solvent drips into the boiling chamber below. The same instrument and solvent can be used for the treatment of osteological material.

Penecke (1929) developed a technique to clean dirty Coleoptera specimens by dipping the previously relaxed specimens in diluted hydrochloric acid for a very short period and immediately transferring them in aqueous solution of sodium bicarbonate. The chemical reaction generates gas on the surface of the specimen and lifts off the particles of dirt. The specimen should remain in the sodium solution for at least 15 minutes in order to assure a complete neutralization. After this period, it should be rinsed in distilled water.

SYNOPSIS OF TREATMENTS FOR EACH INSECT ORDER

The following synopsis has been extracted, with the kind permission of Courtenay Smithers, from his book, *Handbook of insect collecting* (*Collecting, preparation and preservation and storage*) published by A.H. & A.W. Reed, Sydney, Australia.

"The synopsis includes the generally used techniques for dealing with the insects of each order. In the larger orders there are often families or sections of the order for which specialists have devised special techniques. These are not included as they usually apply to insects not collected by the general collector.

"In this synopsis killing methods are marked D and W to indicate which is to be used if dry (D) or wet (W) preservation respectively is to follow. The preservation methods to follow in each case are then indicated by D or W according to whether the killing agent used is appropriate for dry or wet preservation. That is, if the specimen was killed by a method marked D it should be preserved by a method marked D. M indicates a microscope slide preparation of the whole insect, which is made from a liquid-preserved specimen. Where 'killing jar' is given as the killing agent it should be understood that any of the vaporous killing agents can be used; exceptions to this are mentioned and in some cases a specially recommended killing agent is mentioned separately. In each case an indication is given to which stages in the life cycle the technique is appropriate.

"Insect eggs are preserved in 70%–80% alcohol."

Order Thysanura

Killing:	W	Drop into 75% alcohol.
Preserving:	W	In 75% alcohol with a little glycerin added.
	M	Mount in Euparal or Canada balsam.

Order Diplura

Killing:	W	Drop into 75% alcohol.
Preserving:	W	In 75% alcohol with a little glycerin.
	M	Mount in Euparal.

Order Protura

Killing:	W	Drop into 75% alcohol.
Preserving:	W	In 75% alcohol with a little glycerin.
	M	Mount in Euparal or Canada balsam.

Order Collembola

Killing:	W	Drop into 75% alcohol.
Preserving:	W	In 75% alcohol with a little glycerin.
	M	Mount in Euparal.

Order Ephemeroptera

Killing:	D	Killing jar (adults).
	W	Drop into 70% alcohol (adults and nymphs).
Preserving:	D 1.	Mount on card (adults).
	D 2.	Pin through centre of thorax and set with all wings spread (adults).

	D 3.	Double mount, using minuten pin (adults, small species).
	W	In 70% alcohol (adults and nymphs).

Order Odonata

Killing:	D 1.	Killing jar (ethyl acetate not recommended) (adults and nymphs).
	D 2.	Inject ammonia (commercial strength) into thorax by hypodermic needle (adults).
	D 3.	Drop into nearly boiling water (nymphs).
	W 1.	Drop into 70% alcohol (nymphs).
	W 2.	Drop into Oudeman's fluid (nymphs).
	W 3.	Drop in nearly boiling water (nymphs).
Preserving:	D 1.	Pin through centre of thorax and set adults with all wings spread. (If desired, before pinning, internal organs of abdomen can be removed by cutting along ventral side of abdomen from near base to near apex and extracting contents. Abdomen should then be stuffed with cotton wool dusted with borax).
	D 2.	Mount on card (nymphs).
	W 1.	In Oudeman's fluid (if killed in that fluid) (nymphs).
	W 2.	In 70% alcohol (nymphs).

Order Plecoptera

Killing:	D	Killing jar (adults).
	W	Drop into 70% alcohol (adults and nymphs).
Preserving:	D	Pin through centre of thorax and set with all wings spread (adults).
	W 1.	In 70% alcohol (adults and nymphs).
	W 2.	In 2% formalin (adults and nymphs).

Order Grylloblattodea

Killing:	W	Drop into 70% alcohol (adults and nymphs).
Preserving:	W	In 70% alcohol (adults and nymphs).

Order Orthoptera

Killing:	D 1.	Killing jar (adults and nymphs).
	D 2.	Expose to fumes of benzol (adults and nymphs).
	W 1.	Drop into 70% alcohol (adults and nymphs).
	W 2.	Drop into nearly boiling water (adults and nymphs).
	W 3.	Expose to fumes of benzol (adults and nymphs).

Preserving:	D 1.	Pin through prothorax and set with wings of one side spread. Before pinning, stuff abdomen of large specimens as in Odonata (adults).
	D 2.	Pin through prothorax (nymphs).
	D 3.	Double mount on minutens (small species).
	W	In 70% alcohol with a little glycerin (adults and nymphs).

Order Phasmida

Killing:	D	Killing jar (adults and nymphs).
	W	Drop into 70% alcohol (adults and nymphs)).
Preserving:	D 1.	Pin through mesothorax and set with wings of one side spread. (Stuffing of abdomen can be carried out as for Odonata) (adults).
	D 2.	Pin through mesothorax (adults and nymphs).
	W	In 70% alcohol (adults and nymphs).

Order Dermaptera

Killing:	D 1.	Killing jar (adults).
	D 2.	Expose to fumes of benzol (adults).
	W	Drop into 80% alcohol (adults and nymphs).
Preserving:	D 1.	Pin through right elytron and set with wings of left side spread (adults).
	D 2.	Double mount on minuten pins (smaller species).
	W	In 80% alcohol with a little glycerin (adults and nymphs).

Order Embioptera

Killing:	D	Killing jar (adults).
	W	Drop into 70% alcohol (adults and nymphs).
Preserving:	D	Pin through thorax and set with all wings spread (adults).
	W	In 70% alcohol (adults and nymphs).

Order Blattodea

Killing:	D 1.	Killing jar (adults and large nymphs).
	D 2.	Drop into near-boiling water (adults and nymphs).
	W 1.	Drop into 70% alcohol (adults and nymphs).
	W 2.	Drop into nearly boiling water (adults and nymphs).
Preserving:	D 1.	Pin through centre of thorax and set with wings of one side spread (adults).
	D 2.	Mount on card (small nymphs).
	W	In 70% alcohol (adults and nymphs).

Order Mantodea

Killing:	D	Killing jar.
	W	Drop into 70% alcohol.
Preserving:	D	Pin through thorax and set with wings of one side spread.
	W	In 70% alcohol (adults and nymphs).

Order Isoptera

Killing:	W	Drop into 75% alcohol (adults and nymphs).
Preserving:	W	In 75% alcohol (adults and nymphs).

Order Zoraptera

Killing:	W	Drop into 75% alcohol (adults and nymphs).
Preserving:	W	In 75% alcohol (adults and nymphs).
	M	Mount in Euparal or Canada balsam (adults and nymphs).

Order Psocoptera

Killing:	W	Drop into 75% alcohol (adults and nymphs).
Preserving:	W	In 75% alcohol (adults and nymphs).
	M	Mount small species in Euparal or gum chloral (adults and nymphs).

Order Mallophaga

Killing:	W	Drop into 80% alcohol (adults and nymphs).
Preserving:	W	In 80% alcohol (adults and nymphs).
	M	Mount in Euparal (adults and nymphs).

Order Siphunculata

Killing:	W	Drop into 70% alcohol (adults and nymphs).
Preserving:	W	In 70% alcohol (adults and nymphs).
	M	Mount in Euparal (adults and nymphs).

Order Hemiptera

Killing:	D	Killing jar (adults of Heteroptera and larger Homoptera).
	W 1.	Drop into 70% alcohol (adults and nymphs of all families).
	W 2.	Drop into lacto-alcohol (aphids and scale insects).
Preserving:	D 1.	Pin through scutellum and set with wings of one side spread (larger species, adults).
	D 2.	Pin through scutellum (smaller Heteroptera and Homoptera, adults and nymphs, nymphs of larger Heteroptera).

D 3. Double mount on minuten pins or card point (smaller species, adults and nymphs).
W 1. In 70% alcohol (adults and nymphs).
W 2. In lacto-alcohol (aphids, scale insects, adults and nymphs).
M Mount in gum chloral (aphids, scale insects, adults and nymphs).

Order Thysanoptera

Killing:
W 1. Drop into 70% alcohol (adults and nymphs).
W 2. Drop into mixture of 50% methyl alcohol and 0.5% ethyl acetate (by volume). Preserve specimen within three days (adults and nymphs).

Preserving:
W In 70% alcohol (adults and nymphs).
M Mount in Canada balsam.

Order Neuroptera

Killing:
D 1. Killing jar (adults).
D 2. Chloroform (adults).
W 1. Drop into 70% alcohol (all stages).
W 2. Drop into Carnoy's fluid (larvae).

Preserving:
D 1. Pin through thorax and set with all wings spread (adults).
D 2. Mount on card (larvae).
W In 70% alcohol (all stages).

Order Mecoptera

Killing:
D Killing jar (adults).
W 1. Drop into 70% alcohol (all stages).
W 2. Drop into Carnoy's fluid (larvae).

Preserving:
D Pin through thorax and set with all wings spread (adults).
W In 70% alcohol (all stages).

Order Lepidoptera

Killing:
D 1. Killing jar (must be dry) (adults).
D 2. Kill with ammonia (adults of very small species).
D 3. Chloroform or ethyl acetate (larvae).
D 4. Inject with Carnoy's fluid (pupae).
W 1. Drop into Kahle's fluid (larvae).
W 2. Drop into boiling water (larvae and pupae).
W 3. Inject with Carnoy's fluid (larvae and pupae).

Preserving:
D 1. Pin through thorax and mount with wings of both sides spread (adults).
D 2. Mount on minutens, set on miniature setting board and double mount (very small species, adults).
D 3. Pin or mount on card (pupae).
W 1. In 70% alcohol away from light (from boiling water, Carnoy's fluid or Kahle's fluid) (larvae).
W 2. In 70% alcohol (after injection with Carnoy's fluid or from boiling water) (pupae).

Order Trichoptera

Killing:
D Killing jar (adults).
W 1. Drop into Pampl's fluid (adults, larvae and pupae).
W 2. Drop into Pamle's fluid (adults, larvae and pupae).

Preserving:
D 1. Double mount on minutens, after setting (small species, adults).
D 2. Double mount on minutens, after setting (small species, adults).
W 1. In 70% alcohol (adults, larvae and pupae).
W 2. In 3% formalin (from Pample's fluid) (adults, larvae and pupae).

Note:
Larvae can be preserved in their cases or the latter removed and mounted on card for dry preservation. It is preferable to retain the case in liquid with the larva.

Order Diptera

Killing:
D 1. Chloroform (adults).
D 2. Killing jar (adults).
W 1. Drop into 80% alcohol (larvae, pupae).
W 2. Drop into KAA for two hours (larvae).
W 3. Drop into boiling water (larvae).

Preserving:
D 1. Mount on cards, specimens glued on their sides (adult Tipulidae).
D 2. Mount on card points (small species, adults).
D 3. Pin through top of thorax (adults, larger species).
D 4. Double mount on minutens from above (adults of smaller species).
D 5. Pin through side of thorax with minutens, double mount (adults, small species).
D 6. Pin from below with minutens, double mount (adults, small species).
D 7. Store in cellophane envelopes (adult Tipulidae).
W 1. In 80% alcohol (from 80% alcohol or boiling water) (larvae and pupae).

	W 2.	In 95% alcohol from KAA (larvae).
	M	Mount in Canada balsam (small species, adults).
Note:		Several special techniques are used for some families of this order.

Order Siphonaptera

Killing:	W	Drop into 70% alcohol (adults, larvae and pupae).
Preserving:	W	In 70% alcohol (adults, larvae and pupae).
	M	Mount in Euparal (adults and larvae).

Order Hymenoptera

Killing:	D	Killing jar (adults).
	W 1.	Drop into 70% alcohol (adults and larvae).
	W 2.	Drop into Carnoy's fluid (larvae and pupae).
Preserving:	D 1.	Pin through thorax and set with all wings spread (large adults).
	D 2.	Mount on card points (small adults).
	D 3.	Pin on minutens and double mount (adults).
	W	In 70% alcohol (from alcohol and Carnoy's fluid) (especially small species, larvae and pupae).
Note:		Bees of all kinds are usually preserved dry with the mouth and legs set in an extended position.

Order Coleoptera

Killing:	D 1.	Killing jar (adults).
	D 2.	Ethyl acetate fumes (adults).
	D 3.	Drop into boiling water (adults).
	W 1.	Drop into Carnoy's fluid (larvae).
	W 2.	Drop into 70% alcohol (larvae).
	W 3.	Drop into Pample's fluid (pupae).
Preserving:	D 1.	Pin through right elytron, near front (adults).
	D 2.	Mount on minuten pin, placing minuten as in D 1. double mount (adults, smaller species).
	D 3.	Mount on cards (small species).
	D 4.	Mount on card points (adults, small species).
	W	70% alcohol (from Pampl's fluid, Carnoy's fluid or alcohol) (larvae and pupae).

Order Strepsiptera

Killing:	D	Killing jar (winged males).
	W	70% alcohol (males, females in their hosts).
Preserving:	D	Mount on card points (winged males).
	W	70% alcohol (males, females in their hosts).
	M	Mount in Canada balsam or Euparal (adults).

DISPLAY

The methods of collecting and killing insects apply to both taxonomic collections and public displays. The methods pertaining to setting and pinning do not apply to specimens required for display unless they are to show the methods involved in setting up a scientific collection. Displaying insects in a lifelike manner means pins cannot be stuck out from the bodies of specimens, although in a few cases specimens may be pinned during the setting and drying process, but the pins must be removed and the hole camouflaged afterwards. For some reason or another, specimens from a scientific collection may have to be used. This necessitates the removal of the pin. Sometimes this is not an easy matter as the insect's juices corrode the pin so that it sticks. One way of removing the pin is to place the specimen into a relaxing jar until the pin can be easily removed, or it can be removed using an electric depinner (Fig. 8.53). This device is a high current, low voltage, heavy wound thick wire transformer of about 2 volts. The insect pin acts as a fuse when placed between the two transformer wires and glows red hot, which in turn burns through the corroded material. The actual time taken should be minimal or a hole will burn through the top of the specimen, giving an undesirable appearance.

The presentation of insects in mini dioramas projects a better understanding of habitat types, camouflage, behaviour, life cycles, and growth.

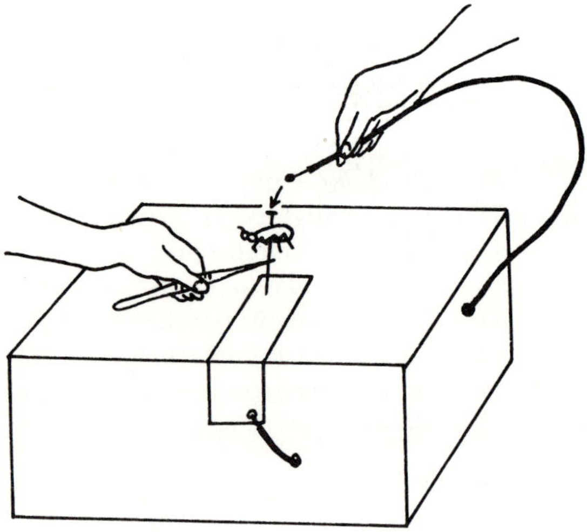

Fig. 8.53. A depinning device.

Small displays can be made, which effectively depict a particular subject, using large and easily identifiable specimens. Ants, although small in size, can be displayed to show the variety of nests and their construction. This involves moulding and casting nests, cut away, to show the internal structure above ground and tunnels under ground, and removing nests from trees such as those made by the green tree ants. Tebble (1980) moulded termite mounds with latex. The termite mouds chosen for casting were of the "magnetic" type *Amitermes laurensis*, an Australian species. Three were selected, two 2.5 metres high, and the other 4 metres high.

LATEX MOULDING TERMITE MOUNDS

Preparation for spraying

An area 30 cm wide around the base of the mound should be cleared to prevent loose debris being blown onto the rubber surface during the spraying operations.

The spray gun should be set between 275 kPa and 620 kPa depending on the surface and penetration required. Rubber is decanted from the tin into ice cream containers via a wire mesh filter to remove any lumps. It is advantageous to have two or three pots for the gun, preferably of large capacity (for example, 1 litre) so that the spraying operation can be more or less continuous. Should spraying be discontinued, even for a short period, the gun must be cleaned in fresh water.

Procedure

Spray the latex over the entirre mound surface, using as many coats as necessary to produce a good skin thickness. Usually four applications are sufficient.

Starting from the top drape the precut cotton stockingette down the surface of the still tacky mound. Using a dry brush, press the material onto the surface, making sure that it conforms properly to the rubber surface. It is sometimes necessary to cut and patch small sections to achieve this. When the stockingette is firmly in place it can be wet with a thick coat of latex. This coat is better brushed on as the thickness can be built up more quickly and minor faults in the stockingette can be more easily seen and corrected. When this has dried, a final coat can be sprayed on.

Constructing the fibreglass jacket

Chopped strand mat (15 g m^{-2}) is rolled out, cut to length, and backed with 50 μm polythene sheeting, leaving a 10 cm margin of fibreglass on either side for overlap. This can then be rolled up ready for use. These rolls should then be tacked in position on top of the mound. Begin at the edges, folding the glass around so as to have a clean surface for subsequent cutting. Once the rolls of cloth are correctly positioned, with the polythene sheeting on the inside, they can be unfurled and fastened into position with 40 mm nails fitted with thick cardboard washers. These fastenings should be prepared in advance, cutting 30 mm wide strips of cardboard into 30 mm sections, leaving just enough material to prevent the individual washers from separating. Drive a nail through the centre of each square leaving strips of tear-off fasteners ready for instant use.

When the glass is firmly in place, it can be wetted out. This has to be done in alternate vertical strips 30 cm wide around the perimeter of the mound. If too big an area is wetted out at once, the weight of the resin will cause the glass to sag away from the surface. Once these strips have set, the alternate areas can be wetted out. If the glass sags too much during the wetting out process, more fastening can be added.

With this stage completed, consideration can be given to the number of separate pieces required to remove the jacket. This should then be marked onto the jacket with a waterproof pencil. Make sure that there is an adequate number of reference marks to accurately realign each separate piece. Once the marking has been completed those pieces still *in situ* can be reinforced with wooden or fibreglass reinforced paper straps to prevent excessive warping. The jacket can then be separated using an angle grinder. As each separate piece is removed, the reference marks must be transferred to the latex mould to facilitate subsequent repositioning. With the jacket removed, the latex sections can then be cut using a sharp knife, scalpel or scissors. As each rubber section is pulled off it should be laid directly into its jacket and taped into place. This will prevent the rubber from becoming damaged or distorted.

Transporting the moulds

To prevent damage and distortion, the rubber moulds are transported laid inside their respective jackets, packed with grass, and wrapped in polythene. Hessian cloth can then be tightly wrapped around and sewn firmly into place using a bag needle and string. This results in a well-cushioned mould which is water, dust, and abrasion proof.

Lay up in fibreglass

In moulding, no separator is used so as to reproduce as much of the original surface as possible. The rubber moulds are placed into their jackets and carefully realigned with the reference marks. Where necessary, super glue is used to hold the rubber in place. Any loose particles of dirt are then blown away; care must be taken not to remove any material from the rubber.

Coloured gel coats can now be sprayed onto the rubber, two or three different colours being employed to match the original surface colours as accurately as possible. A coloured resin paste is then applied over the entire surface to eliminate small surface irregularities, thus reducing the chances of blisters and air bubbles on the finished cast. The last layer is 458 g m^{-2} chopped strain mat of fibreglass and the cast is left to cure. Once removed from their jackets, the separate pieces are joined together with glass mat from the inside and a coloured polyester paste between the joints. This can be trimmed with a knife while still green. Finally, the joints can be dressed with coloured powders suitably matched to hide the jointing.

CONSOLIDATING TERMITE MOUND SECTIONS

Sections of termite mounds were collected and prepared by the authors using the following method: A suitable, small size mound was selected and thick sections were cut from it with a two handled timber saw. The sections were fumigated in order to kill the termites within and several coats of consolidator sprayed on. Consolidator consists of 160 g Paraloid B72 (an acrylic polymer obtained from the Rohm Hass Co.) dissolved in 1600 mL toluene. The substance is applied with a small spray gun and allowed to dry between the successive coats.

HABITAT DISPLAYS

Displays which show insects going about their business are always very popular with the public. Most people find such displays more comprehensible than a large collection showing rows and rows of pinned insects. The specimens can be shown in small, diorama like settings, depicting various habitats and their inhabitants. Displays which deal with aspects of insect life usually hidden from casual observers are most educational (Fig. 8.54).

Camouflage displays are popular as the interested onlooker must search in order to find the respective individuals.

Crosscut sections of insect damaged timber, underground nests, and tunnels all make very interesting displays and show the public something that is rarely observable.

Since there are very many ways of displaying insects, interested parties should search the wide range of literature available to ascertain which method is appropriate for their needs.

However, some information can be provided to assist in the setting and preserving of insects solely for display purposes. Butterflies with open wings are set and dried using the butterfly board with paper strips across the wings.

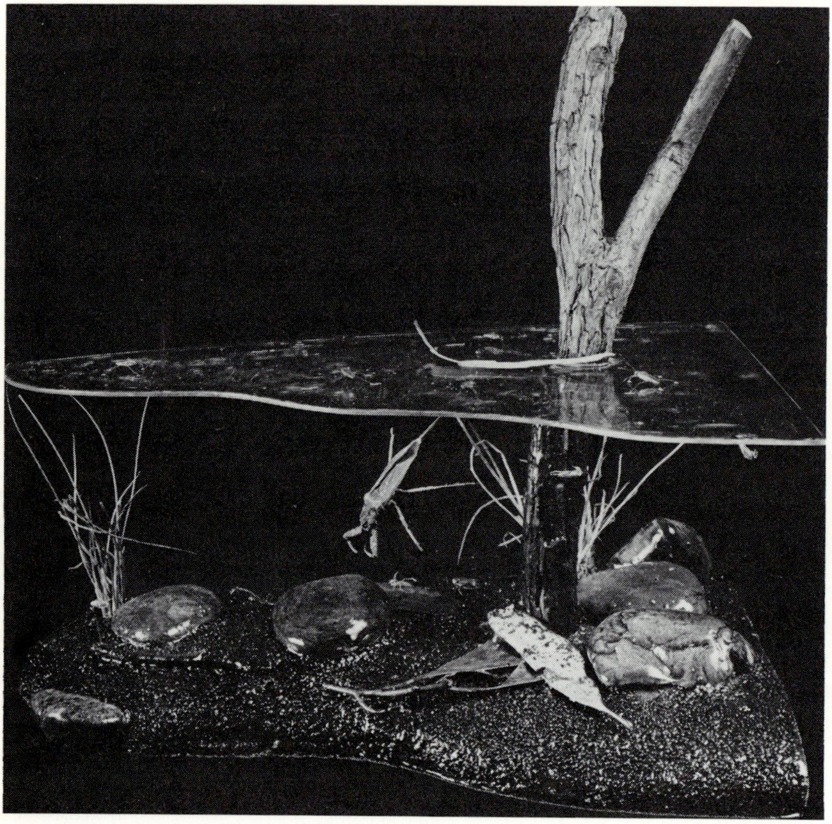

Fig. 8.54. Water habitat display. Preparation M. Dingley. Photo K. Lowe. Courtesy of The Australian Museum.

The legs are not set, nor is the specimen pinned through the thorax. When dried, the ventral surface of the body is glued to its final position within the display area. Specimens that are required to have their wings closed require a little patience and care for they are most likely to be observed from the side, thereby necessitating properly aligned legs and body. The wings should be pushed together and held in place, using small sticky labels until they set, or a couple of small drops of water soluble glue can be placed on the veins of the dorsal surface, so that, when pushed together, the wings stick together. The butterfly is then laid on its side and the body and legs set upon a twig, pins holding the legs in position. The specimen is allowed to dry in this position until required.

All winged insects can be set using the above technique, but care should be taken when using glue on transparent wings.

Beetles mounted with spread wings (in flying position) sometimes present a special problem. Due to the humidity of the surrounding air, the wings may partially fold in time, thus giving an unnatural appearance to the specimen. This can be prevented if the wings are injected with nitrocellulose lacquer. Just before setting, the wing muscles should be injected with the lacquer and, if the specimen is large enough, some lacquer can also be injected into the main veins of the wings. If the beetle is small, the muscles (and also the veins, if possible) can be punctured with a needle dipped in lacquer. The body cavities of the insect can also be injected. This will render it more durable and longer. The lacquer can be removed with a solvent.

It is sometimes necessary to reinforce fragile specimens. This can be done by spraying the set and dried insects with clear lacquer. Inralac is excellent for the purpose. This lacquer was developed by the International Copper Research Association. It is used to coat brass surfaces, and available from various commercial suppliers.

Insects normally seen walking (even though they have wings), such as cockroaches and bugs, should be set and dried using pins as props in such a way that they hold the specimen off the ground, and mount their legs in a natural position without actually putting a pin through any part of the body.

It was mentioneed earlier that large fleshy bodied insects require removal of the viscera and subsequent filling with cotton wool. Freeze drying produces better results and, in most cases, colour is not completely lost.

Larvae have special problems because of their soft skin and fleshy contents. Large specimens may be dried in warm air after removal of the viscera, or they may be moulded using dental impression rubber and cast in polyester resin prior to painting. They also can be freeze dried.

CASTING THE INTERIOR OF AN ANT NEST

There is no better method of demonstrating the extensiveness and complexity of an ant nest than by preparing a cast of all its chambers and passages. The description below, by Mr H. Plowman, is a full account of such an undertaking. The work was carried out by the Queensland Museum, Brisbane, Australia, preparation section, lead by Mr Tebble. Early work in this field was undertaken by G. Ettershank of Monash University, Melbourne, Australia, who successfully cast meat ant nests using molten lead. One intriguing feature he discovered was that each entrance to a colony leads to an independent system of underground galleries, that is, an ant cannot enter one hole and later come out another.

"Catalysed plastic resins present an obvious alternative to lead casting. The resin used in this experiment was a water extended polyester pigmented with standard polyester pigments, using a different pigment colour in adjacent entrances to demonstrate the separation of gallery systems. This resin was catalysed with standard methyl ethyl ketone peroxide (MEKP) by approximately 1%. A total of 220 kg of resin was used in this nest.

"This particular resin adapts well to casting because of its slight flexibility. This enhances excavation and the subsequent handling of the cast. It cures in a damp environment, and even under water, which is to be expected in a subterranean environment. Before attempting to cast the gallery system, the ants in the nest were killed using chloroform administered from a squeeze bottle. This was done by inserting into the entrance holes a series of small plastic tubes, which were held in position with pieces of Plasticine; the chloroform was squeezed down the plastic tubes, which were then sealed off. The ants were killed to stop them burrowing from their own gallery system into an adjacent gallery in order to escape the flow of resin, although whether or not they would have done this is not known. Certainly, the chloroform killed all activity in the nest and no breakthrough was evident in the cast.

"The first step was to make a latex cast of the surface of the nest. Before the latex was sprayed on, all entrance holes were plugged with cotton wool to prevent latex running down them. The surface was sprayed with three to four coats of latex, giving a good coverage, and ensuring that all ironstone pebbles and twigs were adequately covered. The surface was then covered with cotton stockingette and a further coat of latex applied by brush. This was followed with a layer of Visqueen. A fibreglass jacket over the top finished the casting of the nest surface.

"After the latex was peeled off the surface, all entrance holes were once again plugged with cotton wool. Those holes selected for pouring had these plugs removed and small plastic funnels inserted. These were held in position with coloured plasticine. Those holes to have the same coloured resin had similar coloured plasticine at the base of funnels.

"The decanted resin could now be mixed with the selected colour pigments and lightly catalysed. All of mixing was done in 2 litre ice cream buckets using 1 cm^3 of catalyst. This gave ample time to pour the resin into the funnels. With four preparations being poured at once, the rate at which resin was consumed was considerable. There seemed to be no bottom to those gallery systems. The best way to pour the resin was to fill the funnels completely and wait for the resin to slowly flow into the galleries. This process was repeated until no further flow was evident. The funnels were left standing until the resin started to cure and go 'cheesy'. The funnels were then removed.

"The nest was left to stand overnight to ensure that all resin was cured prior to excavation. An interesting point is that no release agent is necessary with this type of casting because the inside surface of the gallery system is coated with a smooth layer of salivary secretion which acts as a natural release agent. Thus, soil releases freely from the resin cast. The excavation was 5 m ç 4 m ç 1.5 m deep and took four men three days to complete, using picks, shovels, a jackhammer, an aquajet, awls, and screwdrivers.

"The starting point was a circle around the nest 0.5 m from the nearest gallery system. This was to ensure that the vertical progress of the excavation did not interfere with galleries poured with resin. A vertical and horizontal trench was dug to approach the edges of the galleries; once these were encountered they were followed down until the base of their systems was reached. The first 0.5 m of the excavation was easy as we were digging into loose sand. The remaining metre was into a consolidated clay which made picks useless. The remaining part of the excavation was continued using a small jackhammer run from a generator. This valuable piece of equipment made comparatively light work of the heavy clay.

"So the hole grew bigger and bigger until there was a column of earth laden galleries isolated at the centre of a large hole. At the outside edge of this column the numerous outer edges of the galleries were seen. As the excavation got underway, it became obvious that there was a mass of upper level galleries just below surface level. Out of curiosity it was decided to clean away the sand which surrounded them. Because of the compactness of these galleries and the delicate connections between them, it was extremely difficult not to damage them while removing the earth.

"The only practical means to achieve this was to spray the surface with the aquajet. The jet had a long nozzle which allowed access into those areas. The results were quite amazing as the spray from the gun readily cleaned away volumes of earth, revealing the mass of coloured galleries. Great care had to be taken using this tool as it was extremely powerful; it is extremely easy to blow off delicate gallery chambers. Also, much water in a given spot saturated the clay and sections of the nest would collapse from the side of the mound. All sections of the upper galleries revealed after spraying were bound together with string and fibreglass tape in an attempt to consolidate them and prevent damage.

"Consideration was then given to the construction of a suitable frame to allow for the support and removal of the gallery system back to the Museum. The frame was constructed using a weld mesh top, which was not only strong, but also had numerous anchor points, Dexion sides, and legs with feet to prevent the weight of the nest forcing them into the water sodden clay. This frame was bolted together to ensure its strength was sufficient to carry 220 kg of resin. This frame was then lowered over the top of the galleries and, when in place, was approximately 100 mm above the nest surface.

"The freshly cleaned upper surface of the nest was secured to the weld mesh top using lengths of 8 gauge wire which were looped under and around those areas which needed supporting. The sides of the gallery system were excavated using awls and screwdrivers and, as fresh areas of galleries were exposed, they were supported by tying off one end of 8 gauge wire to the weld mesh and looping the other end down, around the under galleries which it seemed fit to anchor securely. All wires were fastened to the resin cast using fibreglass tape. Extra supports which seemed necessary were added. The removal of earth away from the galleries and vertical shafts, and the subsequent wiring of some to the weld mesh top was a tedious task, but one which had to be done with great care, considering that these support wires were to replace the support provided by the earth, and that this large mass had to be lifted up and out of a hole and onto a truck for transportation back to the museum.

"The excavation continued towards the centre of the nest and freshly exposed galleries were supported. Any inaccessible areas were cleaned with the aquajet. After the excavation was completed, a few gallery systems, whose bases were not exposed, were broken off at ground level. The vehicle used on this trip was a Toyota fourwheel drive with overhead gantry. This vehicle was reversed to the edge of excavation, and the gantry boom was extended. A timber beam was added to the steel boom to give greater support for the block and tackle.

"The Dexion frame, which was now completely supporting the meat ant nest cast, was roped to the block and tackle and pulled up and out of the hole. This was loaded onto the back of a truck to be transported to Brisbane. Loose bases of gallery systems hanging from the frame were bound with fibreglass tape to prevent them from swinging freely while in transit".

ARACHNIDS

Arachnids form the second largest class of the Phylum Arthropoda after insects. These animals are almost all land dwellers, with a few exceptions such as waterspiders, and are usually aggressive, predatory creatures.

COLLECTING

Most arachnids can be collected with methods similar to those employed for collecting insects. Many of them are nocturnal in their habits or lead secretive lives, under logs, rocks or in the leaflitter on the forest's floor. For effective collecting, one must be familiar with the habits of species to be pursued. Ground dwelling arachnids can be collected by singled collecting either at night with a torch or during the daylight hours from under logs, rocks, bark and other similar hiding places. A sweep net or a beating tray can be beneficial to the collector when capturing specimens from the vegetation. Pit traps, described earlier, can be employed with great success. Various selectors can be employed to collect specimens from leaf litter, decaying wood and soil, in the same way as described earlier. Members of the Order Araneae (spiders) which spin webs but often hide nearby can be easily captured by using a tuning fork. If the fork is hit so that it vibrates and is then held close to the spider's web, the spider will run from its hiding place and move towards the tuning fork. However, a spider is very quick to determine whether the stimulus is prey or not and will move rapidly back to its hiding place. If the collector is quick enough, a small fish tank net or jar can trap the spider and break the web. The jar or a small container can then be used to confine the spider. If a tuning fork is unavailable, then a captured live insect thrown into the web will cause the spider to advance, as does tapping very faintly with a twig. Tapping around the burrow entrance of the wolf spider (Lycosidae) and some other ground dwelling spiders, or prodding down the burrows, will cause the occupants to surface. Alternatively a knife can be quickly thrust slantwise into the ground to intersect the burrow below the spider and so cut off its escape.

Ordinary small glass bottles without a lid can be used to trap those spiders which are seen on flat surfaces such as brick walls, smooth barked trees and the internal walls of homes. The bottle is held open end towards the spider and moved slowly towards it. When the bottle is very close it is then pushed into the flat surface, encasing it. If the spider does not want to move into the bottle it is forced to by sliding a piece of paper or thin cardboard between the bottle and wall. When the open end is sealed the bottle and paper can be moved to an easier position to allow the bottle to be upturned, after which the paper is flicked using the index finger until the spider falls down and into the bottle.

Collecting webs

Two dimensional webs are best collected using the technique devised by Knudsen (1972). It is very simple but effective and is as follows. A clean, undamaged web is chosen and sprayed with an oil based black paint from an aerosol can without forming globules. Spray the web evenly until it becomes black throughout. A stiff white cardboard of ample size is coated on one side with water soluble glue thinned with water in the ratio of three parts glue to two parts water and thoroughly mixed. The painted surface of the board is then carefully maneouvred towards the blackened web until the web sticks onto the board. The supporting strands are cut with a knife and the board and web are allowed to dry.

Collecting ticks and mites

Although a great number of mite species live in the soil and leaf litter from which they can be collected by the usual techniques employed by entomologists seeking insects in these habitats, a great number of species of this order are parasitic and live on a wide variety of hosts. Some mites are free living. Many are strictly host specific.

It is virtually essential that the vertebrate host be trapped alive, and kept alive in as nonstressful in a state as possible, until it is collected. The trapped or shot host should be quickly placed in a closeweave calico bag with a draw string closure, which should be tightly closed. On no account should two specimens, even of the same species, be placed together. Important data to obtain are the sex and age of the host, and the locality. A label can be tied to the outside of the bag. The bag and contents (if necessary together with several others to save time) are then placed in a plastic bag, a wad of cotton wool soaked in chloroform is dropped in, and the bag is sealed.

When it is certain the host is dead (at least half an hour) a bag at a time can be removed and the contents tipped into a white enamel dish. The calico bag should then be everted over the dish and searched thoroughly for specimens which may be caught up in the seams and in the material. The actual specimen must also be searched thoroughly. Animals should be well combed for free living parasites.

Inner ears, scrotal sacs, armpits and inner thighs should be searched for trombiculid mites, with any specimens being removed by carefully snipping the skin around them. Ticks also should be removed still attached to the skin to prevent damage to their mouth parts. Nasal passages should be searched for nasal mites. Birds should be treated the same way, but try to prevent too much damage to feathers by heavy combing. Blowing against the lie of the featers will often uncover parasites. With birds, ear sockets, nasal orifices and bare skin around the eyes are particularly subject to parasite infestation.

It is absolutely essential that an accurate identification be made of the host, so a study skin should be prepared or the body wet preserved. It is best to give the vertebrate host a field number immediately it is caught. Field data can then be inserted in the field diary against that number, and the *same* number should be given to all ectoparasites found on that specimen. Then, at all times in the future, an accurate tie-in of host and parasite can be made.

Keep in mind that other vertebrates are host to ectoparasites. Check all reptiles, which are particularly prone to ticks and mites; fish have external and throat lice. Frogs, however, seem to be singularly clean.

Invertebrates can also host parasitic arachnids. The best method for collecting is to drop the invertebrate host into a container filled with 70% alcohol. If only one host animal is kept in each container, the accurate determination of tie-in between host and parasite can be achieved with certainty.

Some ticks are found on vegetation and so Baker and Wharton (1952) devised a drag consisting of a square metre of flannel cloth tacked on one side to a one metre stick. A rope is tied to either end of the stick, forming a loop which the operator holds onto as he or she drags the cloth through the bushes and grass near animal tracks. The cloth is periodically checked for clinging ticks.

Water mites can be collected by seeping a dip net through water plants or through open water. The water samples are placed into a white bottomed tray and examined for specimens which can be picked up with a small paint brush dipped in alcohol.

KILLING, PRESERVING AND STORAGE

Killing is effective using a potassium cyanide or other dry killing bottle (see above). Specimens designated as dry preparations for display purposes are best killed in a killing bottle with vapours of ethyl acetate. It is important that the specimen does not become wet; delicate hair should remain dry and retain its natural appearance.

Landois (1879) prepared spiders in the following way. The abdomen was separated from the thorax, the viscera squeezed out through the anus, and the skin inflated by air through a straw. The inflated skin of the abdomen was dried with warm air and then attached to the thorax. The specimen was pinned out to dry on a block of cork. This method is very similar to that described above to prepare the skins of insect larvae. Naton (1960) killed small spiders in ethyl acetate vapour. Large specimens were killed by pouring the same chemical on them, then placed in acetone for two to 10 minutes (the larger the spider, the longer it should remain in the acetone). This helps to relax the legs. They can then be arranged as desired. After this, the spiders are placed in acetone for four weeks. During this time the abdomens may shrink. To prevent this, the spiders should be fed before they are killed.

Specimens which are to be preserved wet can be killed by being dropped directly into a jar containing 70% alcohol. Another method is to place specimens into Pampl's fixative (see above) for eight to 12 hours and then transfer them into 70% alcohol.

Brown, black and grey colours are maintained after the spiders have been subjected to Pampl's fixative. The storage medium should be 70% alcohol; the specimens should be placed in glass phials containing a label with data such as locality, date of collection, habitat, and the name of the collector recorded on it. The end of the phial is plugged with cotton wool after the preservative has been added. The phial can then be placed in a larger jar with a leakproof top also containing 70% alcohol. Very small arachnids, like ticks and mites, are usually prepared as microscopic preparations.

PREPARATION FOR DISPLAY PURPOSES

Soft bodied arthropods, such as spiders and the like, cannot be air dried because their thoraxes and abdomens may shrivel, and thus lose their lifelike appearances. Very large spiders and scorpions can be eviscerated and their abdomens, as well as their thoracic cavities, filled with cotton wool as described earlier. These specimens then can be pinned out in the required pose on a block of rigid plastic foam or cork and air dried. However, only a few species can be treated this way. Most arachnids require more elaborate preserving and display techniques.

Landois' method can produce lifelike display specimens, as can the technique described by Matausch (1914) by which the abdomens of the spiders are removed and carved in wood. The abdomenless spiders are dehydrated in a series of alcohol baths (40%, 60%, 80% and 96%) then placed in a bath of one part carbolic acid and three parts xylol, and finally in xylol. Small spiders can be dehydrated and degreased in acetone. The prepared spiders are dried, then the wooden abdomens are attached and painted the appropriate colours.

Many arachnids are large enough to be displayed in an acrylic wet box with a white background. Spiders are sometimes displayed with a male and female of the species together to show structure and size differences.

Embedding spiders of any other terrestrial invertebrates in clear plastic is a must if they are to be handled and examined closely by students. The result is a long lasting and durable display. Arachnids, like any other soft bodied animals, need special care when being embedded. Details are given in Volume 2, Chapter 11.

Freeze drying is the best method for producing lifelike specimens (see Volume 2, Chapter 9). Some specimens will have to be painted in order to render them more natural in their appearance. Generally, spiders retain their markings at an acceptable level for inclusion in displays.

WEB MAKING

If a web is needed to be displayed in as natural an environment as possible, what better way to make one than to use the living spider. Branches or bushes are first placed into position permanently so they cannot be moved after the web has been made. This is done inside the display case, and a living spider is placed on the vegetation. The spider may make a perfect web the first time, but if it is not good enough, break it down and let the spider start again. This can be repeated as many times as possible until the spider dies. When the web is complete the spider should be captured, killed and pinned out onto a soft board in the position that it will take on the web. It should be frozen and later freeze dried (see Volume 2, Chapter 9) after which it can be painted if necessary and attached to the web. If the tarsal claws have dried in a hooked position, the spider may hang on the web strands; however, a small drop of glue on each tarsus will secure it permanently.

DIPLOPODA AND CHILOPODA

These animals can be collected with the same techniques as ground dwelling arachnids. Killing and preserving techniques are also the same, although most collectors prefer to drop the freshly collected specimens in 70% ethanol for 24 hours and then transfer them to fresh 70% ethanol for permanent storage. The best method to prepare them for exhibition purposes is freeze drying.

SYMPHYLA AND PAUROPODA

These small arthropods are usually soil or leaf litter dwellers and can be collected by sifting. The sifted leaf litter should be placed in a Tullgren funnel in order to separate the animals from the debris. Specimens are dropped in 70% ethanol, and later transferred to fresh ethanol of the same strength for permanent storage. These small animals are often prepared as microscopic preparations.

ONYCHOPHORA

Onycophora, or velvet worms, can be found in damp, dark places such as under fallen logs, rocks, moss and in cavities in the soil. Killing and preparation methods are similar to those employed for arachnids. These soft bodied animals tend to curl up if killed with a harsh, fast reacting chemical. Therefore, it is practical to use something which has an anaesthetic effect. Ethyl acetate is ideal, but it is important that only the vapour reaches the animal. Its skin must not become wet. The best method is to place the animal on a piece of blotting paper and allow it to settle and extend its body. Prepare a killing jar with some cotton wool and add a few drops of ethyl acetate. Place the blotting paper with the velvet worm on it in the jar and close it. The animal will expire without curling up. It can now be prepared either as a wet preserved specimen, like an arachnid, or freeze dried for exhibition purposes.

REFERENCES

Baker, E. W., and G. W. Wharton: *An introduction to acarology.* Macmillan, New York 1952.

Endrödy-Younga, S.: Collecting methods and material processing of Arthropoda in the Transvaal Museum. *Transvaal Museum Bulletin* 17/1979: 20–26.

Knudsen, J. W.: *Collecting and preserving plants and animals.* Harper and Row, New York 1972.

Landois, H.: Conservierungs-Methoden der Spinnen für der Sammlung. *Jahresbücher Westfal. Prov. Ver. Munster Zool. Sekt.* 1878/79: 42–43.

McKillop, W. B., and W. B. Preston: Color measurement of entomological specimens. *Canadian Entromologist* 113/1981: 255–258.

Matausch, J.: A new method of preparing spiders for exhibition in numerous groups. *Science* 40/1914: 710.

Naton, E.: Ein Beitrag zur Konservierung von Spinnen durch Behandlung mit Aceton. *Zool. Anz.* 165/1960: 329–332.

Penecke, K. A.: Ein Mittel zur Entfernung des erdigen Überzüges von der Oberfläche von Kafern. *Coleopt. Cbl Berlin* 4/1929: 85–86.

Peterson, A.: *Larvae of insects. Part 1. Lepidoptera and hymenoptera.* Edwards Bros., Ann Arbor 1959.

Smart, J.: *Insects.* British Museum of Natural History (Leaflet no. 4), London 1940.

Smithers, C.: *Handbook of insect collecting.* A. H. & A. W. Reed, Sydney 1981.

Tebble, T. P.: Moulding termite mounds using latex. Abstracts of papers presented at the Conference of Museum preparators. Australian Museum, Sydney 1980.

Wagstaffe, R., and J. H. Fidler: *The preservation of natural history specimens.* Vol. 1. Invertebrates. H. F. G. Witherby, London 1970.

Wiegel, K. H.: Die Nikotintotungsmethode und die Behandlung von Lepidopteren. *Nachr. Bl. bayer. Ent.* 7/1958: 35–36; 45–47.

SOURCE MATERIAL

Amateur Entomologist's Society (London): Leaflet no. 28 (revised). *Killing setting and storing butterflies and moths.* London 1972.

Borror, D. J., and D. M. DeLong: *An introduction to the study of insects.* Rinehart, New York 1954.

Carne, P. B.: Preservation techniques for Scorabeid and other insect larvae. *Proceedings of the Linnaean Society of New South Wales* 76/1951: 26–30.

Cooke, J. A. L.: Notes on some useful arachnological techniques. *Bulletin of the British Arachnology Society* 1/1969: 42–43.

DeLong, D. M., and R. H. Davidson: *Methods of collecting and preserving insects.* Ohio State University Press, Columbus 1936.

Oldroyd, H.: *Collecting preserving and studying insects.* Macmillan, New York 1959.

Oman, D. W., and A. C. Cushman: *Collection and preservation of insects.* U.S. Department of Agriculture Miscellaneous Publication 601, Washington D. C. 1948.

Peterson, A.: *A manual of entomological equipment and methods. Part II.* John S. Swift, St Louis 1937.

Ross, H. H.: How to collect and preserve insects. *Illinois Natural History Survey Circular* 25/1934.

Ross, H. H.: *A textbook of entomology.* (2nd ed.) Wiley, New York 1956.

Swain, R. B.: *The insect guide.* Doubleday, Garden City NY 1956.

TECHNIQUES FOR DISPLAY AND SAFETY

9
Freeze drying

FREEZE DRYING has become one of the most convenient ways of preparing biological specimens. Plants and invertebrates adapt well to preparation in this manner. The freeze drying of vertebrates is dealt with in Volume 1, Chapter 9.

FREEZE DRYING PLANT AND INVERTEBRATE SPECIMENS

The freeze drying process is intended to preserve specimens which contain a large percentage of water. To carry out the process, the specimen should be frozen, then placed in a vacuum chamber. The temperature of the chamber is kept low and, when the specimen's ice content sublimes, the process is completed. Since the ice changes directly from solid to vapour state, the usual chemical, physical and enzymatic changes associated with other forms of drying do not take place.

The firm of Edwards & Crawley, of Sussex, United Kingdom, makes various freeze drying apparatuses, some of which are specifically designed for biological specimen preparation.

Since it is desirable that the specimen's temperature be kept low (below freezing), especially at the beginning of the process, the apparatus embodies a heat shield that reduces the energy reaching the specimen from the chamber walls at room temperature.

In one form a vertical cylindrical chamber contains a helical coil, the lower end of which is closed by a flat spiral. The coil is cooled by direct expansion refrigeration in such a way that the spiral and the lower part of the helix are at 45° C, while the temperature of the upper part of the helix averages –25° C. The upper part, therefore, acts as a heat shield and the lower part as a condenser for water vapour. A heat diffuser in the form of a cylinder of metal, blackened on the inside and outside, may be fitted inside the helix, but not in contact with it.

GENERAL PROCEDURE

The specimens are killed in such a way that they die relaxed, and are posed with the means of supports (pins, cardboard strips, or modelling clay; Fig. 9.1). They are frozen in a deep freezer or in the freeze dryer specimen chamber (without applying the vacuum). After freezing, the specimen is weighed (the weight of the supports is taken into consideration) and the weight recorded. The specimen is then placed in the chamber, which is sealed and evacuated.

Liquid nitrogen is sometimes used to freeze the specimens very rapidly and in a required pose, but great care should be exercised to avoid distortion and splitting. It is practical to apply only a small amount of liquid nitrogen to the specimen, thus forming a frozen, hard shell on its surface. If the entire bulk of the specimen is frozen right through instantly, the hard frozen outer shell may rupture when the interior freezes and expands.

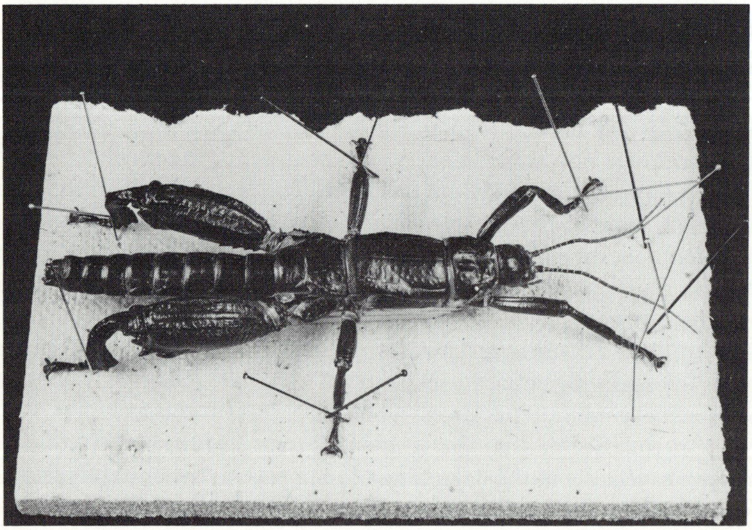

Fig. 9.1. The method of pinning out an insect for freeze drying. Photo A. Pitman. Courtesy of The Australian Museum.

ACCELERATED DRYING

Since the specimen's temperature is kept low during the most of the process, drying by sublimation is slow. Specimens at the top of the chamber (using an Edward Specimen Freeze-Dryer) will dry faster than those at the bottom, as they receive heat by radiation through the lid. The thermally insulating cover, which is usually only removed for brief periods for inspection, can be completely removed and the temperature of the upper section of the chamber further raised by suspending a 60 watt lamp 50 cm above the lid. This will assist the sublimation of the ice crystals from the specimen, while the lower section of the refrigerator coil still provides enough cold for the ice molecules to condense.

The basic principles of freeze drying are discussed in the first volume of this book, along with other aspects of general procedure and the techniques for vertebrate freeze drying.

FREEZE DRYING PLANT SPECIMENS

Most plant specimens freeze dry well, although many lose their original colour. Delicate flowers, although they also freeze dry quite well, lose most of their colour and, after removal from the vacuum, rapidly absorb water from the atmosphere. Plant specimens which behave this way are not suitable for freeze drying.

For study purposes, where lifelike appearance is not of paramount importance, almost all species of plants can be freeze dried. If the specimen is meant to be used in a display, careful consideration of the right method of preparation is necessary. Freeze drying is not always the best technique to employ; however, for some groups of plants it is the most suitable, and, often, the simplest, method.

Marine algae

Marine algae usually collapse when removed from their marine environment. If specimens are to be freeze dried in a natural pose the following method gives good results: Wash specimens in tap water for 30 minutes to remove salt. Place them in a container and cover with fresh water so that they form a normal shape. The container should be placed into the freezer until solid. This is then placed into the freeze drier until the ice sublimates and leaves just the specimen.

Foliage

It is always difficult and extremely labour consuming to reconstruct trees and shrubs for display purposes. The work can be simplified to a certain extent if a freeze dryer is employed to prepare the foliage. If the freeze dryer has a large specimen chamber, freshly cut branches with leaves attached can be treated in one piece and, once prepared, installed on the tree or shrub.

Fruit

Many kinds of fruit which could not be dry preserved in earlier times can now be freeze dried. Fleshy fruits, such as citrus, often have a waxy surface. This wax should be removed by washing the fruit with a suitable solvent. The skin of the fruit should be perforated with a fine pin to assist dehydration. If the fruit is prepared for a display and will be viewed from one side only, the non-show side can be opened and some of the interior removed, thus reducing the mass of water-containing tissue. The removed flesh should be replaced with a filling and the specimen thoroughly frozen before freeze drying. A relatively low temperature is necessary throughout the process (–30° C).

Fungi

Freeze drying is the ideal process for the preparation of fungi. Mushrooms make spectacular exhibition pieces and most of them freeze dry well and reasonably quickly. Unfortunately, dried mushrooms (and also the small fungi species) are readily devoured by insects. Mushrooms can be soaked in saturated borax solution for a few hours prior to freezing and freeze drying in order to render them less palatable to insects. After dehydration has been completed, some species can be soaked in matt lacquer (25% clear matt lacquer, 75% thinners); however, others react unfavourably if brought in contact with the substance. It is appropriate to work at –25° C to –20° C during freeze drying.

Succulents

Here again, some species are more suited for freeze drying than others. Those thickly covered with hair, thorns or scales freeze dry better since, if any shrinkage appears on the specimen's surface, the covering matter hides it. Perforation and the hollowing out of fleshy sections speeds up the process. To minimize shrinkage and distortion, –30° C should be considered as a working temperature.

The freeze drying of plants, as well as their successful lifelike dry preservation, is a field of a specimen preparation where much more research is needed.

INSECTPROOFING AND FINISHING

Most plant specimens are vulnerable to insect damage. If they are soaked in a saturated borax solution, at least their surface receives a protective coating of borax. The soaking, of course, must take place before freezing. A suitable fumigant should be placed in closed display cases containing plant specimens. If the fumigant is periodically replaced as recommended by the manufacturer, reasonable protection is provided to the specimens.

Some specimens become extremely fragile after dehydration. If this represents a problem (for example, through handling) a fine mist spray coat of thinned rubber latex should be applied to the surface. Once the first coat has dried, a further coat can be sprayed on without any detrimental effect. The fine coat of rubber provides some protection and flexibility to the specimen.

Reconstitution is a serious problem with plant specimens. If the nature of the plant allows, it should be sprayed with a fast drying lacquer immediately it is removed from the vacuum in order to seal the surface of the specimen. Unfortunately, many plants react rather unfavourably when sprayed with lacquers or any other solvent based substance. Their surface shrivels (for example, strawberry), or the whole specimen collapses (for example, flowers of sweet-pea).

Reconstitution is accelerated by the presence of salt in the specimen. It is essential that marine plants are thoroughly washed prior to preparation to remove as much salt as possible.

The colour restoration of plant specimens is relatively simple. Providing the surface of the plant can take lacquer, this is the most practical substance to use for recolouring. Usually the lacquer is applied with a spray gun or air brush, or, if the nature of the plant specimen allows, it can simply be dipped in a wash of colour. Water based paints are not suitable for colouring freeze dried specimens.

If lacquer is not compatible with the specimen, a fine coat of rubber latex should be sprayed on the surface first to act as a sealer. Although the rubber latex is a water based substance, if it is applied as a very fine mist, the small amount of water in it normally will not reconstitute the specimen. If the work is carried out in a well ventilated place, drying is practically instant and a seal is established quickly.

FREEZE DRYING INVERTEBRATE ANIMAL SPECIMENS

Minute invertebrates

Many of these can be freeze dried successfully. It is important that the specimens are killed with the appropriate killing agents and that they die in a relaxed manner. If the wrong chemicals or methods are used the animals die in a cramped position; they may retract their extremities, or their entire bodies, into their carapaces, shells or similar. Relaxed animals, especially marine species, should be washed in distilled water to remove as much salt as possible in order to minimize reconstitution after the freeze drying has been completed. The recommended method of fixing is either 3% formalin or 7% ethanol.

Specimens which have been stored in ethanol, or in ethanol and glycerol, can also be freeze dried, providing that the alcohol (and glycerol) is removed. If these substances are not thoroughly removed their presence hinders proper freezing. The specimen must be completely frozen when the vacuum is applied. If a separate freezer is used for the freezing prior to drying, it often happens that small specimens thaw during the transfer from the freezer to the freeze dryer. This, of course, is undesirable; in order to prevent it the best method is to freeze the specimen in the freeze dryer without applying the vacuum. Once the specimen has frozen the vacuum can be applied and drying carried out. This procedure seems to be practical also when freeze drying the larger invertebrates.

Hower (1962; 1979) described a technique of preparing Ostracods for scanning electron microscope examination: the specimen was narcotized with a mixture of chloral hydrate and menthol, and 2% osmium tetroxide in distilled water was used for fixing. (However, if the specimen was of salt water origin, the fixing solution had to be isotonically balanced to avoid distortion due to osmotic pressure.) After fixing, the specimen was thoroughly washed in filtered distilled water and cleaned further by immersion in an ultrasonic cleaning bath. The cleaned specimen (still in the distilled water) was pipetted onto a round cover slip able to fit the scanning electron microscope stub, and frozen with liquid nitrogen. The frozen specimen was then placed in the freeze drying chamber and freeze dried at a constant -30°C. After drying, the chamber, still a vacuum, was heated to room temperature, or the specimen was transferred to a desiccant jar in order to prevent rehydration from condensation on the cold cover slip. The above method could be applied to other minute invertebrates.

Larger invertebrates

The freeze drying of various invertebrate groups where the individuals of the species are large enough to be examined by the naked eye, or with the help of a hand-held magnifier, is more complex. Most of these are suitable for display, either as exhibition specimens or as teaching aids. It is obvious that a dry specimen, on a suitable mounting base, is more practical for these purposes than a wet preserved one.

Killing and fixing: As has been outlined above, the specimen should be anaesthetized first in order to relax it, and then killed. Specimens often need mechanical supports to attain a lifelike pose. Once this pose is achieved the animal should be fixed (formalin is the most widely used preservative), insectproofed, and washed, if necessary. After this it is thoroughly frozen, then placed in the specimen chamber and freeze dried. Sometimes, the specimen must be killed instantly and fixed at the same time. This is best achieved either by plunging the specimen in liquid nitrogen, or by pouring the liquid nitrogen on it. If this method is employed, the quick frozen specimen must be placed immediately either in a freezer or in the prechilled specimen chamber.

Coelenterates: These soft bodied animals present a problem when a lifelike pose is desirable in the prepared specimen. Scyphoza (jellyfish) and some Anthozoa (sea anemones and corals) are best relaxed, then fixed and placed in distilled water where they will attain a lifelike pose by themselves (although not all species and not every specimen). The specimen and the surrounding water can then be frozen and placed in the freeze drier and dried at a constant –30° C.

Cephalopods: Squid, cuttlefish and octopuses can be killed by being immersed and then frozen in fresh water. Once the specimen has died, the ice block is thawed and the animal taken out of the water. Cephalopods are often coated with a heavy layer of mucus. This should be removed by immersing the specimen in alum water. If the body has slightly collapsed it can be filled with cotton or Dacron wool through a small incision on the underside. If the arms seem to have collapsed they can be filled with 10% formalin, and thickened with Aerosil (silicone dioxide), using a syringe and heavy needle. Once the desired "plumpness" has been achieved, the specimen should be posed and fixed with 10% formalin. Fixing is done by injecting the animal as well as immersing it in the fixative. After fixing, the specimen should be removed from the fixing solution, rinsed in water, and insectproofed with Edolan. This is followed by freezing and freeze drying.

The specimen can be given specific poses by inserting armature in its body. A small, suitably shaped balsa wood block can be inserted through an incision on the ventral side of the body. Zinc coated wires, with tapered points on each end, are pushed into each arm. The points toward the body are anchored in the balsa block. The block is pinned to a suitable mounting board, thus fixing the specimen to a stable base. The arms can now be positioned in any configuration by bending the wires.

Arthropods: Many larger arthropods have hard exoskeletons which render them ideal for preservation. Those which are completely encased in such "armour" (for example, crabs, or beetles) can be quickly and efficiently vacuum dried at room temperature. Fixing is necessary, especially with large specimens, to stabilize the large mass of tissue within the exoskeleton. Insectproofing is also important and should be carried out either before freeze drying by injecting the specimens with Edolan in water, or after the

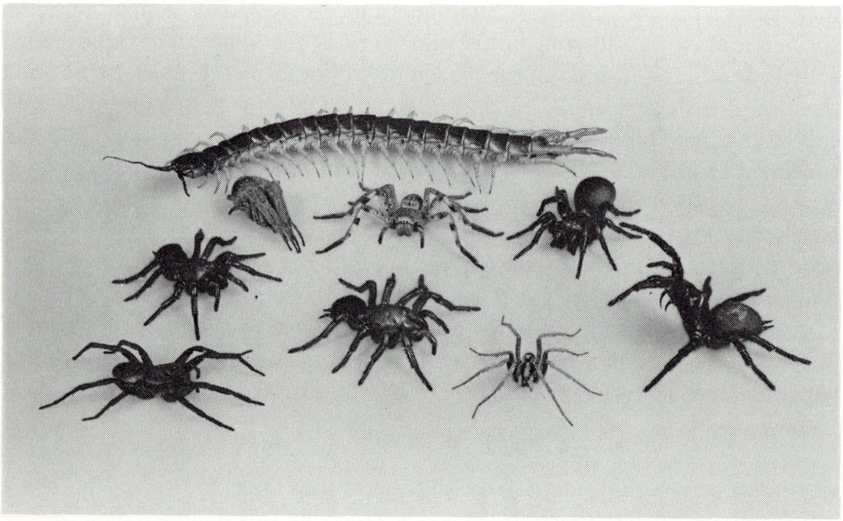

Fig. 9.2. Freeze dried arachnids look lifelike. Photo A. Pitman. Courtesy of The Australian Museum.

process has been completed by injecting the specimen with Edolan in ethanol. Most other arthropods with semirigid exoskeletons or even those with soft bodies can be freeze dried successfully (Fig. 9.2), but, unlike those mentioned above, these must be thoroughly frozen and dried on a low temperature.

Crustaceans: The larger species of Malacostraca (prawns, lobsters, crabs, and so on) should be eviscerated in order to reduce the mass of freeze drying matter. The specimens can be fixed in 10% formalin prior to posing, since the animals remain flexible enough in the fixing solution. If the shell of the animal has been opened for evisceration, internal insectproofing will also be thorough. The more soft tissue removed from the interior of the animal, the sooner the freeze drying will be completed; the less edible matter remaining within the exoskeleton, the less attractive the finished specimen will be for insect pests.

The fixed and rinsed specimen should be soaked in an Edolan and water solution and parts which were not opened during evisceration (for example, legs) should be injected with the chemical. The accessible sections of the interior (for example, the hollowed abdomen) should be rubbed with borax. This not only provides added insectproofing but also facilitates the drying. If necessary, the body cavity should be filled with cotton or Dacron wool, and the specimen can then be posed.

Posing is relatively simple. Using a mounting board of appropriate size, the specimen's legs and other extremities are arranged in a lifelike manner

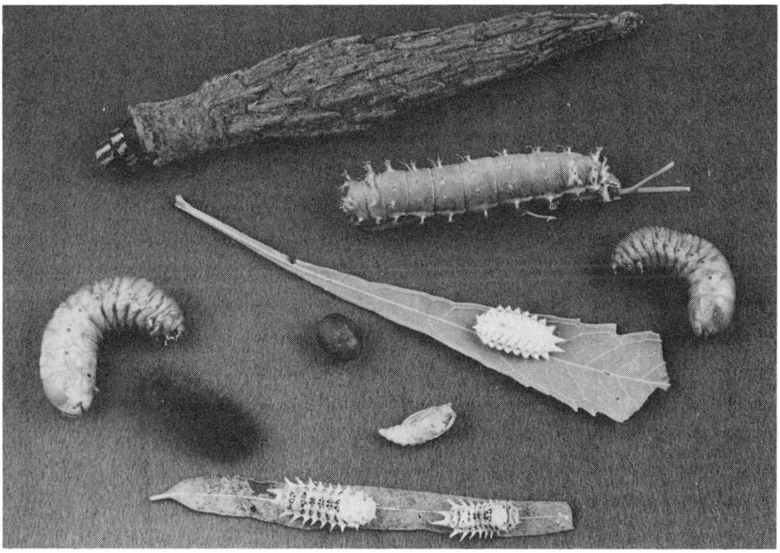

Fig. 9.3. Freeze drying is the best method to prepare most insect larvae and pupae. Photo A. Pitman. Courtesy of The Australian Museum.

and held in position with pins, paper strips, and so on. If the animal is to be posed walking, with its body somewhat elevated, a small block of wood should be placed underneath, thus creating the impression that it is supported on its legs. After freeze drying, this block can be removed.

Smaller specimens can be posed and their position fixed with the help of small lumps of oil based modelling clay. Slabs or blocks of rigid polyurethane foam, soft composite boards, and natural cork are suitable as mounting boards. Polystyrene foam is not suitable as it expands in vacuum, thus distorting specimens.

Arachnids: Specimens can be killed by freezing or with piperonyl butoxide spray. Frozen specimens should be thawed before being posed. Larger arachnids can be posed on mounting boards and their legs held in the required position with pins. The smaller ones are best posed with the help of small lumps of Plasticine, as described above. The freeze dried specimens can be insectproofed by immersion in an Edolan–alcohol solution.

Insects: Soft bodied insects can be freeze dried with success and, in most cases, the prepared specimens are very lifelike. Fresh specimens are the best for freeze drying. Where colours can fade or change easily, the method of killing must be chosen carefully. As an example, many Acrinids (Orthoptera) turn reddish if they are killed with chloroform or ethyl acetate, or dropped in alcohol. These animals must be killed in a cyanide jar or in the freezer in order to preserve their original colour.

Freeze killing is best for a number of other groups also, but it can prove to be a cumbersome method if many insects are collected at once. The transportation of large numbers of live insects from the field to the laboratory can be very difficult, if not impossible; therefore, a compromise should be worked out. If colour preservation is not a problem the insects should be killed with ethyl acetate and transported to the laboratory in sealed jars. The vapour of the ethyl acetate keeps the specimens in a relaxed state. It is a good practice to pose the specimens immediately on return to the laboratory and freeze them.

Alcohol or ethanol preserved specimens can also be freeze dried, but with varied success. The alcohol must be washed out, for the reasons described above, then the specimen can be posed in the usual way and frozen. If the animal becomes rigid in the alcohol it can be relaxed before being posed by being soaked in a weak (1%–2%) carbolic acid in water solution.

Insect larvae should always be freeze killed (Fig. 9.3). After killing, the specimen can be thawed, or semithawed, and posed. Often it is not necessary to pose the animal, since it dies in a more or less natural posture. If the specimen is to be placed in a collection it can be placed live in a tight fitting glass vial and frozen. This ensures a uniform, straight form. Very large larvae can be perforated on the ventral surface to speed up drying. A no. 1 insect pin should be used for this work and, if the specimen is to be placed in a scientific collection, the perforation should be noted on the data label.

Most insect specimens can be insectproofed after the freeze drying process has been completed by being soaked in an Edolan–alcohol solution.

REFERENCES

Hower, R.O.: Freeze-drying biological specimens. *Smithsonian Institution Information Leaflet* no. 324 1962.

——. *Freeze-drying biological specimens.* U.S. National Museum, Washington 1979.

(See Volume 1, Chapter 9, for other references and source material.)

10

Plastic display box manufacture

THE PREPARATION OF good quality boxes to contain wet specimens requires acrylic plastics, and special machinery and tools. This section is aimed at the preparator who has access to such matcrials and aids.

ACRYLIC PLASTICS

Acrylic plastics belong to the group of plastics called thermoplastics which become soft when heated and hard when cooled. Acrylic, used in the production of specimen containers, is clear, transparent, and has a glossy smooth surface.

TYPES OF ACRYLIC

Acrylic is normally available in sheet, tube, and rod form.

Sheet

Sheets come in various sizes, the most widely used sizes being 1905 mm × 1390 mm, 2540 mm × 138 mm, 2760 mm × 2600 mm, and 2760 mm × 1905 mm. Clear or coloured sheets are available with thicknesses ranging from 1.5 mm–50 mm. Boxes of average size containing less than five litres of fluid usually use a thickness of 3.2 mm for the walls. While there are no set guidelines for the required thickness of acrylic for a certain sized box, containers holding large volumes of fluid will obviously require thicker walls.

Tubing

Tubing is normally used to make round boxes of small size as it comes in only a limited range of sizes. Such boxes generally have a wall thickness of between 3 mm and 12 mm and have diameters of between 9.5 mm and 152 mm. There is a greater pressure exerted on the ends of boxes made with tubing so they are more useful for containing dry specimens.

Rods

Some techniques for securing certain specimens to the box require the use of an acrylic rod. Useful diameters range from 6 mm–200 mm.

PROPERTIES OF ACRYLIC

Although acrylic has many known properties, only a few need be described here. For more information, the ICI technical service notes, listed in the references, will be useful.

A. Acrylic scratches easily and so manufacturers protect acrylic sheeting with a layer of paper on both sides. This paper must stay on the acrylic during machining, and should be removed only when glueing pieces together. If the acrylic has been stored for a long time, the adhesive becomes hard and the paper very difficult to remove. A liquid compound (X55—manufactured by Ajax chemicals) dissolves the glue and allows the paper to be removed with ease. Xylene and alcohol, although they are eventually harmful to acrylic, can also be used to remove the hard adhesive and paper. A clean, soft rag should be dampened with the solvent and rubbed over the surface of the paper. After removing the paper the acrylic should be washed with warm water and liquid detergent.

B. Acrylic reacts with various chemicals such as alcohol, xylene, acetone, acetic acid, ethylene dichloride, methylene dichloride, and chloroform. This reaction can be used to glue the acrylic together.

C. Acrylic sheet is relatively even in thickness enabling precise machining to be performed. This evenness helps to prevent distortion.

D. Cutting acrylic is easy. It can be scored and snap broken as well as sawn using various types of saws.

E. Polishing the edges with a buffing wheel produces a clear finish as does the application of a flame, solvent, or a strip of wood impregnated with Brasso or a similar brass polish.

F. Acrylic becomes pliable at a temperature of around 120° C and is soft enough to bend into shapes at temperatures between 150° C and 170° C.

GLUES AND CEMENTS

Three types of materials can be used to bond acrylics to each other: solvents; cements; and acrylic chips dissolved in solvent.

SOLVENTS

Chemicals such as chloroform, ethylene dichloride (EDC), and methylene dichloride, when applied to the surfaces of the joints, fuse the areas together, after which the solvent evaporates.

CEMENTS

These are manufactured materials consisting of the polymer dissolved in the monomer. A hardener is added before the solution is applied to the surfaces to be joined, resulting in a softening of the surfaces and polymerization of the cement to form a bond.

ACRYLIC CHIPS DISSOLVED IN SOLVENT

Acrylic chips are dissolved in a solvent to make a viscous liquid similar in consistency to motor oil. Between 5%–10% of this glue is made up of acrylic chips. When the solution is applied to the surfaces to be joined, the solvent softens them and they then fuse together. The dissolved chips slow down the rate of evaporation of the solvent, allowing a longer time for the surfaces to soften before they have to be joined together. The ideal solvent is ethylene dichloride (EDC).

MACHINERY

The following tools are needed to produce a wide range of acrylic containers, and they should not be used for working with wood or metal.

CIRCULAR SAW

Tungsten carbide tipped saw blades are the best. These give a superior finish to cut surfaces and allow the material to be fed through at a faster rate than a normal blade. A 150 mm diameter blade having two to three teeth per centimetre is a good all round blade. Other high speed steel saws should be hollow ground or have a slight set to remove swarf, which produces binding and overheating (see ICI technical service note — Machining). The number of teeth depends on the thickness of acrylic to be cut. Acrylic 3 mm thick requires from three to four teeth per centimetre up to one to two teeth per centimetre for acrylic 13 mm thick and over.

Chipping of the surface occurs when the saw is showing signs of bluntness and when the acrylic is pushed through the saw blade too quickly.

BENCH CIRCULAR SAW

The height of the blade should be just higher than the thickness of the acrylic. However, if the height is too low the acrylic will melt slightly because the blade does not cool quickly enough; also, the acrylic will jump upwards producing an irregular cut. The rate of feed of the acrylic through the blade also affects the smoothness of the cut.

RADIAL ARM SAWS

The blade is adjusted to cut the acrylic just below the undersurface.

OTHER SAWS

Acrylic can be cut with most saws which have a slight set to remove the swarf, but these are not generally used for box making because they cannot cut straight edges. These saws may be used when small, irregular shapes are required, such as when certain specimen supports are needed.

DRILLS

Drills which have a slow spiral and wide flutes make the cleanest holes. Crazing (stress lines) occurs if the drill is allowed to heat the acrylic. Therefore, correct drilling speeds and a coolant are necessary. The drill must run true and the work to be drilled should be clamped firmly. The speeds at which different size drills should revolve are: 1.5 mm diameter — 7000 rev./min; 6 mm diameter — 188 rev./min; and 12 mm diameter — 900 rev./min.

Other sizes will need to operate at corresponding speeds if absolute accuracy is required, but, if only an approximate finish is needed, the speed can be taken from those listed. The drills should be used only for acrylic work and not for drilling wood or metal. They will last longer, but, more importantly, they should have their cutting edges ground to produce a greater angle and less depth. This means that the point of the drill should be less than the

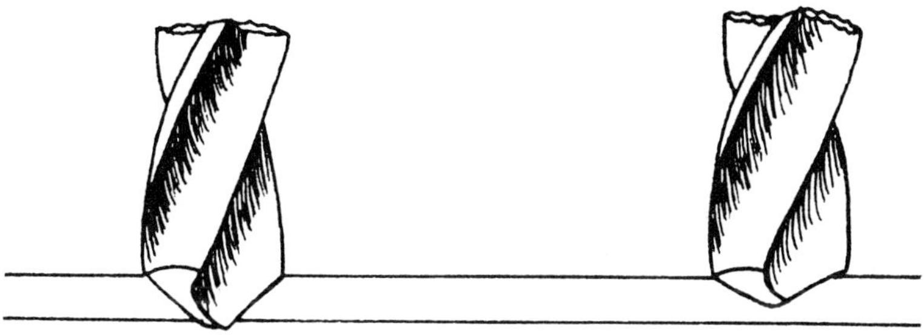

Fig. 10.1. Two drill bits cutting through a sheet of acrylic. The left one is a normal drill which would tend to crack the acrylic. The drill on the right has had the cutting edge ground down to produce a greater angle, less depth and hence no cracking of the acrylic.

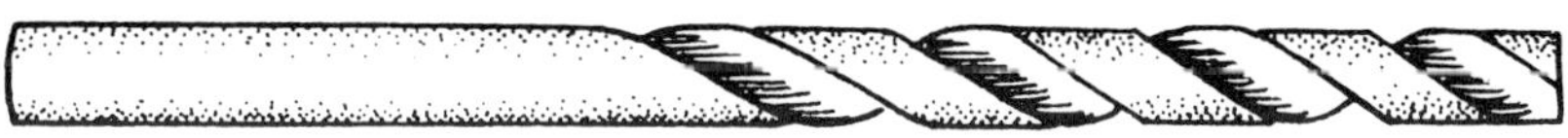

Fig. 10.2 A reamer made by grinding flat the cutting edges. The drill should be the same diameter as the head of the nylon screw.

thickness of the acrylic. If the point of the drill pierces the undersurface of the acrylic before the top of the cutting edge has entered it, the acrylic will crack. This is really critical when very thin acrylic is worked (Fig. 10.1).

A reamer with the same diameter as the head of the nylon screw is used to make a flat bottomed hole the depth of the screw head (Fig. 10.2). The reamer is only used after the hole to be tapped has been drilled.

ROUTER

The only other cutting tool required is the router. A high speed woodworking router is suitable, but only the best and strongest should be purchased.

A table should be constructed so that the router can be fixed to the undersurface with the cutter protruding through a hole. A cutter with a diameter of 13 mm and a ball race on top is all that is required (Fig. 10.3). The inclusion of the ball race above the cutter enables the acrylic box to be pushed against the cutter without it gouging the sides. The table should be covered with felt or laminex to have a smooth scratch resistant surface on which to slide the boxes.

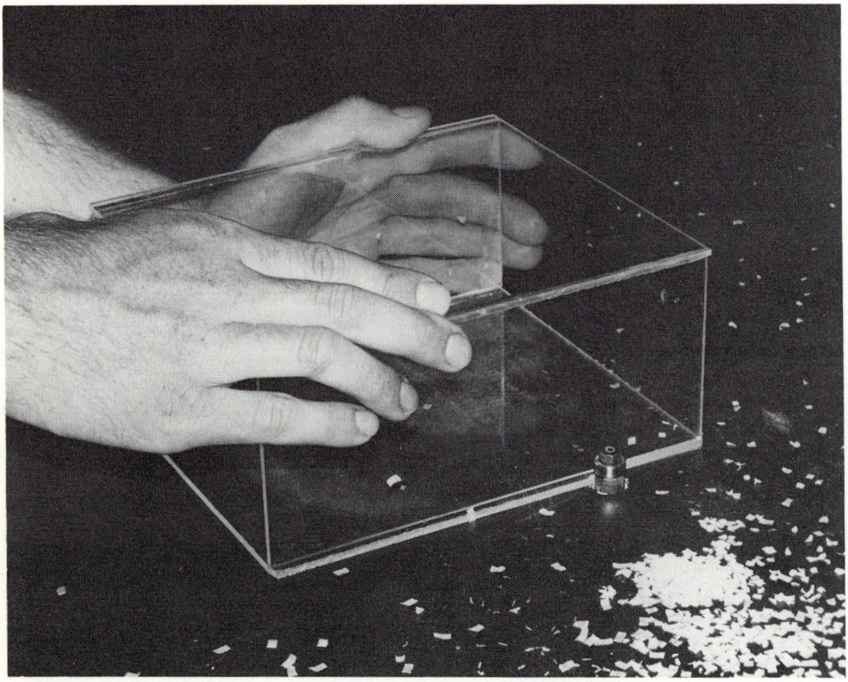

Fig. 10.3. The box is pushed against the ball race of the router cutter to remove the waste. Photo M. Dingley.

TAPS

The acrylic containers should be made so that a nylon screw can be inserted into the bottom. Therefore a good set of taps is essential.

GRINDING

Grinding edges on a disc sander makes the job of cleaning and contouring edges easier. A 250 mm diameter disc should turn at about 3000 revolutions per minute. Larger sanders are a luxury item and are better than the smaller ones and if you are lucky enough to possess one its speed should be slower.

BUFFING WHEEL

Final polishing of the edges is attained using a buffing wheel. One mounted on a stand or wall is best (Fig. 10.4). If large and heavy boxes are made, then the buffing wheel should be attached to a flexible shaft so that the wheel may

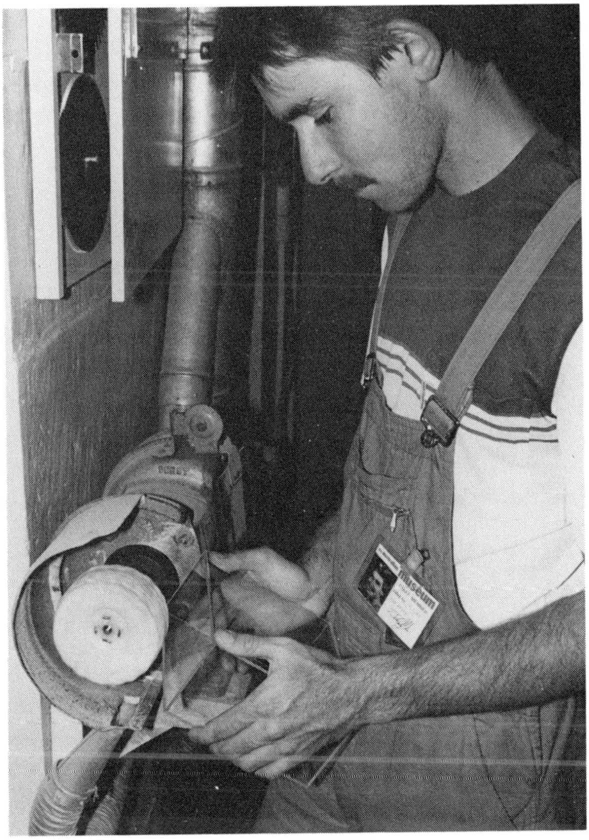

Fig. 10.4. Polishing the edges of an acrylic box using a buffing wheel. Photo M. Dingley.

be taken to the box if it is too heavy to manoeuvre. The buffing wheel should be made of calico. The average size is 150 mm diameter, with a rotation of about 1400 revolutions per minute. Bar wax should be applied sparingly to the rotating wheel, and the box fed to the surface of the wheel. Too much pressure will overheat and leave burn marks on the acrylic. It will then have to be sanded back and repolished. Bar wax can be purchased from hardware stores. A clean wheel should be used to put a final and superior shine onto the box.

HOT AIR OVEN

After machining or sanding, the edges of acrylic have molecular disturbance and this can introduce a stress factor which will become evident during and after glueing unless the box is first annealed in an oven. The oven should be accurate to ±2° C, be able to reach 150° C and have air circulation. The oven should be loaded when cool and allowed to rise in temperature at a rate not exceeding 18° C per hour. When a temperature of 87° C–93° C is reached, it should be maintained for a time according to the formula:

$$\text{time (hours)} + \frac{\text{thickness of acrylic in mm}}{3}$$

The oven should cool at a rate of 12° C per hour.

TOOLS

Tools useful in working with acrylic include: ethylene dichloride; wet and dry paper (grades 280, 400 and 600); glass syringes; fine wire-fuse or Nichrome; cheese head or tapered nylon screws 3 mm–6 mm; pipettes with rubber teats; sewing needles and nylon fishing line; acrylic antistatic cleaner; felt squares 500 mm × 500 mm (on which to place boxes to prevent scratching); and a screwdriver; preserving solutions (see Volume 1, Chapter 7); and a small artists paint brush. The following pieces of apparatus may be made in the workshop and are useful in their applications.

Turntable

This is used in the laboratory, as well as in the teaching museum. It enables the box to be viewed from all sides without the problem of scratching from too much handling. It can be made quite easily from pieces of wood, acrylic, cork sheet, and ball bearings. An acrylic base has a circular groove cut into it to house the ball bearings on which the turntable top rotates. The groove is cut on a lathe, but, if one is not available, the base can be designed to hold a bearing assembly. The acrylic square is mounted in the turntable top and should have a circular groove cut into it to match the base if the bearing assembly is not used. The base is held to the turntable top with a large bolt.

Specimen measuring stand

Specimens should be accommodated in a box so that there is not much wasted space around them; as well, too little space will make the specimen appear cramped. The odd shapes of most specimens make a measuring stand a useful and easily made piece of equipment.

A wooden base approximately 400 mm^2 is scored at 10 mm intervals. Two uprights are made with a groove placed vertically on one side of each upright

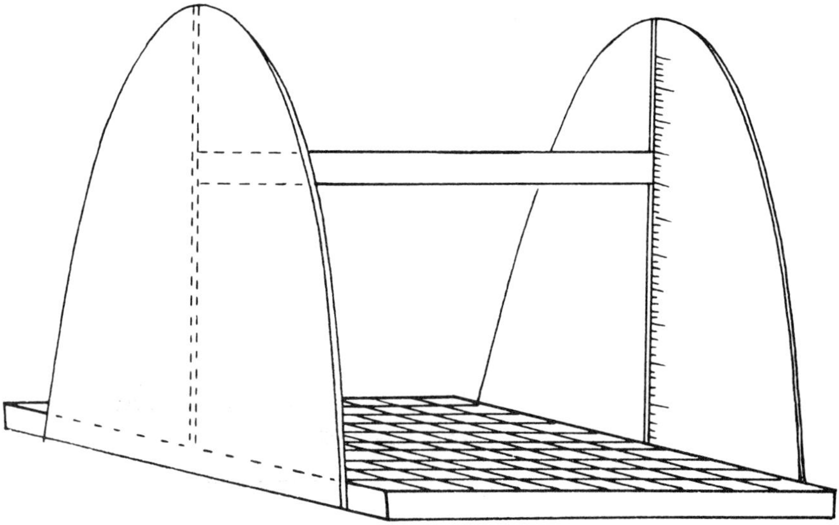

Fig. 10.5. The specimen measuring stand allows the internal measurements of the box to be made. The specimen is laid on the base and the bar is lowered to allow for the correct depth to be calculated.

to house a cross beam which can be raised up and down (Fig. 10.5). The uprights should each have a scale (in centimetres).

To measure a specimen, place it near the left hand front corner, allowing enough space on either side to prevent it being cramped when in the box. The 10 mm squares on the base can be read to give the length and breadth. The cross beam can be lowered for measurement of height. All measurements are then recorded as the internal dimensions of the display box.

Acrylic glueing holder

This holder (Fig. 10.6) allows two sides to be glued onto another surface at right angles and at the required distance apart. Exacting work is required to make this apparatus, otherwise problems occur causing errors in the squareness of the two sides during glueing.

The apparatus consists of two inverted U-shaped jigs or holders which slide along two parallel lengths of brass tubing. The dimensions of the holders may be changed for individual requirements. Two of these holders will be useful when long sides have to be glued; the sides can be held parallel all along the entire length of the surface to be glued.

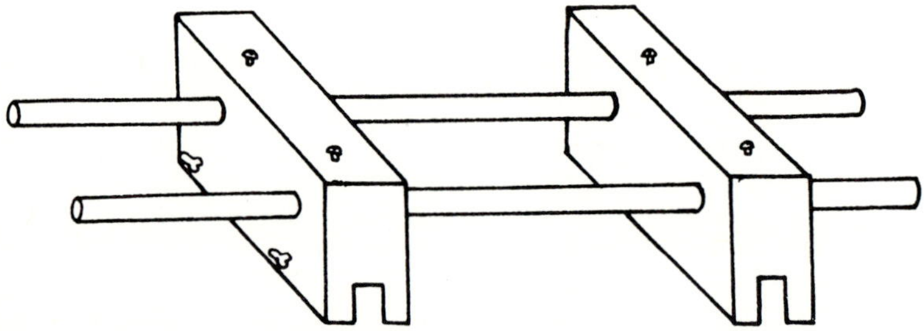

Fig. 10.6. The acrylic glueing holder is made from two acrylic jigs held parallel to each other by means of two brass tubes. The distance between the jigs can be adjusted for different sized boxes.

Heated glass table

The softening of acrylic can be hastened if the glass on which it is softening is heated slightly. High humidity slows down the rate of solvent action; a warm glass table dries the immediate area producing a more efficient glueing process.

A table 1200 mm × 900 mm is constructed with an inlaid sheet of asbestos. A length of heating element should then be wound over the asbestos. An aluminium sheet should then be put in place to distribute the heat more evenly. Onto the aluminium, place a sheet of glass which has had the undersurface painted black. The paint allows the operator to see the solvent evaporate so that he or she can apply more if required. The glass top should be totally level so that the solvent does not flow away from the acrylic edges. The temperature should be raised to about 40° C. The inclusion of a thermostat will enable users to regulate the temperature to suit their own requirements. A 6.4 mm hole can be drilled through the glass and the table to allow a tube to be attached to a negative pressure source. During the glueing of the top onto the container, the box is inverted so that the open end rests over the hole and the solvent fumes can be sucked out. The same procedure can be followed when the base is glued on.

Clamp for holding sides of the box in position

When the top is being glued onto a box containing a specimen and mounting fluid, the sides of the box will bow outwards due to the pressure of the fluid. Unless a square box is made (which requires two clamps), a single clamp is all that is needed to straighten the sides before glueing the top in position.

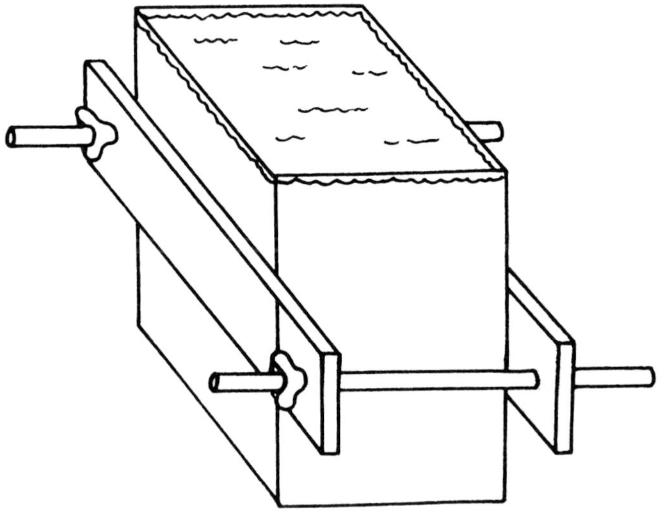

Fig. 10.7. The clamp is mounted to prevent the two sides of the box from bowing outwards whilst it has been filled with the storage medium.

Two pieces of straight, clean, thick timber are cut to length (two or three different sized holders may be made to cover all sizes of boxes) and two holes drilled through each piece of timber near the ends to accommodate two long bolts. Washers are placed in position and wing nuts are used to tighten the clamp and straighten the sides (Fig. 10.7). The timber should be thick enough so that it will not bend when pulled together.

Acrylic bending machine

Some workers prefer to bend a single sheet of acrylic to form four or two sides and then add a top and a base. A disadvantage of this method is that some distortion is evident at the corners. The advantage is that a stronger box is formed compared to the glueing process, but, as containers made this way have lasted many years without leaking, it is really a matter of individual preference.

The acrylic bending machine can be made in the workshop; it is primarily two flat tops hinged so that the front section can be lifted upwards to produce any angle up to 90°. The area which brings the two flat tops together should have a gap approximately 10 mm wide so that the heating element, situated below this gap, will radiate the heat through the opening (Fig. 10.8). The gap

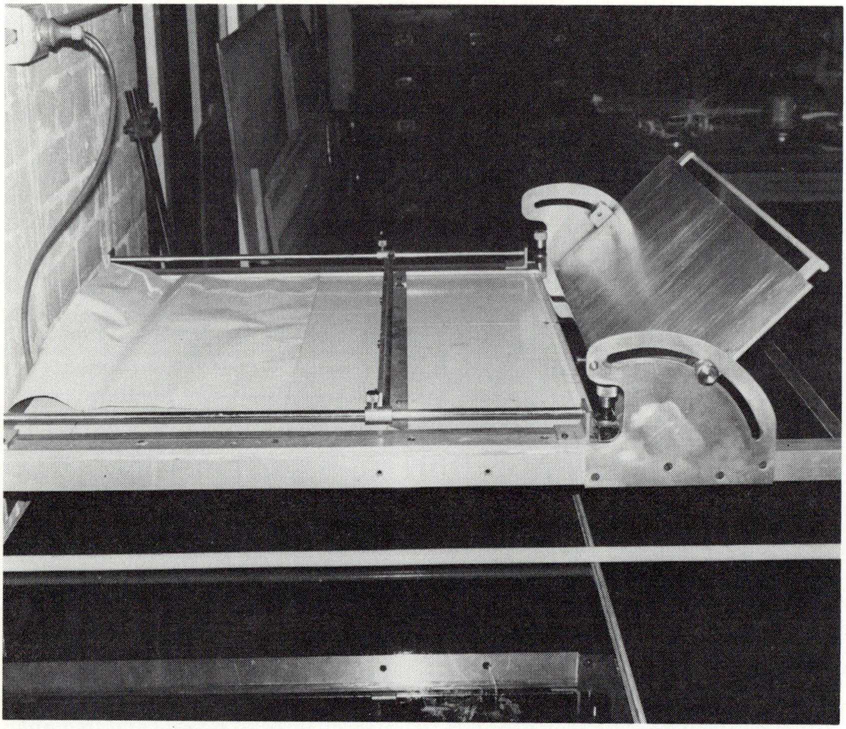

Fig. 10.8. An acrylic bending machine made in the workshop of the Department of Veterinary Anatomy, University of Sydney, Australia. Photo M. Dingley.

may be increased when thicker acrylic is to be bent and a larger surface area needs to be heated. The heating element should have a variable thermostat. The overall size of the machine will be controlled by the size of acrylic to be bent. The width of the top and the element should exceed the width of acrylic so that an even heat is applied. A machine with top surface measurements of 500 mm wide and 800 mm deep is regarded as more than sufficient for most boxes.

MAKING THE BOX

Measurements should always be internal ones. A record must be kept of these measurements, along with a diagram of the specimen showing the viewing face and the proposed technique of fixing it to the box. Information on such aspects as wide bases and coloured backgrounds should also be recorded.

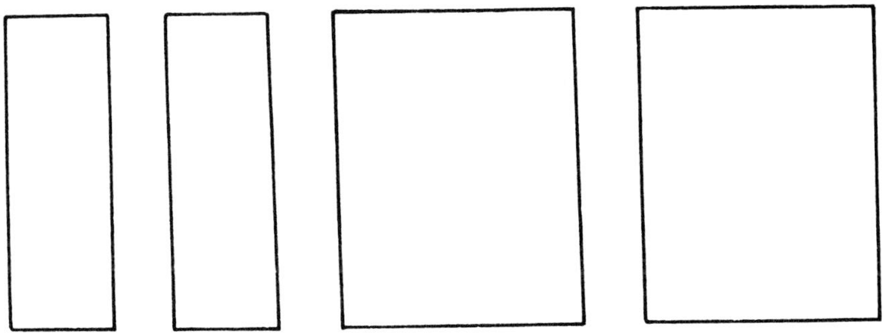

Fig. 10.9. Showing how to cut the four sides of a box keeping the height constant and sawing the faces to the required sizes.

CUTTING THE ACRYLIC

When using power machinery always wear protective goggles. Four strips of acrylic should then be cut with its width the same as the height of the box. The two ends are cut to the exact measurements but the front and rear faces are made wider (Fig. 10.9) so that when the two ends are placed in position there will be an overlap that can be removed later with a router.

After cutting, all pieces should be annealed in the hot air oven before glueing commences, but this process can be omitted if a hot air oven is not available. Just before glueing, the backing paper can be removed and the acrylic wiped with a clean rag dampened with Xylene to remove the minute traces of glue adhering to it. The pieces should then be washed in warm water and liquid detergent, and dried.

The two side pieces should now be clamped into the acrylic holder and the lower edges placed across thin fuse or Nichrome wires on a level, glass-topped table (Fig. 10.10A). A glass syringe or pipette should be filled with EDC which should be carefully squirted under the edges. The solvent quickly spreads between the edges and the glass and immediately starts to soften the acrylic. Four minutes is a reasonable softening time for 3.2 mm acrylic on a cold surface. More solvent may have to be applied if evaporation takes place. While waiting the acrylic to soften, peel off the protective paper from one face of the front piece, and lay this piece flat on the table. When the acrylic has softened, pick up the holder, flick off the wires and lay the two edges gently onto the other surface. Be very careful as the joint at first will be slippery and will want to move (Fig. 10.10B). Hold the surfaces to prevent this slippage for 30 seconds and then place small 50 g weights to secure a strong bond. Leave the box in this position for at least an hour before proceeding to the next stage.

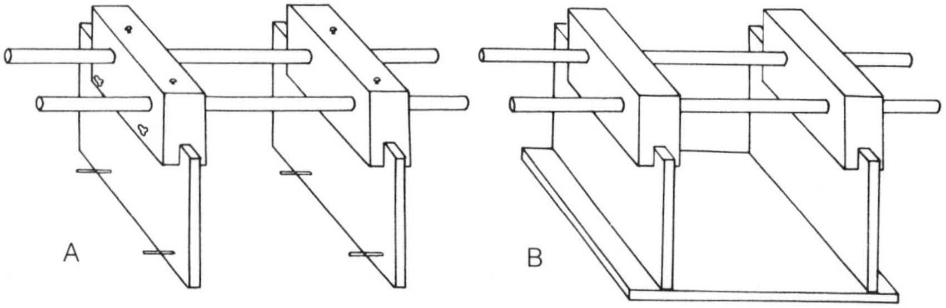

Fig. 10.10. A. Two sides of the box are gripped in the acrylic glueing holder and placed so that the edges are resting on short Nichrome wires. The solvent is run under the acrylic to soften the edges. **B.** After softening the edges, the wires are flicked off and placed onto one of the other sides of the box. It stays in this position until the joints are firm.

In the next stage, the holder can be removed by unscrewing the bolts, and the partially mounted box inverted so that the two free edges are resting on wires placed on the glass top (Fig. 10.11). The solvent should be applied for the prescribed time. Peel the paper off one face of the side to be glued and place this piece on the glass top. After the acrylic has softened, pick up the box and flick off the wires before joining the edges together. Hold it for 30 seconds before applying the weights. The weights should be evenly distributed. To aid this, a few lengths of wood, 13 mm acrylic, or glass plate, long enough to overlap the sides of the box, can be placed in position. When put in position, the weights should have an even distribution. Do not place the dissolved acrylic away from the edges allowing bubbles to enter the joint. This may produce a weakness which could lead to leakage of the mounting fluid.

The joints should bond together strongly before they are machined so at least two hours should elapse before the weights are removed. The overlapping edges can be cut off using a router. The box should be pushed right up to the cutter. As the cutter has a ball race on top of the shaft, the side of the box will not be damaged by the cutter. Push the box smoothly and at a constant speed away from you until the edges have been shaved. Repeat the procedure for the other edges. The two open ends can then be sanded so that they will be level and ready for the top and base.

To sand the ends, grade 280 wet and dry paper should be stuck onto a glass plate with double sided tape to prevent it moving during the sanding process. It should then be placed where water can flow over it to remove the acrylic dust. The end of the box should then be moved in a circular motion on the paper until the edges have been ground flat and even. To check that they are flat, cover the end of the box with a piece of flat acrylic so that it overlaps

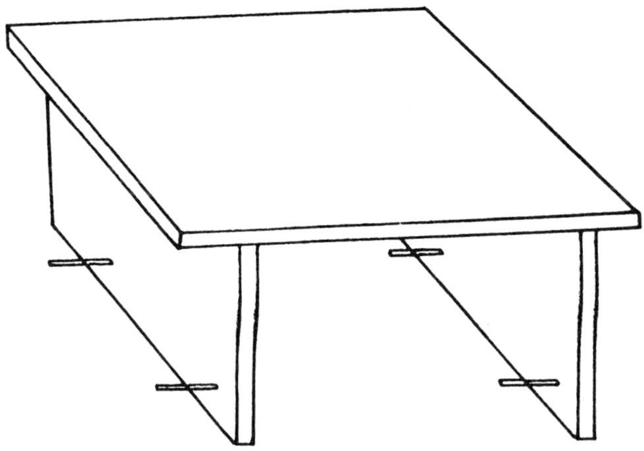

Fig. 10.11. To glue the last side in position the two edges are placed again onto Nichrome wires and softened using a solvent.

the sides. Hold the middle of this, and, tap the top all round the touching surfaces. A sharp sound denotes uneven edges which will need resanding. Once all edges are even the container can be annealed in the hot air oven for the prescribed time before the top and base are bonded in position.

The finished box will have a nylon screw mounted into the base so that fluid can be topped up. This base will therefore have to be 6.4 mm thick to allow sufficient depth for the screw. The screws in the base must have a really good fit or leakage will occur. Some manufactured boxes use round headed screws and fit them to the top. This is very unsightly. Countersunk screws in the base of the box are almost invisible to the observer, especially if situated in the rear corner.

GLUEING THE TOP

The base and top sections should be cut at the same time. They should be about 3 mm larger all round the box size. To glue the end, place the edge to be glued onto four wires and flood the area with EDC as previously described. Once the acrylic has softened, place the edge onto the top section and hold for 30 seconds. Invert the box. Place the open end over the hole in the hot table so that the heavy solvent fumes can be extracted. If the fumes are allowed to float near the joints, there is surface distortion. Place small weights in position and allow the joint to fully harden before removing the overlapping edges with the router. The container should then be washed with detergent and warm water to remove finger prints and glue from the backing paper. Use cotton wool to apply the detergent, then rinse several times in tap water followed by a final rinse in distilled water. Let the water drain from the box until dry but if there are streaks when dry, wash the box again and rinse several times in distilled water.

GLUEING THE BASE

The base is fitted in place once the specimen has been inserted into the box, but it can be done in different ways depending on the type of fixing method employed. The most common way is to fill the box to within 3 mm–4 mm of the top with mounting fluid and glue the top into position using acrylic chips dissolved in EDC. The clamp for holding the sides in position is secured to the box and the top edges are cleaned of oil and dirt using a clean rag moistened with alcohol. Place a lamp nearby so that there will be a good light source. Using a pipette, apply a ridge of the glue all round the edge. Allow the glue to cover the entire surface. Wait a minute or two, then apply another layer; do not apply too much because the glue will run down the inside of the box when the base is lowered into position. Place the base onto the box and allow it to settle under its own weight before applying additional weights. A clear, thick, and flat piece of glass or acrylic, large enough to cover the top, is placed onto the surface and the weights added. Depending on the size of the container, these weights may weigh up to 4 kg; they should be added if it seems really necessary. The weights are lifted after half an hour so that the glue joints can be checked for bubbles and weak points. If the surfaces were flat and enough cement was applied there should be no bubbles; if there are bubbles they cause weaknesses. If there are many bubbles, the base may be removed and the procedure repeated using a new base. If there are only a few, a scalpel should be inserted into the joint and glue removed from the immediate area. More cement can be added using a very thin tapered pipette. Specimens for which this method can be used are soft bodied animals, and anatomical specimens, such as livers and spleen, which cannot dry out and have to be supported in the mounting fluid. Containers which are tall and thin require a wide base to reduce the chances of being knocked over. The base should be cut to allow enough of a rim all around so that the box has to be tilted through an angle of 45° before it topples over. The edges of the base should be polished before it is glued to the box. Care must be taken during the glueing so that the box does not slip off centre.

Another method is employed when specimens are very sturdy, have their surface patted dry, and are able to withstand air support. Some which fall into this category are muscle dissections, hearts, and bone structures. The specimen is mounted in the box without mounting fluid and the whole container is inverted so that the open end is facing down onto the table. The same method is used to glue the base as was used to glue the top. The fumes can escape through the screw hole in the base. After allowing the surfaces to be in contact for a few minutes, the box should be inverted so that the hole is uppermost. Weights can be added so that they leave the hole uncovered. While the surfaces are bonding, the mounting fluid can be fed into the box until the specimen is covered. The fluid helps to support the specimen. Leave overnight before filling the box completely with fluid. When it is full, remove excess acrylic using the router.

SANDING AND POLISHING

The edges which have been routered smooth should now be sanded using 280, 400 and 600 grades of wet and dry paper to prepare them for polishing with the buffing wheel. The coarsest wet and dry paper (280) should be fixed to a wooden block with double sided tape, masking tape, or staples. The block can be pushed back and forth along the routered edges. Water should be on hand to pour over the box and paper to remove acrylic dust so that it will not clog the paper. When all edges have been sanded, repeat the procedure using 400 and 600 grades of paper.

If the base was sealed using acrylic chips dissolved in EDC, and this has dribbled down the outside of the box, it can be removed during sanding. When polishing with the buffing wheel, be careful not to create burn marks (melted acrylic) and also be careful that the box is not knocked out of your hands. When the edges have been polished, the box can be cleaned with detergent and warm water and dried using a soft, clean cloth. An antistatic compound should then be applied to the acrylic. Such compounds include Cirrasol (ICI) and Kunststoff.

SPECIAL ATTACHMENTS

There are many ways in which to support specimens in a box, but the preparator must always think of a support which will be the least noticeable to the viewer. This cancels out a lot of nasty looking methods where, for example, specimens are tied to a plate using thick nylon and the nylon is passed around the specimen instead of through it. In time, the specimen will become loose and move off centre, with the result that it will fall to the bottom of the container. A display jar in which the specimen has been allowed to rest on the bottom without any form of fixing should never be considered. Specimens should be centred correctly, and a clean overall appearance should be achieved. When this has been done, it shows that the preparator has taken the trouble to perfect his or her technique, and it is most evident in the completed display.

The following notes and diagrams are followed by the types of specimens that have been fixed by the authors. This is only a guide for preparators to understand the basic fundamental principles involved and be able to use their own imagination when a problem arises.

Backing plate

This plate is normally a clear sheet of acrylic to which the specimen is attached. The idea is to mount the specimen onto it and then slide it into the box. It may be necessary to use a coloured backing plate to enhance certain characteristics of specimens. The use of white or black backing plates is most favoured. The use of coloured plates should be restricted. The term backing plate is misleading because it can be used when placed at the front of the box or even on the front of a specimen. The plate is cut and the edges polished to just fit the box. On one face, five small acrylic spacers are cemented; one near each corner and one in the middle (Fig. 10.12A). These spacers keep the plate away from the container allowing the fluid to reach all spaces within. If the plate does not have these spacers there will be no gap and bubbles will be trapped and be almost impossible to remove. To stop the plate from tipping forward, four small acrylic nibs are made and cemented onto the two ends of the box. They are mounted near the base and top allowing enough room in which to slide the plate. The nibs are made from short (3 mm–4 mm) lengths of 1.6 mm acrylic rod, with one end rounded and polished and the other end cut flat (Fig. 10.12B). The flat end is soaked in solvent to soften it, and is then placed into position. The various specimen supporting fixtures are fixed to the backing plates.

The simplest method of fixing is to use nylon fishing line, the breaking strain of which is determined by the size of the specimen. The specimen is laid out on the plate and positioned. Areas capable of hiding the nylon and supporting the specimen are located, and the underside of the backing plate is marked. This can only be done on clear acrylic; if coloured acrylic is used, the mark has to be made on the same side as the specimen. At least two (but it is better with three) areas are marked, although some specimens may require more to give optimum support. The marks should then have two 1.6 mm holes drilled; this is sufficient to allow a needle to pass. A needle should be threaded with the fishing line and passed through the specimen so that the loose ends of the nylon are above the specimen. The nylon must be pulled reasonably tight and a double loop knot made to prevent it slipping before knotting. The line will cut through specimens which are soft and

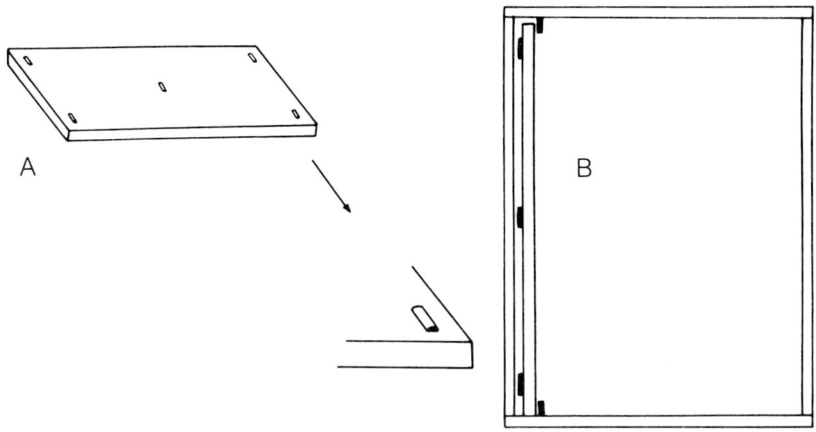

Fig. 10.12. A. This figure shows a backing plate having five spacers glued in position. **B.** The backing plate mounted in a box (side view) showing how the spacers keep the backing plate away from the box and also the nibs to prevent the plate from tipping forward.

friable, so a small plastic ring should be inserted. This ring can be a circle of acetate sheet (cut from a clear sheet with an appropriately sized cork borer). A ring of 5 mm diameter is often sufficient. Two small holes should be made close together in the ring. The needle and nylon are passed through the undersurface of the plate and through the specimen where the acetate ring is threaded onto the nylon and knotted firmly. The ring prevents the nylon being pulled through and, as it is clear, it will not be noticed. Soft specimens such as fish, worms and pathological slices are suitable for this type of fixing (Fig. 10.13).

If bisected hearts and heads are to be displayed, they may be fixed directly to the inside of the box (or onto a backing plate) using special lugs. The front plate is made the same way as the backing plate, but is fitted to the front of the box. The lugs are made from 4.8 mm thick acrylic and the size varies with that of the specimen. However, lugs 10 mm × 5 mm × 4.8 mm allow a needle hole to be drilled through the side. The minimum number of lugs needed to hold a specimen in position is three; the lugs should be positioned so that they are in a triangular formation for greatest strength. Bone has to be drilled to allow the needle to pass and the lugs are fixed to the flat surface of the side to be viewed. Acetate rings are used if they are necessary. When the lugs are securely tied, they are rested on the plate and adjusted so that they are flat. The specimen is coated with gelatine (see below) and allowed to solidify. To glue the lugs onto the plate, they are again positioned while any fluid lost from

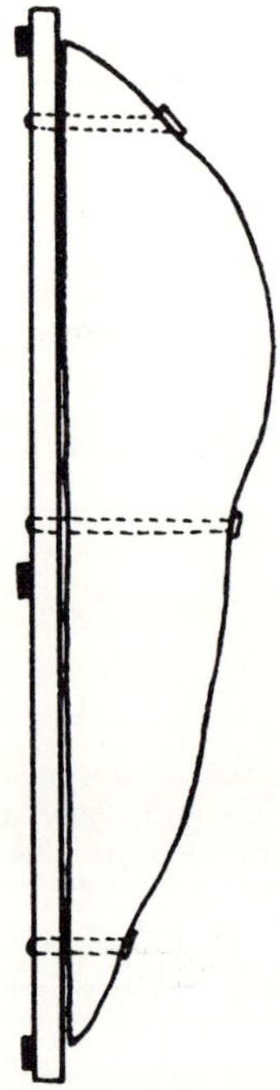

Fig. 10.13. Side view of a soft specimen tied to a backing plate using nylon fishing line (dotted) and acetate rings.

the specimen should be mopped up. The lugs are lifted one by one with a long thin knife. This also lifts part of the specimen. Acrylic chips dissolved in EDC are pipetted onto the plate and the lug is lowered (Fig. 10.14). When all lugs have been glued, the specimen should be covered with paper towelling soaked in mounting fluid to prevent it drying out whilst the cement is hardening.

Rods

The last method of attachment uses acrylic rods inserted into the specimen either from the back or base of the box. They can be glued to a backing plate or directly onto the inner surface of the box. The specimens must have holes made into them to take the rods. When rods are inserted from the base, the base is glued onto the box last and at least three rods are needed for mounting into the rear of the specimen. If muscle tissue has to hold the rods, holes can be made using a cork borer of smaller diameter than the rod. If bone has to be spiked, it should be drilled with a hole the same diameter as the rod.

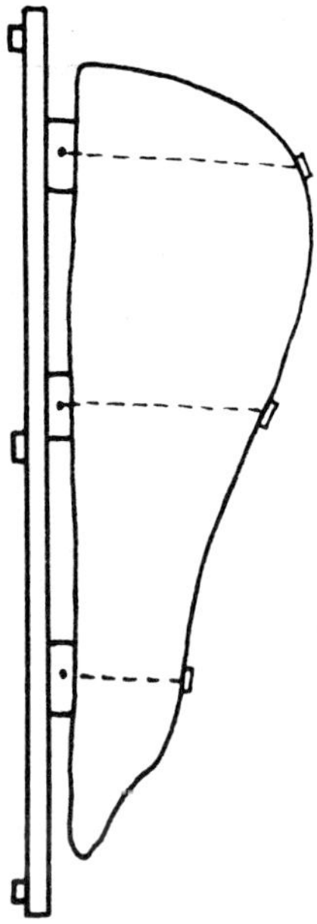

Fig. 10.14. Side view of a specimen such as a bisected head showing the lugs glued to the plate, nylon fishing line (dotted) and acetate rings.

Perpendicular holes are necessary. The rods should be inserted once they have been cut to length and sanded flat so that they are all equal in length when holding the specimen (Fig. 10.15). The rods also should have the end that will be stuck into the specimen sanded to form a four sided point; this will grip the tissue better than if it was rounded. The rods are glued by raising each rod separately and applying acrylic cement such as Tensol No. 7. When withdrawing the pipette after applying the cement, be very careful not to let any fall onto the plate.

Specimens suitable for rod fixing include lungs, livers, and spleens. These are best attached to a plate, while limbs, showing muscles and bones, and brains may be glued directly to the inner surface of the box.

SPECIAL HEAT FORMED BOXES

It has been shown above that boxes are made by cutting and cementing together sides, tops, and bases, or by using a bending machine. A method has been devised by Mr R. Scott-Child at the Australian Museum in which small

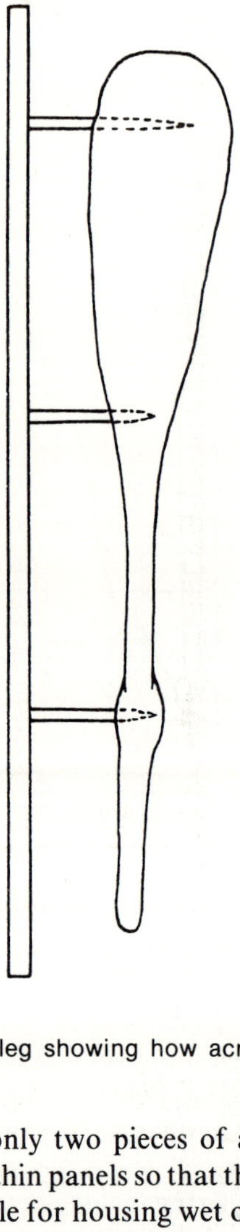

Fig. 10.15. Side view of a dog leg showing how acrylic rods are used to spike the specimen.

boxes can be made using only two pieces of acrylic. They were originally designed for mounting into thin panels so that they could be used in travelling exhibitions. They are suitable for housing wet or dry specimens, but they are limited in depth because of their insertion into panels. Because it takes a long time to make the jig, boxes should only be made when large numbers of the same size are required.

MAKING THE JIG

A piece of wood of equal thickness to the depth of the panel has a hole cut through it equal to the width and height of the box and is sandpapered very

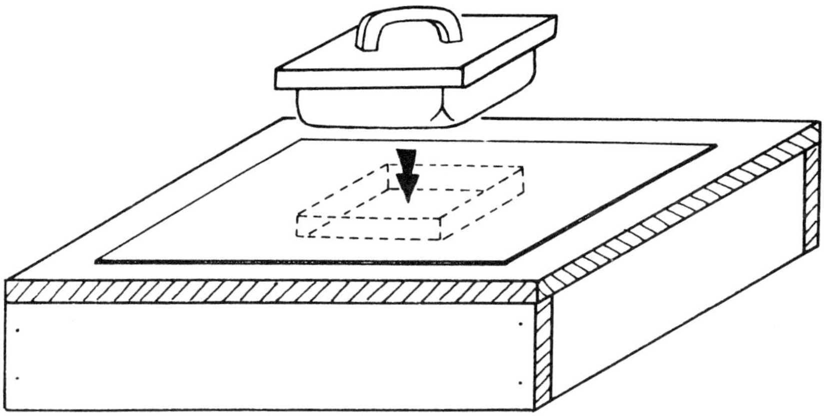

Fig. 10.16. A jig to show how heat formed boxes are made. The male block, with the handle, is pushed down onto heat softened acrylic and into the female mould.

smooth. This is the female mould. If the acrylic to be used is 1.6 mm, a block of wood the same thickness as the depth of the panel is cut so that its dimensions are 1.6 mm smaller all round. This part (the male mould) is also sanded very smooth. The male mould is made smaller because it will push softened acrylic through the female mould. A handle and two extensions should be attached to the back of the male mould. The extensions fit into two uprights mounted onto the female mould so that they guide the handle up and down in exactly the right place. Remove the protective paper from a sheet of clear acrylic and soften it in a hot oven set at 150° C. When soft, it should be laid over the hole of the female mould and the male section pushed downwards, forcing the acrylic with it (Fig. 10.16). Allow the acrylic to cool for five minutes, then remove it from the mould and place it in a dish containing cold water until it cools completely.

The edges will be convoluted and should be removed using a band or circular saw so that the depth will be the exact width of the panel. The edges are sanded flat using grade 280 wet and dry paper. A backing plate 3.2 mm thick should then be cut to overlap the box, with holes drilled through the overlap so that it can be screwed to the back of the panel. The specimen is mounted onto this plate and the formed box is then cemented onto it.

The display panel should now have a hole cut into it to fit the box, which is pushed in from the rear and screwed into position. If the box is to contain liquid, a screw is inserted into the back plate as close to the top as possible, and the mounting fluid is passed through this opening.

SPECIAL CONSIDERATIONS

DISSECTION

Anatomical preparations are useful in some cases where special features are to be displayed and the need for a good dissection is paramount. All dissection work should be carried out slowly, methodically, and knowledgeably. If mistakes begin to creep in, leave the work alone, do something else for a while, and then return refreshed. During dissection, loose ends become evident. These are more apparent when the specimen is immersed in fluid so they have to be removed by cutting them with curved scissors. Friable specimens and those with loose ends (those which cannot be removed entirely) must be coated with a layer of gelatine to prevent cloudy solutions and to stick the loose ends down. The gelatine is made up as follows: a screw-top jar has 15 g powdered gelatine added to 100 mL cold distilled water and is left overnight in an oven set at 40° C. The water soaks through the gelatine and dissolves it. When the gelatine is completely dissolved, one or two crystals of thymol should be thoroughly stirred in to act as a fungicide. The gelatine is applied to the surface of the specimen once all dissecting and fixing work has been completed and the specimen is ready for insertion into the box. The gelatine makes the specimen very slippery and difficult to hold; it should not be handled too often after application. At room temperature, the stock gelatine is solid so for use it needs to be remelted. This is done by placing the jar into a beaker of hot tap water. A pastry brush is ideal for brushing the gelatine onto the specimen; it should be applied so that an even layer covers the entire surface. The gelatine should be allowed to cool, then the specimen should be submerged in a bath of distilled water. A lamp should be shone onto the specimen to see if any bubbles adhere to the surface. If bubbles are present they can be removed by remelting the gelatine using steam. A steamer can be made in the workshop by soldering a metal tube onto the outlet cover of a camping kettle (Fig. 10.17). A rubber tube is attached to this. A water trap may be inserted somewhere between the nozzle and kettle, but is not essential. Water is placed into the kettle and is heated to boiling point; steam flows out of the nozzle. This piece of apparatus is very basic, but melts the gelatine quite adequately. The steam should be applied to the bubbles for only a second or two so as to melt the gelatine and not cook the specimen. When all the bubbles have been broken, the gelatine should be allowed to cool again, and the whole surface of the specimen can be sprayed with a fine mist of formaldehyde (40%). This converts the gelatine to an irreversible gel and prevents it from being remelted, as well as setting it more quickly. The gelatine may also be sprayed onto the specimen if it is found that brushing it on deposits a thick layer.

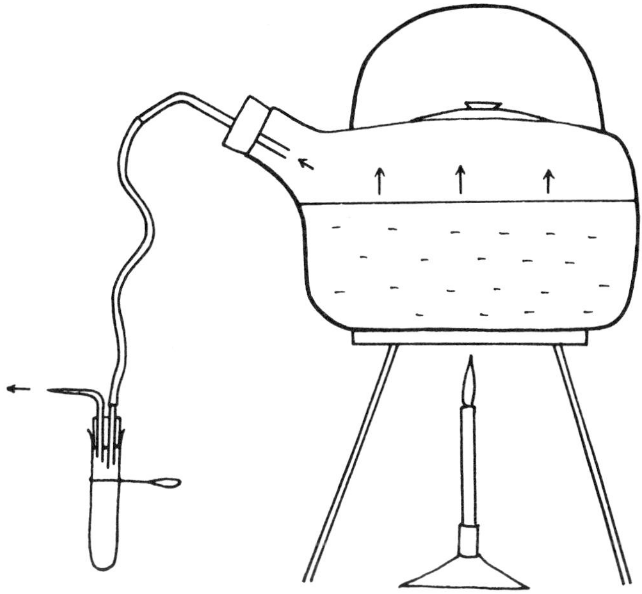

Fig. 10.17. A steamer made from a camping kettle. The water trap near the outlet is not necessary but if one is used, a handle is necessary as it gets very hot to hold.

CLOUDY SOLUTIONS

Various reasons have been put forward for cloudy solutions, but it should be remembered that prevention is better than a cure.

With friable specimens cloudy solutions may be due to maceration of poorly fixed material, leaving small fragments of tissue suspended in solution.

With fatty specimens, fat seeps out of tissues and deposits itself onto the surface of the specimen. In time, it floats to the surface and becomes trapped under the top of the box, forming a pale yellow sludge.

Remedies

First, do not use a fatty specimen unless it is an important one, and then always coat it with a layer of gelatine. For friable specimens and those which have become pigmented, gelatine also helps to some degree. If the solution becomes cloudy it should be filtered and changed. If the mounting fluid already contains sodium hydrosulphite, air must be excluded or oxidation will take place, causing the colour to fade. Therefore, the fluid (when filtering) must always cover the specimen. To filter the fluid, another hole must be drilled through the bottom of the box (if the screw has been inserted into the base), and the screw removed from the original hole. A pump with a variable speed control is connected to two lengths of tubing. The outlet tube is placed so that the open end is almost touching the bottom of the box (while it is inverted this is usually the top) and the input tube is set just inside the box. If the sodium hydrosulphite was added to the original fluid, it must be added to the fresh fluid. A very fine filter should be inserted in the outlet tube. The filter can be compressed paper or a system of grades of activated charcoal. The process of filtration proceeds until the solution in the box remains crystal clear. *At no time should air be allowed to contact the specimen if the specimen has been in contact with sodium hydrosulphite.*

REJUVENATION OF DRIED SPECIMENS

Museum storage containers used to consist of glass jars. The lids were normally of a push-in fit, or were just glass plates which rested on the rim of the jar. They were sealed with a variety of sealants to prevent evaporation of the fluid. Over the years, the sealants broke down due to temperature changes or because they were not handled properly. Some jars had their lids replaced after people removed specimens for research. Evaporation could then take place. The specimen could then dry out, leaving a hard and sometimes unrecognizable mass.

Cleave and Ross (1947) utilized an aqueous solution of trisodium phosphate which, when applied to dried preserved invertebrates, altered the permeability of the tissues, reconstituting the specimen back to an almost original form. Leeches, centipedes, tiny fish, and small crustaceans were soaked in a 0.5% solution of trisodium phosphate until their form returned. Larger invertebrates and some vertebrate anatomical material needed a higher concentration of around 2%. Better results are obtained using a lower concentration over a long period of time. Also, if the temperature of the solution is warmed to 35°C, better results are obtained. This method varies in its success rate, giving slightly better results for invertebrates which have relatively thin bodies than for the thicker vertebrates.

Banks and Williams (1972) used a surfactant to rehydrate coccoidea and Thysanoptera insects. Specimens were soaked in cold Decon 90 (Decon Laboratories Ltd, Sussex, United Kingdom; or, in America, Contrad 70 from Harleco, Philadelphia, PA, U.S.A.) overnight, or less if the specimens are very small. After rehydrating, the specimens were rinsed twice in distilled water and transferred to a preserving–storage medium.

Thompson *et al.* (1966) rehydrated crustaceans by soaking specimens for 12–24 hours in a mixture of 50% ethylene glycol and 50% distilled water. Specimens were then transferred to 50% alcohol, and then to 70% alcohol. The use of ethylene glycol is being discouraged because of possible health risks; Marhue (1983) substituted propylene glycol, and this gave good results.

SEALING GLASS MUSEUM JARS

If glass museum jars are required to hold specimens or old storage containers need refurbishing, the following list of sealants and their preparations will be useful.

The mounting fluid should cover the specimen and a layer of air should be allowed between the specimen and the top to allow for contraction and expansion of the fluid. It is advisable to drill a hole in the top plate so that air is displaced when the lid is lowered onto the jar. The hole also enables fluid to be topped up when the level drops to an unacceptable level. A glass cover slip may be stuck over the hole, using a silicone sealant, to minimize this evaporation.

Bladders

The earliest type of cover consisted of pig or sheep bladders. The bladders can be obtained from a local abattoir. They are usually dry and need to be soaked in water to soften them before they can be stretched over the opening of the jar and allowed to dry completely. The bladder thus prevents evaporation if the jar is not moved around, but it does break easily if tampered with.

Stockholm tar and red lead

This mixture is made up as follows: add a little red lead (litharge) to Stockholm tar to make a stiff paste. Mix thoroughly so that they combine completely. Spread a thin layer onto the rim of the jar and place the lid in position. Weigh it down and allow it to harden for three to four days.

Bitumen

Sealing lids with bitumen can be hazardous because of the need to heat the lid before it is placed on the jar. The lid is heated in an oven whilst a small amount of bitumen is melted over a Bunsen burner flame. The bitumen is spread over the top of the jar and the lid put in position before the bitumen cools. If the bitumen cools before the lid is placed in position, it should be remelted and the process repeated.

Silicone sealant

When expelled from its container and applied to the top of the jar, silicone sealant forms a good bond. After application, it should harden for 24 hours. It is available from hardware stores.

LIGHTING

'Natural' colour in anatomical specimens relies not only on the addition of sodium hydrosulphite but also on the type of lighting, temperature, and background colours. Temperatures above 27_{1}C tend to alter the colours, producing pale red to grey. Variations in temperature expand and contract the fluid, and this, because of the tendency of acrylic to pass liquids through the walls, causes bubbles to appear so a temperature controlled room is advised.

Some specimens which appear slightly pale can be enhanced by giving the showcase a pale green background. Red and green are complementary to each other, so a green background will highlight the red tissues.

Lighting in the room containing specimens treated with the hydrosulphite method should be of the incandescent type. This form of lighting is strong in the yellow and red parts of the spectrum. Fluorescent lamps, which are strong in the blue area, should be avoided. Even natural lighting does not have much of an effect on colour highlighting. Therefore, it is advisable to have good lighting to enhance even the most subdued red colours.

REFERENCES

Banks, H.J., and D.J. Williams: Use of the surfactant, Decon 90, in the preparation of coccids and other insects for microscopy. *Journal of the Entomological Society* 11/1972: 347–348.

Cleave, H.J., and J.A. Ross: A method for reclaiming zoological specimens. *Science* 105/1947: 318.

Marhue, L.: Techniques to restore dried-up invertebrate specimens. *Syllogeus* 44/1983: 175–177.

Thompson, J.R., M.H. Thompson and S. Drummond: A method for restoring dried crustacean specimens to taxonomically usable condition. *Crustaceana* 10/109 1965: part 1.

SOURCE MATERIAL

Audette, L.G.: A method of constructing plastic jars. *Anatomical Records* 153/1965: 383–388.

Dunker, J.R., and Schluter, O.: Neue Plexiglaskästen für Feuchtpräparate. *Der Präparator,* 12/1966: 99–105.

Farmer, T.W., *et alii*: A simple method of mounting gross biologic material in plastic boxes. *Science* 114/1951: 236–237.

Gordon, H.: Methods of fabricating museum containers from plastic. *Bull. Int. Med. Mus.* 32/1951: 103–116.

Hackett, C.J., and W. A. Norman: The Dugid-Young technique for making plastic specimen containers. *Medical and Biological Illustration* 2/1952: 191–197.
Hapatzky, H.: Plexiglasstürze für Trockenpräparate. *Der Präparator,* 1/1955: 56–57.
Harris, R.H.: 1979. The conservation of one of the earliest known examples of a fluid filled dissection. *Museums Journal* 79/2 1979: 71.
ICI. *Technical Service Notes* (PX 118 Tensol Cements; PX 199 Machining; IX 120 Cementing; PX 121 Shaping; PX 122 Properties).
Kennedy, A.J.: A modification of methods of fabricating museum containers from plastic. *American Journal of Clinical Pathology.* 22/1952: 196.
Morton, J.E.: 1954. Plastic containers for museum specimens. *Medical and Biological Illustration* 4/1954: 178.

11

Plastics workshop, embedding, and workshop safety

THE TERM plastic is really misconstrued. A dictionary definition of plastic is usually something that is characterized by moulding, readily assumes a new shape, is susceptible to influence and impressionable, pliable, supple and flexible. Many materials fall into this category such as clay, wax, glass and metals which all undergo a change of shape when proper conditions apply. They will also retain this new shape.

In this chapter we will describe some materials used in museum work that have this plasticity, but which are purchased in a liquid form which, by altering certain conditions, produces a solid material. The following list encompasses the more common materials that are used in the areas of museum display; polyesters, epoxies, urethanes, acrylics, silicones, and latex.

These plastics take many forms and are adaptable to many situations within the realms of museum work. They are used in a variety of jobs such as in glueing, fabrication, moulding, casting, and gap filling. Many books have been written on plastics, their chemistry, properties and uses. In this chapter we will describe the uses and properties only.

Plastics are used because they are strong, lightweight and easy to machine and glue. They may be coloured to the desired hue or painted to match certain requirements. They are water resistant as well as fungus and mildew resistant. They can be used to make moulds and casts, and last, but not least, items can be mass produced if required. With all of the following materials described in this chapter, a small sample should be tested from the batch to be used on the day on which it is to be used so that the pot life can be determined. This will depend on the ambient temperature and humidity, both of which can substantially alter the cure time.

Plastics are generally divided into three categories:

Thermosetting plastics: Once these plastics have become hard they will not revert to a fluid, even under the influence of heat or pressure. Examples are polyesters, epoxies, phenolics, and ureas.

Thermoplastic plastics: These become soft upon heating and harden again upon cooling. Examples are polyethylene, polyvinylchloride, polystyrene, cellulose acetate, polyamide (Nylon) and polymethyl methacrylate (acrylic).

Elastomers: These plastics withstand repeated stretching and return back to their approximate original length. Examples include silicones, urethanes, and nitrile rubbers.

PLASTICS WORKSHOP

POLYESTERS

These are thermosetting plastics and are mostly used in casting, moulding, and laminating. They are probably the most widely used plastics in museum exhibition work due to their low cost and the ease with which they are worked. Some of their characteristics include: a shrinkage of around 7%–12%; poor adhesion because of shrinkage; and ability to alter polymerizing times.

Polyester casting resins contain waxes or flexibilizers which minimize cracking and tacky air-exposed surfaces. They shrink volumetrically depending on the amount of catalyst used. Although the shrinkage seems high, in reality it is not, as it is volumetric not linear.

Generally, methyl ethyl ketone peroxide (MEKP) is used to harden the resin although there are other hardeners that are suitable for certain applications. MEKP is a strong oxidizing agent. It should be kept in airtight containers and stored in the dark. If possible, the resin should be kept in a refrigerator. Although artists and craftsmen make large casts, museum workers usually cast small models, embed objects in it (see below) and use it for other useful purposes such as pouring thin layers onto acrylic sheet to reproduce the surface of water. In this case, one of the properties of casting resin that we can take advantage of is that it produces a somewhat uneven or wavy surface, giving an immediate impression of an aquatic environment.

Laminating resins are designed so that they may be combined with layers of fabric, paper, or glass and acrylic matting to produce, from a mould, articles that are thin walled hollow and lightweight. They generally have a higher exotherm than casting resins as they are used in thin layers where heat generation is not normally a problem. Laminating resins can also be used in certain circumstances to produce moulds where flexibility is required, such as when multipiece moulds are made where many casts are required. Adding flexibilizers (plasticizers), such as Benzoflex, or di-butyl phthalate, to casts enables, with the use of heat, the bending and twisting of the cast, as with fins on fish casts. Fine structures which could be easily broken when a plaster mould is chipped away become more resistant to breakage if a flexibilizer is mixed into the resin prior to laminating the inside of a mould.

Types of moulds from which laminating resins can be extracted are silicone rubbers, plaster of Paris, wood, metal, polypropylene, glass, acrylic, and other plastics. Silicone rubber moulds require no mould release or parting agent as it has no adherent properties. Most other materials require the application of a release agent. Mould releases may be liquid (brushed or sprayed into the mould) or hard (wiped with a cloth). They are based on polyvinyl alcohol, polyvinyl chloride, silicones, and waxes and may be purchased from the suppliers of resins.

Polyesters can be thickened using materials called fillers. Two such fillers, asbestos and Aerosil, when added produce a nonflowing mixture that is ideal when the resin has to hang downwards in a mould undercut. Asbestos is no longer used in many workshops because of its association with respiratory disease. Other general fillers include talc (kaolin), cotton floc, and whiting and, for specific applications, sand, silica carbide chips, mica and pumice.

The first resin coat usually has a thick consistency; this is termed a gel coat. The gel coat can be mixed in the workshop using asbetos or Aerosil fillers or it can be purchased from resin suppliers already made up. When the cast is extracted from the mould, the gel coat becomes the surface finish of the cast. Pigments added before the catalyst or painting after polymerization are applied to the gel coat to finish the appearance of the cast.

Hazards of polyesters

Bourne and Milner (1963) summarize the problems of working with polyester resins and give some case histories. The handling of these plastics may cause skin irritations, burns, and allergic reactions, mostly around the eyes and nose. The hardeners are dangerous as they can spontaneously combust in certain circumstances. Also, there have been reports of serious eye problems occurring many years after the hardener had accidentally splashed into the eyes. When handling these plastics and associated components, follow the manufacturer's instructions explicitly to reduce risks. Use plastic gloves when mixing components, and wear protective glasses when injecting resin with a syringe.

EPOXIES

This group of plastics is very useful for the following: they have negligible shrinkage; they have excellent adhesion properties; they are very strong; they have good chemical resistance; and they can be obtained in many degrees of viscosity. Epoxies have been formulated for specific applications and, in most instances, do not produce good results if they are applied to other situations.

The types of epoxies utilized in museum work are adhesives (fast and slow setting, viscous and pastes), casting resins, surface coatings, gel coat resins, modelling pastes, and foaming resins. Colouring pastes can be mixed easily into the uncured resin to produce a multitude of hues.

Liquid resins can have fillers mixed with them to alter their viscosity and such fillers include talc, kaolin (china clay), wood flour, Aerosil (Cabosil), asbestos, metal powders, and sand.

Epoxies are obtained in two components: the resin, and a hardener. They should be measured out according to the manufacturer's recommendations and mixed thoroughly when required. Some epoxies have a setting time of two to five minutes. This is useful for glueing broken objects quickly. Others have setting times of 24 hours. Generally, the longer the setting time the stronger the bond.

Laminating resins show little or no shrinkage. They also allow chopped fibreglass coth and matting to be used when laying up in moulds. Most moulds (except silicone rubber and latex) require a mould sealer and a release or parting agent to be applied prior to the addition of the epoxy. Such sealers include QZ 1 and QZ 5 (Ciba-Geigy) and release agents include QV 10, QZ 11 and QZ 13 (Ciba-Geigy). When casting epoxies in silicone rubber moulds, the mould should be heated in an oven after every five to eight castings to drive off amines that infiltrate the rubber and inhibit further casts.

When handling epoxies, safety procedures must be enforced. Skin contact should be rigorously avoided by wearing gloves, goggles and protective clothing. If there is a risk of a dust cloud forming, breathing masks should be worn. If contact with the skin occurs, the affected area should be cleaned with disposable paper towels and washed with warm soapy water. Any material that is still adhering to the skin should be removed with a resin-removing cream. Do not use solvents. Any contamination of the eye should be treated with an eye wash for at least 15 minutes; medical attention should be sought immediately. If epoxies are ingested, drink as much water as possible and seek medical attention. Waste epoxies should be disposed of in accordance with local, state and federal laws. Working areas should be kept clean and any spillages must be wiped away.

Finally, it is wise to study the complete product data sheets (which can be obtained from the manufacturers) before using any epoxy to become familiar with the characteristics of each resin.

URETHANES (POLYURETHANES)

Flexible urethanes are used everyday in such areas as cushion and mattress foams, and automobile dashboards, Rigid, semirigid, and flexible foams are used to insulate refrigeration components, pipelines, and cavity walls. The type most commonly used in museums at this time are rigid. They are purchased as a two component system. When mixed together, the components increase in volume and become rigid and lightweight. These foams are used in mannikins for large mammal taxidermy and as a build up agent to produce foreground material in dioramas. One interesting item we have produced using urethane is a crusty loaf of bread.

If very large areas are to be covered using urethanes, then specially designed spray machines are used. For most applications, urethanes are mixed in containers ranging in size from paper cups to buckets. Hand mixing is easy when small volumes are required, but larger volumes require an electric stirrer.

Foaming occurs in three stages: (i) a liquid phase to a creamy consistency; (ii) gas generation within the components which enlarge the mass to many times its original volume, producing air cells in the process; and (iii) a short gel period following by polymerization and stiffening of the cell walls.

The two components consist of an isocyanate and a polyol. When mixed together, the urethane foams in size due to either carbon dioxide or fluorocarbon production and can reach densities of about 900 kg m^{-3}.

When using urethanes to produce casts, precise volumes must be used. If too much is mixed, high pressures build up, resulting in the mould breaking and splitting. When using urethanes, always wear protective gloves and clothing and work in an area with good ventilation.

Low density foams can be warmed to produce a faster exotherm whilst high density foams require chilling to reduce the exotherm. Poor mixing generally gives rise to the cells collapsing, internal voids, and incomplete rising of the material, although incomplete rising or filling of moulds could be the result of miscalculation of sufficient volume. If the mixture seems slow in its action it can be speeded up by flicking a few drops of water onto the surface of the mixture which will 'kick off' the reaction.

Urethanes do not part from moulds except silicone rubber, so a mould release or parting agent must be applied. Such parting agents include silicones and waxes. Urethane manufacturers should be consulted for correct advice.

ACRYLIC

Common acrylics are Perspex, Plexiglass and Shinkolite. They are thermoplastics. They can be purchased in sheets, rods, and tubing. Some liquid forms can be purchased in two component systems and are ideal for casting small objects from wet moulds. Specimens embedded in acrylic are in some ways superior to those embedded in polyester resins, but problems occur during the preparation of the materials.

Solid acrylic is available in many colours, and can be translucent or opaque, as well as clear and nonreflective. Rods and tubing are normally clear.

Acrylic has uses in the manufacture of display boxes, architectural models, stands, and other essential items. Acrylic can be printed on or engraved; therefore it can be useful as a label or identification panel.

Acrylic can be machined easily with normal woodworking and metal working tools. It can also be heated to be shaped into many forms. Details are provided in Volume 2, Chapter 10, and readers should turn to this chapter for full information on working with acrylic.

SILICONES

The most commonly used silicones are the room temperature vulcanizing (RTV) rubbers, also known as cold cure silastomers. Two types are used in museums: one component adhesive sealants; and two component systems.

One component adhesives, require no mixing. They are used to fill gaps or joints, for sticking objects together, and between similar and dissimilar surfaces. They are also used for sealing, waterproofing, and weatherproofing. They are commonly bought in cartridges and expelled using a caulking gun. The type used for joining glass to form aquariums gives off acetic acid fumes which can be irritating when used in large quantities.

The two component system requires a catalyst to solidify the rubber. There are many different types suitable for many different uses. Manufacturers are able to supply detailed information on the use and function of silicones. Readers should consult with the customer service department of the appropriate chemical manufacturer to ensure that they receive the best and most up-to-date product information.

LATEX

Latex is a water-dispersed colloid from the sap of certain species of tropical trees. The rubber particles are small globules (1 μm–3 μm) and have a solid content of about 40%. Latex is used in moulding or casting and may be applied by brush, spraygun, rollers, or pouring. It dries at room temperature, although this happens slowly at high humidity.

Latex solution contains ammonia which prevents coagulation and fungal attack. Brushes and sprayguns may be cleaned using clean water or with a 2%–4% ammonia solution.

Latex is very elastic, but its main disadvantage is that it shrinks considerably when it sets and is of little use in moulding when critical detail is needed. Fillers such as Aerosil, floc, and kaolin can be mixed into the latex to thicken it, but they should be moistened first with 5% ammonia solution to allow easy mixing. The latex can also be hardened quickly using a hot air gun or by spraying a fine mist of acetic acid onto the layer to be set.

Latex can be coloured using powders that have been mixed in 5% ammonia solution. Liquid pigments can be mixed into the latex without any preparation.

When moulding large objects such as rocks, thin layers of latex should be applied, each layer setting before further coats are applied. Two or three layers are usually sufficient before applying a layer of fabric to act as a strengthener. Even so, the rubber mould removed from the object will be floppy and unable to keep its correct shape. To overcome this, a fibreglass or plaster jacket should be added before the latex mould is removed from the object. When large moulds are made and strengthened using a fibreglass jacket, it is essential to be able to remove this jacket to pull the latex mould from the original cast. To ensure removal, a layer of polythene or plastic food wrap must be laid over the latex cast before the last layer sets. The plastic must be left in position until the latex sets and holds the plastic tight. The first coat or layer of the polyester jacket is applied to the plastic and allowed to set. Fibreglass reinforcement is applied until the required thickness is attained. The jacket can then be removed from the latex mould. Before casting from the latex mould, it should be laid into the fibreglass jacket to form its original shape.

Solid casts are not usually made from latex as it takes too long for the rubber to set within the mass. However cranial endocasts (Hildebrand, 1968) can be produced; that is, thin and hollow casts are made of the brain cavity in skulls so that they may be extracted without causing damage to the skull.

One useful property of solid latex is that it expands up to 1.5 times if immersed in paraffin oil. If a small cast can be made from latex and expanded in paraffin oil. A mould can be made from this expanded cast and recast using latex. This cast can then be expanded, the process continuing until the desired size is reached.

HOT-MELT COMPOUNDS

These compounds, based on polyvinyl chloride, melt upon heating and solidify upon cooling. They are not used as often as other mould making materials due to the high temperature of melting (about 120°C). They may be used as moulding or casting compounds, but care has to be exercised if they are used with Plasticine for the high temperature will melt the Plasticine.

The most common hot-melt compound is Vinamold which is available with different colours, melting points and hardnesses (for different applications). It is manufactured by Vinatex Ltd, Surrey, United Kingdom, who will supply the compound or give details of suppliers.

The compound is rubbery when solid. It is cut into small pieces which are placed in a double boiler for melting. The double boiler should contain an oil so that the high temperatures can be reached. The melting is done on a hot plate and in a fume cupboard. To avoid premature cooling on the object to be moulded, this, too, should be heated. This also helps avert the problem of specimens cracking when covered with the very hot compound.

The compounds are reusable, but have taken second place to the easy-to-work with silicone rubbers.

EMBEDDING BIOLOGICAL SPECIMENS IN CLEAR PLASTIC

The reader should now be aware of the need to display specimens in a way that allows them to be studied closely without breakage and disfigurement. Some specimens can be displayed in a liquid preservative or by being embedded in plastic (worms and spiders), but some specimens (for example, very fragile, dry plants) have to be embedded to prevent damage.

The steps involved in embedding are the same for all biological material, with some deviations for certain types of specimens. In simple terms, all material should be fixed in position, preserved, dehydrated, then embedded in the plastic. Different specimens, however, are preserved differently, and the method of dehydration will also differ. The following steps have been used by the authors many times, with a very high success rate.

THE PLASTIC

Plastics can be divided into three major categories, as indicated in the previous section. Some special information regarding use of plastics for embedding is given here.

The polyester resins are the most widely used as an embedding material for they are cheap, clear when hard, sufficiently stable in light, and they harden quickly. A disadvantage is that polyester resins shrink volumetrically; this can produce enough stress in a fragile object to break it.

Polyesters

Polyester resins are described in detail in the previous section. Here, details relevant only to embedding are discussed.

When the hardener has been added and mixed properly into the resin, a chemical reaction begins. The temperature of the resin increases and it goes through a gel stage before hardening to a hard, transparent block. Generally speaking the molecules join to form macromolecules which sink to the bottom and lay upon each other to form a solidifying mass.

Polyester resin contains two ingredients which aid in solidifying or polymerization; a solvent, and a promoter. The solvent, styrene monomer, has the polyester particules dissolved in it and the promoter allows polymerization to take place at room temperature and within a specified time. During polymerization, when the molecules join to form macromolecules, some of the styrene evaporates, producing a shrinkage in volume of approximately 7% but this shrinkage also depends on the amount of hardener added, the room temperature, and the thickness of the layers.

The resin used to embed biological material should have four prerequisites: it should contain a stabilizer to minimize discolouration and deterioration by ultraviolet light; it should have a low exotherm of between 40°C–60°C to prevent cracking, warping, shrinkage, and production of bubbles from within porous specimens; and it should have a slow even cure after gelling over a period of several hours.

The Australian Museum uses a product which fills all of these prerequisites. However, the product data sheets recommend the use of 1% methyl ethyl ketone peroxide (MEKP) at 25°C while we have used concentrations down to 0.25%, with 0.5% as an average. When thick layers are poured, less hardener should be used. When many layers are to be poured to produce a block, each layer should contain the same amount of hardener.

Once the resin has solidified into a hard, transparent mass, it cannot be returned to a liquid state. If a valuable specimen has been embedded, and the resin on polymerizing shows internal cracks or bubbles, there is a method which can be employed to retrieve it. After the block has hardened it can be sawn and sanded to leave a thin layer of plastic around the specimen. This should then be immersed in a jar containing acetone for several days or weeks (depending on size) until the plastic has dissolved away. The acetone may need to be changed if a large amount of plastic must be dissolved. The skeleton of a sea urchin crumbled into tiny fragments after treating with this method and the only explanation was that the hard resin inside expanded after immersion in the acetone, breaking the calcified skeleton.

EQUIPMENT

Moulds

Craft and hobby shops sell various shapes and sizes of polyvinyl chloride moulds suitable for many types of embedments. They come in square, rectangular and round shapes, but are limited in size. In some cases, this means use of more plastic than is really required. Glass and polished metals such as aluminium provide good nonadhering surfaces. Two aluminium L-shaped pieces can be manoeuvred to allow a range of sizes (Fig. 11.1). A piece of glass can be used as the base. The base and edges of the aluminium must be sealed to prevent resin seaping under the contact points. This can be achieved with Plasticine or by melting paraffin wax and brushing it into the cracks to provide a seal. If a machine room is available, then clear acrylic sheeting may be used to make moulds. A saw with a tungsten tipped blade is all that is required to cut the sides. Fig. 11.2 shows how to place the sides together to form the correct size mould. Ethylene dichloride or chloroform painted around the joints will keep the mould in shape and prevent the liquid resin from escaping. Paraffin wax, melted and painted around the joints, will be suitable if the other solvents are not available.

Plastic water pipes and the like make ideal moulds if cylindrical shapes are required. If round moulds are used, then several more coats of mould release should be applied than for rectangular moulds, otherwise the hardened block will be exceptionally difficult to remove.

Mould releases

Unless thermoplastic moulds are used, all other types of moulds must be coated with a release agent to enable the solid block to be expelled from the mould easily. Glass and aluminium have good nonadhering properties, but if they are scratched or dirty removing the block becomes harder unless mould release is applied to the internal surfaces.

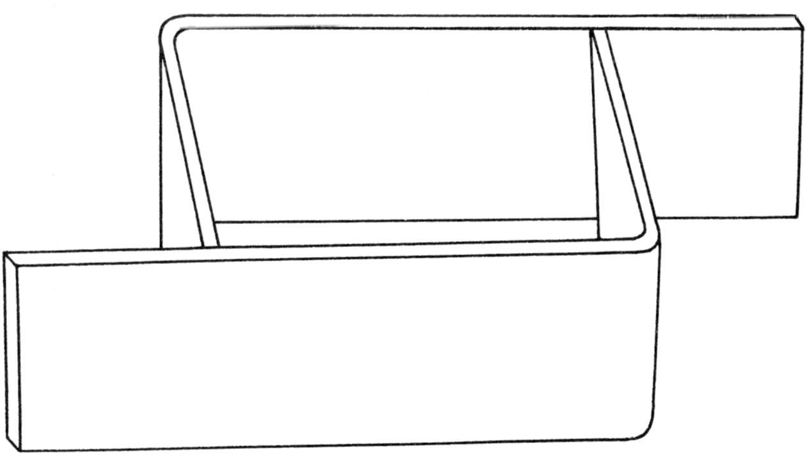

Fig. 11.1. These aluminium L-shaped sections fit together to produce a mould. Joints should be sealed using melted paraffin wax.

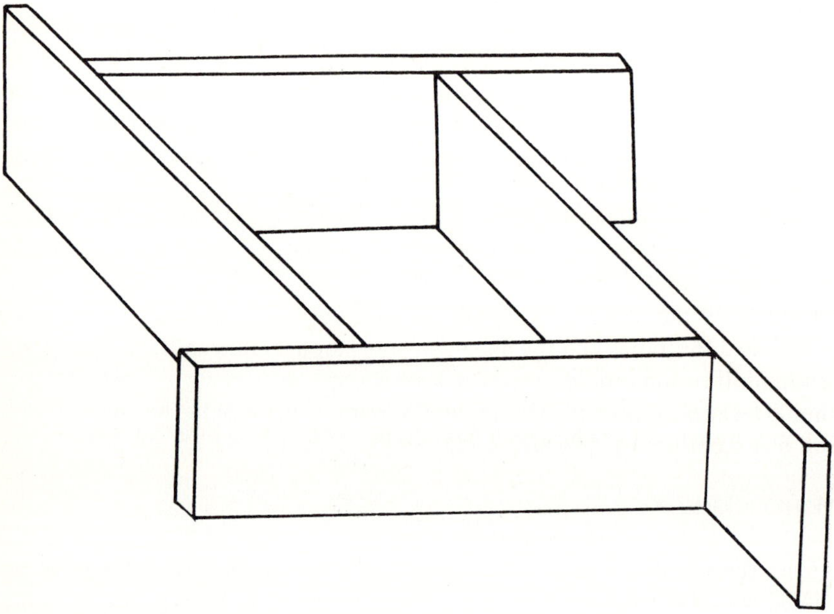

Fig. 11.2. Any size mould can be prepared by cutting four equal sized strips of clear acrylic and cementing them together using melted paraffin wax.

There are several mould release agents available on the market, some of which may be purchased from the suppliers of polyester resins. Hard waxes, such as floor or furniture wax, are suitable, as are the aerosol sprays used for cooking utensils or furniture polish. Polyvinyl alcohol (PVA) release, which is coloured to show where the surfaces have been coated, may be used with success. The release is applied to the mould using a cloth. The coating must have dried before the resin is poured in.

Mixing containers

There are basically two types of containers used for mixing the resin and hardener; disposable, and nondisposable. Disposable containers consist of the paper cups used for serving hot fast food. Do not use waxed paper cups; these have a coating which is dissolved by the styrene in the resin and produces a slightly opaque look in the block. Other unsuitable disposable containers are plastic yoghurt and butter containers as the styrene dissolves them. They may be used if the mixture is poured out immediately, but not if the plastic is left in the container for any length of time. There are many nondisposable mixing vessels. Nonflexible containers, such as glass and ceramic, should be turned upside down after the resin is mixed to allow the contents to drain as much as possible. Two methods will remove resin from the container:

1. Acetone may be poured into the container while the resin is still in its liquid form. The acetone will dissolve it, whilst paper towelling can be used wipe the sides clean.
2. Allow the resin to solidify, then place the container in the freezer where the different expansion and contraction coefficients will allow for easier extraction of the resin.

Other reusable containers include plastic bottles, such as those which contain orange juice (but which have to be cut down to be able to stir in the hardener). The resin can solidify within the container, which is flexible enough to break the hardened resin so it can be removed.

Measuring devices

There are three ways to measure the quantities of resin and hardener: volume by volume (v/v); weight by volume (w/v); and weight by weight (w/w). For those who do not have access to a balance, v/v measuring is recommended. If a plastic, graduated measuring device is unavailable for mixing, the following method should be followed. Pour water to the desired level into the mould and transfer this to a graduated device (kitchen measuring cup). Record the volume. Dry the mould thoroughly and pour in the same amount of resin. Using a graduated eye dropper, add the required amount of hardener and mix thoroughly using a wooden or glass stirring rod.

For those who do have access to a balance, measure the quantities w/w. When using a balance, zero the weight of the container, if possible, or weigh the empty container first and deduct this from the final weight.

When measuring the components, take one gram of resin and hardener to equal one gram or one millilitre of water. These weights are not exact, but the authors have never encountered any problems due to this approximation.

CALCULATING QUANTITIES

Two mathematical formulae are all that is required to be able to calculate all the volumes required to produce rectangular or cylindrical embedments.

Whichever formula is used, the final quantity of resin should be increased by 10 percent to allow for a slight discrepancy between weight and volume (that is, the approximation that one gram of resin equals one millilitre of resin), and also to allow for wastage of resin around the sides and bottom of the mixing container.

For rectangular blocks, use the formula, length x breadth x height to calculate the volume of resin needed. Measure all sides in centimetres as this will make the conversion to volume easy. On average, pour layers 1 cm thick until the techniques have been mastered.

Example

To calculate the volume of a rectangular block 8.5 cm long, 4 cm wide and a layer 1 cm deep use the following:

$$\underset{8.5}{\text{length}} \times \underset{4}{\text{breadth}} \times \underset{1}{\text{height}} = 34\ \text{cm}^3$$

If you take 1 cm^3 to equal 1 gram and add 10% for wastage the final volume is calculated as:

$$\frac{34 \times 10}{100} = 3.4$$

The weight of resin required is 34 + 3.4 = 37.4 grams. This figure can be rounded to 40g. To calculate the amount of hardener required to polymerize 40 g of resin using a concentration of 0.5% apply the following formula:

$$\frac{\text{wt of resin X \% of hardener}}{100} = \frac{40\ \text{g}}{100} \text{ x } 0.5\%$$
$$= 0.2\ \text{g}$$

For cylindrical blocks, the formula πr^2h is used where π is taken as 3.142, r is the radius of the end, and h is the height. If a cylindical mould has a radius of 5 cm and a layer 1 cm thick is required, then the amount of resin required is:

$$3.142 \text{ x } (5)^2 \text{ x } 1 = 78.55\ \text{g}$$
$$\text{add } 10\% = 86.4\ \text{g}$$

The amount of hardener needed at a concentration of 0.5% will be:

$$\frac{86.4 \text{ x } 0.5\%}{100} = 0.43\ \text{g}$$

COLOURED BACKGROUNDS

In certain cases, it may be advantageous to have a coloured base layer. The reasons for this may be:

1. A specimen may have an external surface which will be lost and unable to be distinguished clearly in clear resin as is the case with a white sea slug (Nudibranch). A black or blue background would produce a good contrast for it to be seen.
2. Sometimes the finished embedment may need to be displayed on a wall panel. With a coloured base, bolts can be fixed to the rear of the embedment so that they will be undiscernible.

Coloured pigments are available in various colours and can be obtained from the resin supplier. The pigment is very concentrated and only a small amount is needed to colour the resin. Too much pigment can inhibit the setting time. The pigment should be thoroughly mixed into the resin before the hardener is added. This allows for the correct colour shade to be reached, which, when matching particular colours, may take some time.

The coloured base layer may be poured, first but this results in an upturn of the colour at the corners due to the meniscus. This is quite noticeable after the block has been polished (Fig. 11.3A). The alternative is to pour the base layer last, on top of the specimen already embedded in clear layers (Fig. 11.3B). Both methods have their advantages and disadvantages and each specimen requires careful thought before work is commenced. Pouring the base layer last requires the specimen to be embedded upside down with the viewing surface at the bottom of the mould. This can pose some problems as porous specimens with under cuts or concavities may hold air bubbles which will lead to an unaesthetic appearance. Also, the top layer should be thoroughly tack free (not sticky). Otherwise, when the coloured layer is poured, some mixing of the colours will occur, causing whirls and streaks.

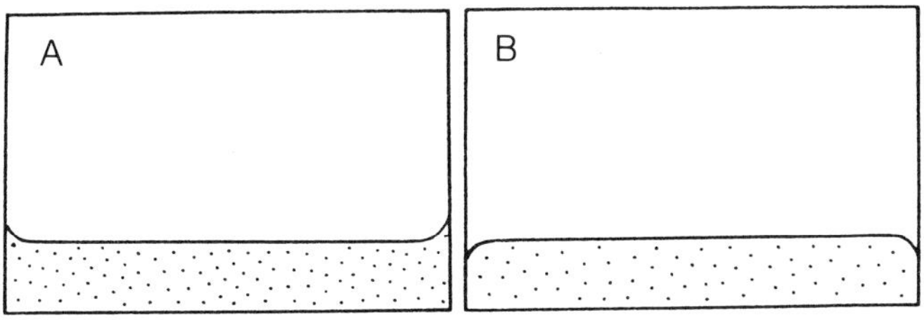

Fig. 11.3. A. The coloured base (stippled) has been poured first which produces an upward edged meniscus in the final block. **B.** The coloured base here has been poured on top of the already poured clear layers producing a downward meniscus.

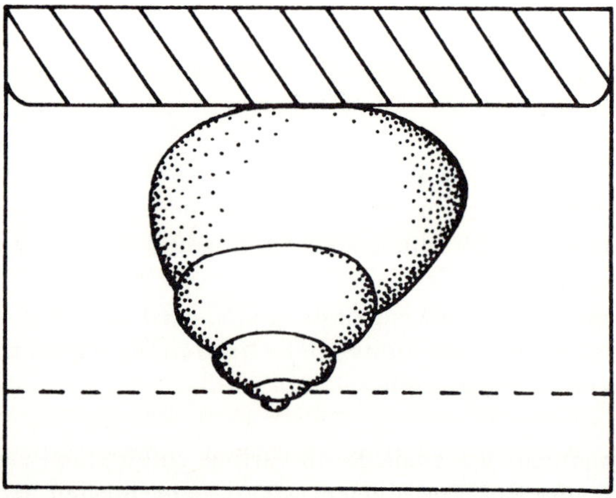

Fig. 11.4. This is how a hollow shell should be embedded, open end up and the base layer poured last.

Examples which retain air are conical shells (Fig. 11.4). These are better embedded with the apex pointing down. If the shell is positioned with the wide, open end down, a large air bubble will be trapped within; upon heating (during polymerization), this will expand and crack the block.

TOP SURFACE LAYER

The curing of the polyester resin is inhibited by air; the top 0.5 mm of the block, if exposed to the air, will be tacky. This clogs the sandpaper that is used later to smooth the surfaces. Many workers have devised different ways to exclude the air from the surface.

If the resin is poured right up to the top of the mould a piece of glass (with a layer of mould release applied) can be laid on one edge of the mould and gently lowered until it completely covers the resin and mould. There must not be any air bubbles trapped under the glass. If there are, the glass must be lifted and more resin added.

Cellophane or cling wrap can be laid over the surface of the resin while it is still liquid; however, this produces an uneven top surface.

A layer which has had hardener added to form a 3%–4% concentration can be poured over the top and this will polymerize quickly to produce an almost tack free surface.

Styrene monomer poured on top of the last layer while it is still in a liquid phase will float and exclude air at the same time. This produces a tack free, and flat, even surface. It is a useful method to use when specimens with a large surface area are to be only half embedded (Dingley, 1980).

LABELS

Labelling specimens has been given much prominence elsewhere. Labels may be handwritten on heavy white paper using India ink, or typed, preferably on an electric typewriter. If these types of labels are to be used they should be soaked in alcohol for a few minutes to rid them of dirt and grease. They can be removed from the alcohol with forceps and dried, then placed into styrene monomer. They should be left in this solution for a minute before embedding. The immersion in the styrene monomer provides easier penetration of the paper by the resin; this prevents air bubbles forming.

Labels made from plastic can be prepared, but they require special machines to make them, or they have to be sent out to a company which specializes in this type of artwork. In either case, the labels should be typed in black onto white paper. As many labels as possible should be typed onto paper with dimensions of 20 cm × 25 cm in order to reduce costs. This the master. It is sent for processing. The final product is known as an Arcright Transpositive. An Artiscope machine, used by many exhibition departments in museums, can perform the necessary steps to produce these labels easily and quickly. The lettering can then be cut into the right size labels and soaked in alcohol, as described for paper labels, then embedded in the plastic.

PROBLEMS

Silvering

The term is loosely used to describe a place where the polyester resin shrinks away from the specimen upon polymerization to produce an effect which masks out detail and which has a silver appearance. This shrinkage from the specimen produces a gap which is silvered by internal reflection and can be quite disastrous. Reasons for the nonadherence and, hence, shrinkage from specimens is due to several factors.

1. A wax coating such as that found on some insects (beetles, grasshoppers, centipedes, millipedes) and some plants cannot be penetrated to give a bonding contact.
2. Air trapped within specimens can expand them during exotherm forcing the cured resin to move away from them.
3. The exotherm within the resin can pressure some specimens to warp and distort.

4. Some extremely smooth objects which cannot be penetrated by the resin cannot form a bonding surface, so the resin shrinks away from them.

Before embedding, the likelihood of silvering occurring has to be evaluated. If there is a chance of silvering occurring, treat the specimen as though silvering is definite. Hard objects such as fir cones or smooth operculi from snails will have the chance of silvering lessened if the amount of hardener is reduced, even down to a concentration of 0.1%.

Those specimens which have a wax coating can be immersed in acetone for 24 hours; also, their surface may be scraped with a scalpel. Allow the specimen to dry thoroughly, then immerse it in styrene monomer for 24 hours before soaking in resin without hardener.

Specimens which contain air, such as coral, soft biological material and bones, will require a more specialized procedure. There are two methods; use of either will depend on whether or not special equipment is available.

1. If the laboratory has a vacuum chamber, such as those used in a histology laboratory, it will make the task much quicker. After all of the required steps involved in the embedding technique, up to the embedding technique, have been followed, and the specimen has been soaked in resin without hardener added, place it into the vacuum chamber. Switch on the vacuum pump and adjust the volume of air being extracted so that the rate of air being extracted is low. Increase the rate until bubbles can be seen coming out of the specimen. Hold the extraction rate there until less bubbles are emitted. Maintain this procedure until lemonade-like bubbles appear out of the resin. Reduce the vacuum rate until there are no more bubbles. Switch off the vacuum pump and allow the chamber to slowly return to atmospheric pressure. The whole procedure must not take longer than 15 minutes. As the styrene solvent is extracted from the resin, the resin will become very viscous and will not be thin enough to penetrate the specimen. If a very fragile specimen is to undergo this treatment it may be wise to thin the resin beforehand with styrene monomer to produce a 50:50 mix.

2. If a vacuum chamber is not available, specimens can be infused at a very slow rate by simple immersion in jars containing various viscosities of resin. The first jar can contain a mixture of resin and styrene monomer in the ratio of 2 parts of resin to 8 parts styrene monomer with, successive ratios of 4:4, 6:4, 8:2, and finally pure resin. Leave the specimen in each jar for 12 hours and do not allow the specimen to drain during the transferring procedure. This method may be made more efficient by placing the specimens in jars and then tumbling them in a machine similar to those used by lapidary workers.

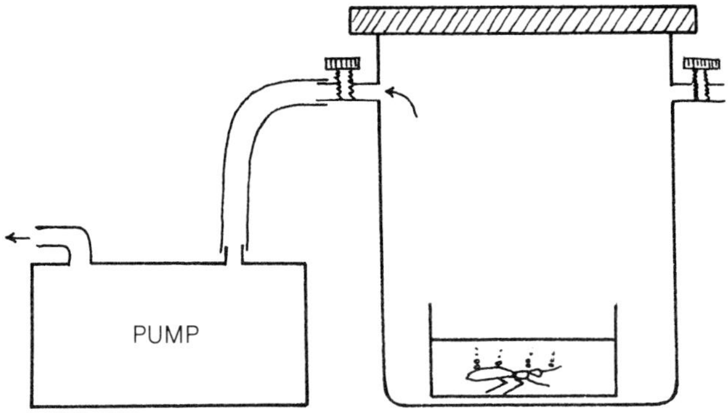

Fig. 11.5. A vacuum chamber is a useful tool for removing air bubbles from within specimens.

Cracks

Cracks occur because of internal pressures that have built up during the exothermic reaction, warping, distorting, or some movement of the specimen. They can also occur due to the failure to apply enough mould release to the sides of the mould, resulting in the resin sticking to the mould, and, upon shrinking, splitting the softer, inner region of the block. In all cases, less hardener should be added and more mould release applied. If a small concentration of hardener is mixed with the resin, the polymerizing time is increased with the result that cracks can be seen appearing before the block has cured to a hard mass. The block will have the consistency of a pencil eraser and cracks can be rectified using a syringe. Fill the syringe with resin and hardener and penetrate the block with the needle, pushing it down into the lower region of the crack. Depress the plunger slowly to exude the resin and withdraw the needle until the resin has filled the crack and the needle has been completely withdrawn. This method is an excellent way of filling cracks without having to dissolve away the block to retrieve the specimen. Other workers have waited until the block has fully cured before drilling a fine hole down into the crack and filling it using a syringe, but the results are not as favourable.

Bubbles

These occur when the specimen has not been infused with resin sufficiently, or when the bubbles have been deposited onto the specimen at the time the

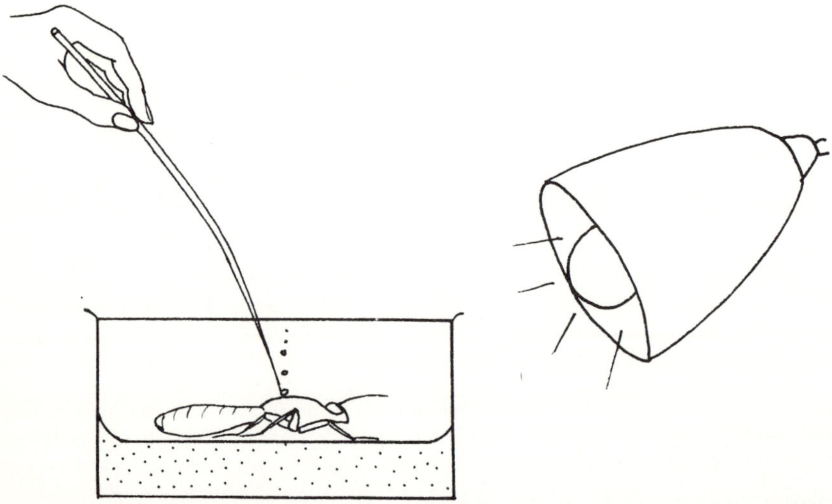

Fig. 11.6. Surface air bubbles attached to specimens are removed using a probe. The light source is to enable the operator to clearly see the bubbles adhering to the specimen.

resin and hardener were mixed and not allowed to disperse. If a few bubbles remain stuck to the specimen, prise them off with a pin or needle and force them to the surface (Fig. 11.6). If a lamp is shone close to and over the mould, it highlights any bubbles, allowing easy removal.

Mirror under the block: Lutz (1969) describes the use of a mirror to check for bubbles underneath a specimen during embedding. The base of the mould must be clear and should be raised off the bench using wooden blocks or similar. A mirror can then be placed under the mould and positioned in such a way that the undersurface of the block can be observed. When the specimen is positioned on the base layer and the second layer is poured, trapped air and small bubbles can be seen and removed using a pin or probe.

Cloudy appearance on the surface of each layer

This is due to a temperature change; cold during pouring, and warm after polymerization so that moisture condenses on the surface of the layer. This can happen if the mould is moved from one room to another (which should not happen), or if the layer was poured in the evening and polymerized overnight. Unless many layers have to be poured, try and pour all layers in one day to avoid this problem.

Cloudy appearance around the specimen

Cloudy, white globules or streaks around a soft biological specimen are the result of failure to remove all moisture. There is no remedy except to follow the dehydrating procedures for each specimen explicitly before embedding.

Whirls and aberrations

Once the block has been polished, whirls or streaky aberrations may be observed. These are due to insufficient mixing of the hardener and resin. They are not normally seen from the front so in most cases the problem is not noticed. The remedy is to mix the components thoroughly.

Hairs, specks and foreign objects

These enter the liquid resin from the air or are dropped from the preparator. Do not perform any embedding procedures in a dusty environment, and, when mixing components, tilt the mixing vessel slightly to decrease the risk of foreign matter falling into it. When the mixing vessel is allowed to stand to allow bubbles to disperse, cover the vessel with a piece of smooth cardboard, glass or plastic. Follow this procedure when layers have been poured into the mould, but leave a very small gap on top of the mould to allow fumes to escape. When looking for trapped bubbles using a lamp, foreign objects can also be observed. They, too, can be removed with a pin or needle probe.

EMBEDDING PROCEDURE

Just prior to embedding, a test mix should be made to check the time between mixing and gelling. The final mixing and pouring should be completed within minutes of this experimental procedure as ambient room temperature affects polymerization times. This test also allows the preparator to determine if the batch of resin is still stable and suitable for use.

Dry specimen

Many objects fall into this category, including rocks, minerals, fossils, bones, dried insects, and shells. To begin embedding this type of material, first place the objects into a warm oven. Leave them in the oven for 24 hours to remove any moisture, then switch off the oven and allow the temperature to drop slowly, lessening the chance of cracking the objects. Shells, if heated and then rapidly cooled, can crack and break. Place the objects into acetone for 15–20 minutes to dissolve any oil or grease. Remove the objects from the vessel

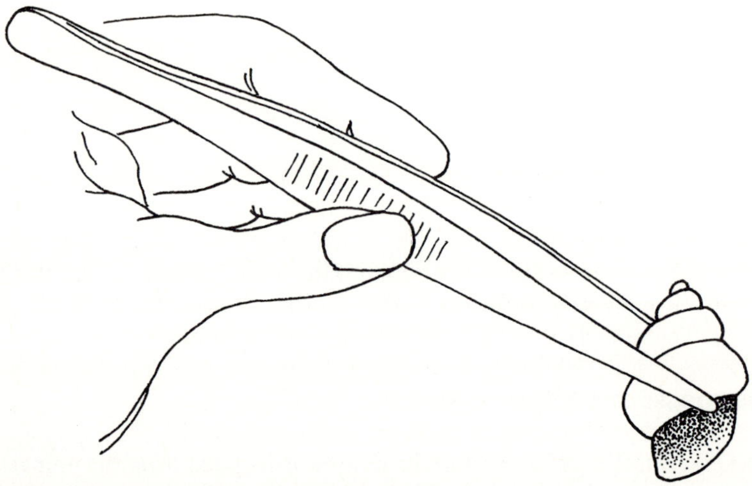

Fig. 11.7. Forceps should be used to handle specimens. Fingers can transfer grime and oil onto specimens resulting in poor embedments.

using forceps (as fingers contain oil) and allow to dry completely (Fig. 11.7). Porous specimens, such as insects, coral and some minerals, will take a long time to dry after soaking in acetone so allow ample time for complete evaporation.

Specimens are transferred into resin in order to 'wet' the surface, allowing some penetration of resin, expelling some air and aiding in reducing silvering. Very porous specimens may be soaked in styrene monomers for 30 minutes initially and then transferred to a 50:50 mixture of styrene monomer and resin for one hour; finally, they should be placed in resin for 24 hours. If light weight specimens such as insects float during the infusion technique, weigh them down. Insects, such as beetles, grasshoppers and cicadas, would benefit from having mixed resin and hardener (1%–2%) injected into the abdominal cavity; the resin either sets within or the specimen is put into catalysed resin and then into a vacuum chamber.

In cases where specimens are heavy, two layers may be sufficient for embedment. If the specimen floats, or if a coloured base is required, then three layers will be the minimum required. Of course, the number of layers also depends on the thickness of the specimen. For example, a bird skeleton measuring 160 mm × 300 mm × 120 mm required 10 layers of resin. This large number of layers was needed to overcome the large amount of heat that would have been produced if the block had a smaller number of thicker layers.

If a layer is poured and allowed to cure, then another layer is poured on top, the result will be a definite line of distortion at the junction. This is only seen from the side and may or may not be unsightly depending on whether the block will be handled. The authors believe that the more unobtrusive this line, the better the finished specimen. The following procedure should be employed to achieve the faintest of lines. For clear blocks, the base layer should be poured into the mould and, later, tested for gelling with a probe. Relatively light specimens can be placed onto this gelled layer and allowed to adhere for ten minutes while the resin hardens a little. Another layer can then be poured. The addition of the second layer before the first has hardened allows for partial mixing of the two layers, thereby lessening the line of distortion. Heavier specimens should be placed on the gelling layer after it has become a little more firm so that it can take the weight without breaking. The second layer should be poured before the first layer has set, and this procedure followed with other layers. When pouring the resin into the mould always pour from one corner or edge. This pushes foreign matter away from the previous layer to the side and it also moves in and around the specimen, pushing any trapped air away. If very large blocks are to be poured and the room temperature is hot, then, to overcome a high exotherm, the amount of hardener is reduced. After pouring a layer, the mould is placed into a refrigerator or into the sink where cold water can circulate around it. Water should not come in contact with the resin, so the mould must be kept covered at all times.

Oncc thc block has appeared to have set after using hardener at concentrations of 0.5% or less, wait for a week before removing it from the mould as the resin may still be rubbery. If it is removed from the mould while it is still rubbery, the stresses that are built up during removal may crack the block. If higher concentrations of hardener are used, removal from the mould may be attempted sooner.

SANDING AND POLISHING

This is the most time consuming and labour intensive part of the whole procedure of embedding, especially if no mechanical equipment is available. If it has to be done by hand, the following information will help.

As the polyester resin shrinks on hardening all surfaces will be slightly uneven and will have to be sanded flat and finally polished. A coarse grade garnet paper (no. 80) should be used to grind all surfaces flat. Check all angles and flatness using a square. The block is sanded using a circular motion, altering the grip often so that an even pressure is applied over the block (Fig. 11.8). An assembly line procedure can be used. Decreasing grades of sanding paper (180, 280, 400, 600, 1200) may be stuck onto a glass plate with masking tape or double sided tape. The larger the number, the finer grade of abrasive.

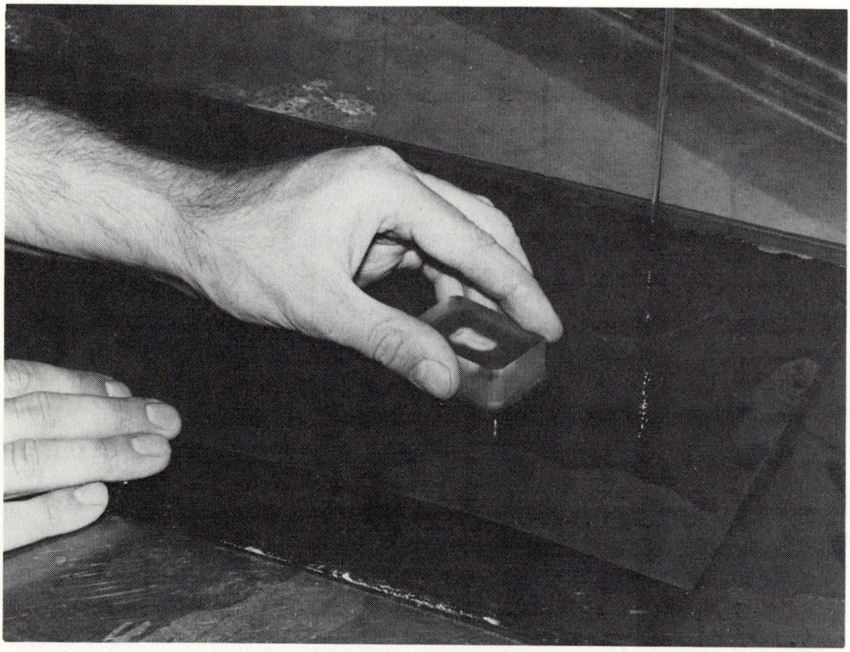

Fig. 11.8. Various grades of wet and dry silicon carbide paper are used to produce smooth surfaces prior to polishing. Photo M. Dingley.

The block is rotated on each grade, sanding all surfaces before moving onto the next grade. The polyester dust will quickly clog the abrasive unless water is applied to wash it away. Water can be splashed onto the paper so that the dust washes off. Repeat this on all grades of paper, making sure that each surface of the block is sanded equally. Final polishing is achieved by dipping a soft cloth into a liquid metal cleaner and rubbing each surface until a crystal clear finish is obtained.

To speed up the sanding and polishing, a finishing machine and buffing wheel can be employed. The finishing machine is a belt sander which allows you to course grind the block into its correct shape. It also enables straight edges and 90° angles to be made exactly. The sanding is achieved as previously described using various grades of wet and dry paper. The final polishing is performed on a buffing wheel using a calico cloth wheel and polishing compounds (Fig. 11.9). Hard polishing compounds in large sticks, Brasso or soap type compounds have been used successfully. Do not push the block onto the revolving wheel too hard as it will create burn marks which will have to be removed using 600 to 1200 paper. Pushing the block hard against the wheel also increases the risk of it being knocked out of the operator's hand and the possibility of it breaking.

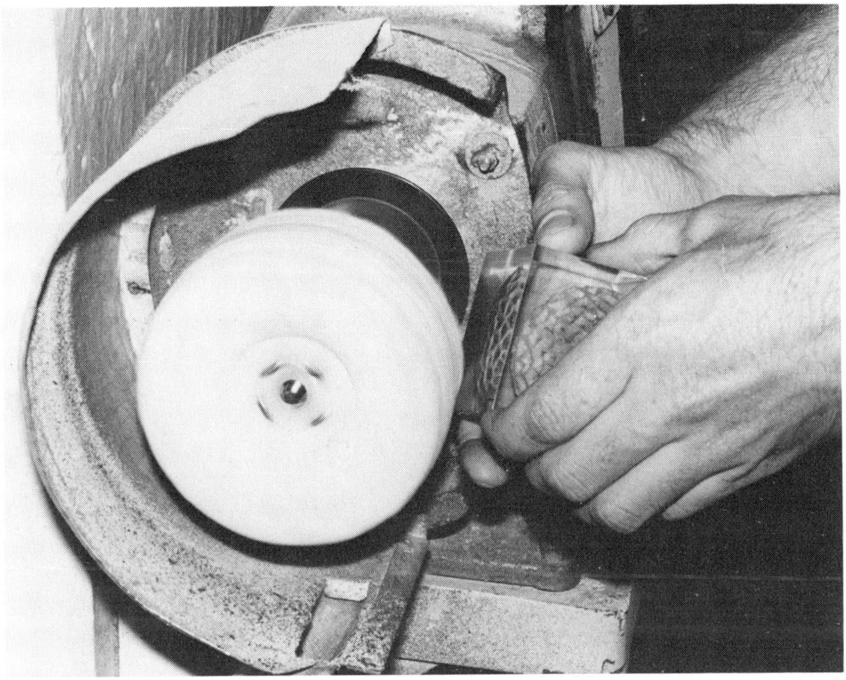

Fig. 11.9. Final polishing is achieved quickly and effortlessly using a buffing wheel. Photo M. Dingley.

Cylindrical blocks

Sometimes the removal of a round block from the mould may be more difficult than necessary, so the mould may have to be cut. The mould may be cut on a band saw if made from plastic. The easiest way to sand the edge of the block is to sand the bottom surface completely flat first. The base can then be stuck onto a lathe plate using double sided tape. It must be held very firmly. The wheel is made to revolve on the slowest speed and various grades of wet and dry paper cut into strips are held against the block as it rotates (Fig. 11.10). The paper will become clogged with resin dust so every now and again the paper should be dipped into a container holding water. Do not have the paper soaking in water, for this may be dangerous with electricity. Wipe up drips quickly to avoid any mishaps. When the edge has been smoothed completely the block can be removed from the lathe and the top surface sanded flat by hand or on the finisher. Polishing can be performed on the buffing wheel. If a buffing wheel is not available, the edge can be polished by replacing the block onto the lathe and a rag dipped in Brasso can be employed to give the final finish (Fig. 11.11).

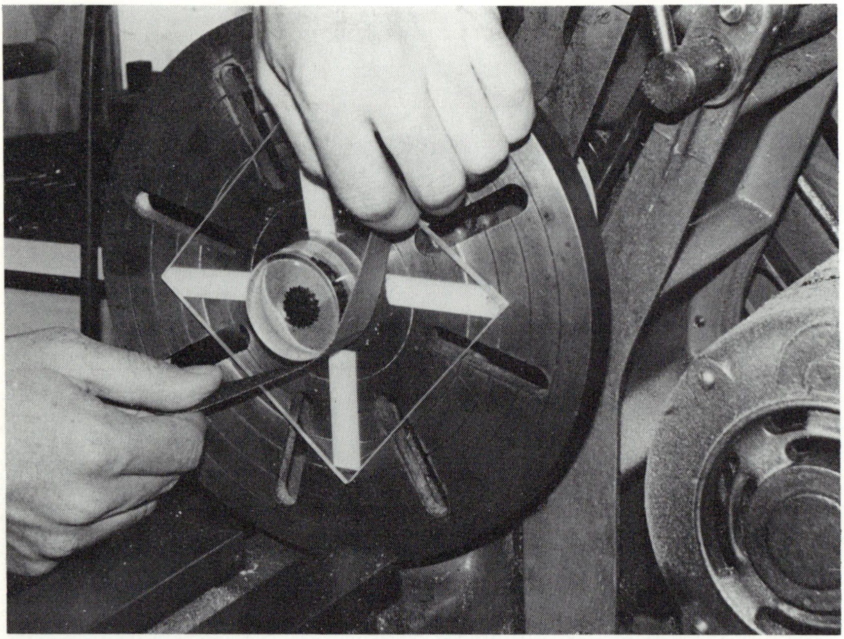

Fig. 11.10. Sanding the edge of a circular embedment is performed using a lathe plate. The base of the embedment is stuck to the plate using double sided sticky tape. The wet and dry silicon carbide paper is held against the edge whilst it revolves. Photo M. Dingley.

WET SPECIMENS

The majority of embedded specimens in a natural history museum fall within this category. They consist of plants, worms, anemones, tadpoles, frogs, and embryos. Some workers may wish to embed anatomical or pathological material. The preparation of specimens from the diverse animal kingdom is well covered in the preceding parts of this work. All biological material has to be preserved and dehydrated and only the basic, general methods will be explained here.

Preservation

Follow the instructions dealing with killing, fixing and preserving as set out in the appropriate chapters of this work.

Dehydration

As water is not compatible with the resin it must be removed from the specimen. There are several methods to achieve dehydration.

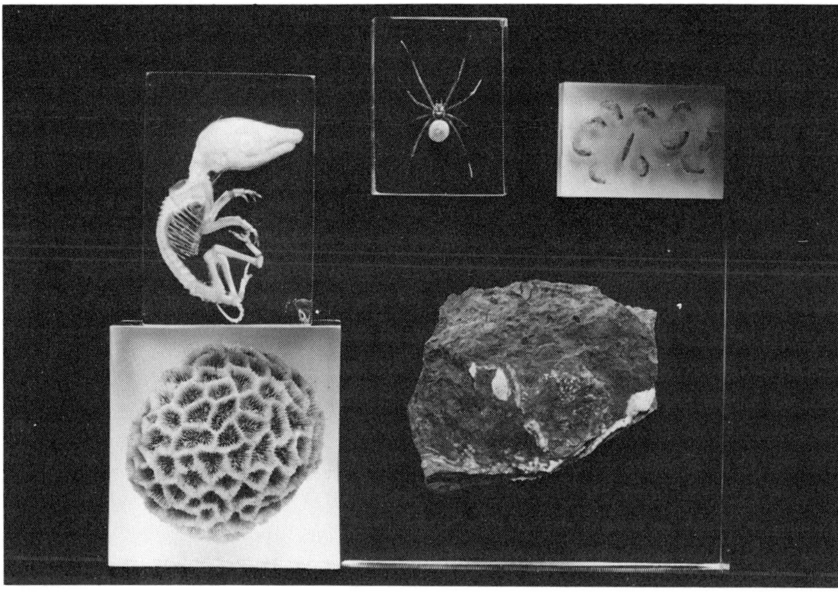

Fig. 11.11. A series of finished embedments. Photo A. Pitman. Courtesy of The Australian Museum, Sydney.

1. Wash formalin fixed specimens in tap water for one hour. Remove and blot dry using a paper towel. Pass the specimen through successive baths of 70%, 80%, 90% and 95% ethyl alcohol, and two baths of 100% ethyl alcohol for 24 hours each. Drain, blot dry, and allow to dry thoroughly in air. Drying in air is only possible with sturdy specimens. It cannot be done with embryological material and the like. Blot such specimens dry, and place them in acetone for a few seconds. The specimen can then be transferred to a bath of resin without hardener or into various mixtures of resin and styrene monomer prior to embedding.

2. Specimens which allow slight penetration of resin, such as tadpoles, frogs and worms, are best dehydrated in glycerine, but it must be anhydrous and kept tightly stoppered, for it is hygroscopic. There are five steps: (i) wash specimens in tap water for one hour, drain and blot dry; (ii) place the specimen into 70% glycerine and water for 24 hours, then pass the specimen through successive concentrations of 80%, 90% and 95% glycerine and two baths of 100% glycerine for 24 hours in each; (iii) several days in 100% glycerine may be necessary for some fleshy material after which the specimen is rinsed in acetone for 30 seconds to remove surplus glycerine from the surface; (iv) air dry the specimen and place it into resin without hardener for four hours, then discard the resin, drain the specimen and place it into fresh resin, repeating this procedure three or four times to allow the excess glycerine to seep out from the specimen; and (v) embed the specimen in resin.

3. Polyethylene glycol 600 is very hygroscopic and specimens placed directly into it will dehydrate. Several baths of polyethylene glycol are needed to remove all water from the specimens. This could take two weeks or more, depending on the thickness and porosity of the specimen. Very fragile embryos become distorted when placed into full strength polyethylene glycol, so successive concentrations of 60%, 70%, 80%, 90%, 95% and 100% are required. Remove the specimen from the glycol and allow it to drain before blotting dry. If the specimen can withstand a temperature of 50° C, place it in an oven to sweat out any remaining glycol. Rinse in acetone for 15–30 seconds and air dry before soaking in resin without hardener.

4. For plants, dehydration may be achieved using a technique which can use one of two different materials. First, make a bed of fine grained sand and lay or stand the plant in question in it. Sift the sand around it until all parts and crevices have been covered and filled. Let this stand for several weeks at room temperature, or place into a drying cabinet. Silica gel or equal proportions of silica gel and calcium chloride may be substituted for the sand.

FLOWERS

Plants and flowers can be embedded after they have been pressed and dried. They will lose their colour in the process unless they are prepared to retain their colours.

Dry and faded plant material can be used for the beginner to get the feel of embedding plants. These specimens are often useful in classrooms to show life cycles or comparative studies of seed types. Whatever the reason for embedding such plants, a high standard should be maintained. For succulent plants and thick material, follow method 4 in the previous section on dehydration. Most other plants can be dried in a plant press until required. When the material is dry it has to be infused with resin to replace the air. Thinned resin can be used first, then pure resin prior to embedding.

Colour restoration

There are chemicals which dehydrate plants and yet enable them to retain their colours. Although they are successful for many different colours they are listed only as a guide. One solution may be sufficient for some flowers, while others may require soaking in two or more. If the list is insufficient, then a little experimentation will be required. Do not treat more than one specimen in any one solution and plants should be immersed for from one to 24 hours depending on the ridigity and fragility of the petals. The temperature should be 21° C. After soaking, remove the specimen, allow it to drain, then place it into a dessicator for a minimum of 24 hours. The specimen can then be embedded.

For the following types and colours of plants and flowers use the solution or solutions noted.

Green, blue and some purples: Solution 1 and some in 2.

Pinks and reds: Solution 3, but some may need soaking in solutions 1, 3, or 4.

Whites: Dehydrate in solution 3 and bleach in solution 5. Some white flowers dehydrate well in solutions 4 and 6.

Orchids: Solution 4.

Orange blossoms and magnolias: Solution 7.

Solution 1:	tertiary butyl alcohol	100 g
	thiourea	1 g
	sodium citrate	2 g
Solution 2:	tertiary butyl alcohol	100 g
	thiourea	1 g
	sodium bicarbonate	2 g
Solution 3:	tertiary butyl alcohol	100 g
	thiourea	1 g
	citric acid	2 g
Solution 4:	tertiary butyl alcohol	100 g
	citric acid	2 g
Solution 5:	tertiary butyl alcohol	100g
	thiourea	1 g
	citric acid	2 g
	formaldehyde	5 g
Solution 6:	tertiary butyl alcohol	100 g
	thiourea	1 g
	citric acid	2 g
	sodium sulphate	2 g
Solution 7:	tertiary butyl alcohol	100 g
	peracetic acid 40%	10 g

Steedman (1976 a, b) showed that drying flowers in silica gel with a mesh size of 30–120 preserves both colour and form, with excellent results. If the specimens are then placed into polyethylene glycol 600 dibenzoate, they will be preserved indefinitely. Also, drying in silica gel at a temperature of 40° C produces better results than at 15° C–25° C.

HAZARDS OF EMBEDDING

The following precautions and hints should be known before starting embedding procedures.

1. Work in a well ventilated area as the materials used are toxic and flammable.
2. Use disposable gloves when handling any of the materials.
3. If any of the components are spilt onto the skin, wash them off with copious amounts of water.
4. If any of the components are splashed into the eyes, wash immediately with water and seek medical attention.
5. Failures do occur but do not despair. Find the mistake and rectify it.

WORKSHOP SAFETY

Safety within a working environment is not as controlled as it should be. Many workers are ignorant of the potential hazards that surround them. This is not always the fault of employers. The areas of concern include: exposure to plastics, chemicals, and dust; personal safety; and zoonoses. Apart from the protection of each individual worker, there is a great risk to others working nearby. An example of this could arise when one person has to use a solvent for a particular job; if there is no exhaust system the worker has to wear a protective mask. Those people who work in the same room can be exposed to those fumes if they are not forewarned of the hazard.

CHEMICALS

Chemicals have the potential for much harm. They may be flammable, explosive, poisonous, acidic, or caustic. Before using any chemical, its potential hazards should be noted either by reading the label of the container or by consulting *The Merck Index* (published by Merck & Co., Inc., Rahway, NJ, U.S.A.). This book lists all the chemicals used in a museum's preparation rooms and laboratories, as well as many others. It is invaluable for the workshop.

When pouring or mixing acids, bases and solvents, always wear protective clothing, shoes, gloves, and safety glasses. If there is no exhaust system, work in the open air, upwind of the fumes, but wear a breathing mask if the more dangerous chemicals have to be used in an enclosed room.

When mixing acids with water, always add the acid to the water slowly and with continuous stirring to prevent a build up of heat. If the water is added to the acid, there is a risk of the acid spurting up and out of the container; or the container, if made from glass, may break due to the exothermic reaction. If acids are spilt onto the skin, wash the affected area immediately under running water. Commonly used acids are: sulphuric (oil or vitriol); nitric (aquafortis); hydrochloric (muriatic); glacial acetic; carbolic (phenol); and formic.

The alkalis generally used in the preparation room and which are caustic to the skin are sodium and potassium hydroxide. They are usually acquired in a pellet or flake form, then water is added slowly, stirring continuously. If spilt onto the skin, an alkali should be washed under running water or neutralized using a weak solution of acetic acid, a container of which should be available at all times. Commonly used alkalis are: sodium hydroxide (caustic soda); potassium hydroxide (caustic potash); and ammonia.

Chemical solvents are used as degreasers, glues, and for painting. They consist of organic acids, ketones, esters and hydrocarbons. A thorough knowledge of their properties and hazards is essential so that safety aspects of their use can be properly understood. The flammability of petroleum solvents is well known. Solvents should be stored in an isolated, brick building with outward opening doors. There should be adequate ventilation and all electrical fittings should be flameproof. The store should house the bulk containers and small quantities should be drawn from these when required. Of course naked flames, acids and peroxides, or smoking should not be allowed near the solvents. Fire extinguishers should be easily accessible near the store and in areas where the solvents will be used. When splashed onto the skin, solvents can produce burns and irritation or itchiness. If large quantities are spilt onto the skin, some may be absorbed through the skin. They can then enter the blood system and cause damage to internal organs. Some cause dermatitis, and irritate the eyes and mucous membranes of the nose and throat. The kidneys and liver suffer from solvents because of their roles as detoxifying organs. Chlorinated solvents and benzene may cause cancer of the kidneys and liver. Some solvents have a narcotic effect on the body, while others cause vomiting, heart attacks, respiratory failure, and, in some cases, death. Some of the more commonly used solvents and their hazards are listed below:

Acetone: highly flammable, causes headaches, irritation of the lungs, fatigue, erythema and, in large amounts, narcosis.

Benzene: flammable, acute irritation of mucous membranes, leukaemia, it is absorbed through the skin, and death may result from respiratory failure.

Carbon tetrachloride: poisoning by inhalation, ingestion or skin absorption. Can cause liver and kidney damage and death.

Chloroform: large doses may cause hypotension, respiratory and myocardial depression and death. Also found to be carcinogenic.

Ethylene dichloride: flammable, causes abdominal cramps, irritation of the lungs and eyes and narcosis.

Methylene chloride: a narcotic in high concentrations.

Toluene: flammable and may cause mild macrocytic anaemia.

Trichloroethylene: has a narcotic effect in high concentrations, but breaks down under heat to produce phosgene which itself causes pulmonary oedema, eye irritation and death.

Other chemicals that are routinely used in museums and laboratories are preservatives. Workers rarely have any knowledge of the hazards that surround such solutions.

Ethanol: flammable at 100%, but lessened upon dilution with water. Causes nausea, vomiting, drowsiness, depression, impaired vision, stupor, incoordination, coma, and death.

Chromium trioxide (chromic acid): powerful oxidizer, irritation, ulceration and allergic eczema of the dermal layer. Causes irritation of lungs and bronchogenic carcinoma from breathing chromate dust.

Formaldehyde solution: pungent odour, vapours irritating to the mucous membranes of the eyes and nose, causes dermatitis, and ingestion may cause severe abdominal pain and coma.

Mercuric chloride (corrosive sublimate): highly toxic, corrosive to mucous membranes. Ingestion causes vomiting, kidney damage, abdominal pain and death after ingestion of one to two grams.

Potassium dichromate: a corrosive poison, destroys mucous membranes and skin ulceration.

Glacial acetic acid: ingestion will cause corrosion of the mouth and gastrointestinal tract, diarrhoea and possibly death.

Osmium tetroxide: causes damage to eyes, skin and respiratory tract.

Insect proofing substances also have debilitating effects upon the body.

Naphthalene: poisoning may occur by ingestion, inhalation or skin absorption. Other effects are nausea, headaches and convulsions.

Potassium cyanide: poisoning by absorption through injured skin or by ingestion.

Paradichlorobenzene: liver damage. Fumes are irritating to the skin, eyes and throat.

PLASTICS

Machining plastics by sawing and sanding and to lesser extents, routing, milling, drilling and turning is hazardous due to the release of dust particles and fumes (from the heat generated). Inhaling fine dust particles is a definite health risk which is even more dangerous when fillers such as silica, talc, Aerosil and asbestos have been used. Silica inhalation can lead to silicosis, and inhalation of asbestos to asbestosis and lung cancer. Other fillers, such as plasticizers, dyes, pigments, promoters and stabilizers, can all be detrimental to the health and well being of the user. They are damaging not only when machined but in their natural state. All plastics will burn and emit toxic fumes so if plastics have to be melted into shape or burnt for any reason this should be done in the open air or in a fume cupboard.

When cutting polystyrene foam, methyl chloride is released in conjunction with unreacted styrene monomer. This cannot be detected unless special equipment is used, for the fumes are almost colourless and odourless. However, they cause dizziness, and have other effects similar to those of ethanol. In large quantities, these fumes will cause death.

Plastics can be solidified (polymerized) by mixing two components, a resin and a hardener. Acrylics consist of a methyl methacrylate monomer, or combinations thereof, and a catalyst. The monomer is a strong skin irritant. Breathing of the fumes can cause headaches, nausea, loss of apetite, and a lowering of blood pressure. The catalyst, benzol peroxide, is explosive, a skin sensitizer, and irritant and damaging to the eyes if splashed into them.

Polyester resins are used extensively in museum preparatorial work as they are cheap, lightweight, and easy to work with. These factors can cause the dangers to be neglected. The resin is a polyester dissolved in styrene monomer and the hardener is a strong oxidizing agent, methyl ethyl ketone peroxide (MEKP). There are other hardeners available, but this is the most common. The monomer is an aromatic hydrocarbon causing irritation to the eyes and lungs, extreme dermatitis, headaches, nausea, loss of appetite, and coma. It is also damaging to the liver, the blood, and the nervous system. The hardener is not without its dangers for it is a strong irritant to the eyes. Blindness can occur years after the hardener has been splashed into them.

Epoxy resins are used mostly as glues, but some putties are used in modelling and taxidermy work. The resin is an irritant, sensitizer and suspected cancer causing agent. It is supposed that it is absorbed through the skin. The hardeners, especially amines, are even more hazardous in small quantities. During the hardening process, a great deal of heat if formed and the vapours that are given off are toxic.

Polyvinyl acetate, polyvinyl chloride and polyvinyl alcohol are all thermoplastics and as such need to be heated to be moulded into various shapes. Vinyl chloride fumes are given off during this process; these fumes can cause hepatitis and liver cancer.

Silicone rubbers are considered safe to work with, but care should be exercised when using the catalysts as they cause dermatitis. If inhaled for long periods, they can cause lung problems.

There are many polyurethane resins, but the most common one comes in the form of a foam, either rigid or flexible. The two liquid phases can produce breathing discomfort and coughing, as well as being irritating to the eyes. Face masks should be worn at all times and the work should proceed in well ventilated areas. Isopropyl alcohol removes the resins from the skin.

Tobacco smoke affects the same parts of the lungs as other air pollutants and the harmful effect of smoking would greatly contribute to the total body burden, so museum workers should be aware of the health hazards of smoking. However, the most obvious danger of cigarette and pipe smoking is the fire hazard. Another hazard occurs when the smoker leaves a burning cigarette on the workbench where it can make contact with various materials detrimental to health. Dirty hands may also introduce foreign elements to the tobacco fumes, especially for those who roll their own cigarettes.

ZOONOSES

Zoonoses are infections and diseases that are naturally transmitted between vertebrates and humans. These zoonoses may be transmitted via inhalation, by skin contact usually through cuts and abrasions, animal bites, parasites, and insect vectors such as mosquitoes. If infected animals are touched and then the infection transferred to the mouth before washing the hands, a lack of personal hygiene is indicated. Live or dead animals may carry potential zoonoses and workers should be aware of the dangers. The great danger of zoonoses lies in the fact that the symptoms are very often mistaken for the symptoms or other more common ailments, or unrecognized. A very short list of the more common zoonoses that can be contracted by preparators and technicians who work in this field follows: acariasis; brucellosis; cowpox; hydatid disease; psittacosis; Q fever; ringworm; salmonellosis; schistosome dermatitis; toxoplasmosis; and tuberculosis.

Working with carcasses is always risky for the preparator, especially if the cause of death is unknown. This is especially true where dead animals have been received after they have been found dead in the forests and on the roadside. Zoo specimens are less risky as they are checked by a veterinarian before they are released from the zoo. However, the problem with zoo specimens is that only the cause of death need be stated. Some latent infection could have been overlooked, only to be contracted by a taxidermist or another museum worker. This is in no way the fault of zoos, but it is mentioned here so that people may be aware of the problem.

Whenever dead animals are used for taxidermic work, skeleton preparation, or any other preparation, surgical gloves should be worn, as should a mask. Preparators should also be injected against tetanus and have the booster shots when required. Most of all, preparators should always wash their hands before eating, and never put fingers in the mouth, or touch other people, without first washing the hands.

Workshop safety remains in the hands of the employee. When pouring and mixing liquids, see to it that protective clothing, shoes, masks and goggles are worn. When deciding upon which chemical to use for a particular job, such as degreasing bones, use a solvent that is least harmful to the user. Make sure that others are warned when volatile chemicals are to be used near them. Always read the labels of storage containers and be aware of the potential hazards before use. When using machinery use protective goggles and never remove the guards. Also, personal hygiene cannot be overstressed, so be very careful, whatever work is being done. Unfortunately, these few short notes cannot identify all of the potential hazards that exist in museum workshops, but accidents and the number of hazards can be reduced if employees think before using chemicals and machinery, and if the work area is spacious and kept clean.

REFERENCES

Bourne, L. B. and F. J. Milner, Polyester resin hazards. *British Journal of Industrial Medicine* 20/1963: 100–109.

Dingley, M. O.: A note on the surface finish of a plastic embedment. *Abstracts of papers presented at the conference of museum preparators (with additional material).* Australian Museum, Sydney 1980.

Hildebrand, M.: *Anatomical preparations.* University of California Press, Berkeley 1968.

Lutz, E. L.: *Handbook of plastic embedding of animals plants and various objects with improved CME D6 polyester resins.* Naturegraph, Healdsburg 1969.

Steedman, H. F.: Flowers in plastics. *Museums Journal* 75/4 1976a.

——. Flowers in plastics 2. The treatment of blocks. *Museums Journal* 76/1 1976b.

SOURCE MATERIAL

Cagle, C. V.: *Adhesive bonding: Techniques and applications.* McGraw-Hill, New York 1968.

Chaffe, P. S.: *Zoonoses considerations.* Reading Park Zoo, Fresno CA n.d.

Christobel, D. C., and D. T. Blackburn: A new procedure for mounting cleared leaves using polyester resin. *Transactions of the Royal Society of South Australia* 99/2 1975: 55–58.

Clydesdale, A.: *Chemicals in conservation.* Scottish Development Agency Conservation Bureau, Edinburgh 1982.

Condon, G.: *The art of flower preservation.* Lane, Menlo Park CA 1962.

De Dani, A.: *Glass fiber reinforced plastics.* John Wiley, New York 1961.

Doolittle, A. K.: *Technology of solvents and plasticizers.* John Wiley, New York 1954.

Doyle, E. H.: *The development and use of polyester products.* McGraw-Hill, New York 1967.

DuBois, J. H., and F. W. John: *Plastics.* Van Nostrand Reinhold, New York 1967.

Edwards, J. J., and M. J. Edwards: *Medical museum technology.* Oxford University Press, London 1959 (Chapter 8).

Halperin, W. E., M. Goodman, L. Stayner, L. J. Elliott, R. A. Keenlyside, and P. J. Landrigan: Nasal cancer in a worker exposed to formaldehyde. *Journal of the American Medical Association* 249/4 1983: 510–512.

Hangay, G.: Occupational health hazards to preparators. National Conference of Preparators and Technicians, Museum of Victoria, Melbourne 1983.

Harden, C.: *How to preserve animals and other specimens in clear plastic.* Naturegraph, Healdsburg 1963.

Heimann, E. H.: *Do it yourself with plastics.* E. P. Dutton, New York 1973.

Hollander, H. B.: *Plastics for artists and craftsmen.* Watson-Guptill, New York 1972.

Horn, M. B.: *Acrylic resins.* Van Nostrand Reinhold, New York 1960.

Johnson, K.: *Cellular plastics: Recent developments.* Noyes Data, Park Ridge NJ 1970.

Jones, A. T.: Museum health hazards. *Kalori* 45/1973: 19–21.

Kerns, J. L.: An improved technique of embedding specimens in transparent plastics. *Anatomical Record* 117/1953: 345–351.

Lassak, E.: Dangerous museum chemicals. *Kalori* 45/1973: 15–18.
Lawrence, J. R.: *Polyester resins.* Van Nostrand Reinhold, New York 1960.
Lee, H., and K. Neville: *Epoxy resins.* McGraw-Hill, New York 1957.
McAnna, M. E.: Embedding in polyester resin. *Kalori* 31/1965: 51–55.
Milby, R. V.: *Plastics technology.* McGraw-Hill, New York 1973.
Miles, P.M.: Preservation of insect and plant specimens in transparent plastics. *Plant Pathology* 1/1952: 65–68.
Moulds, M. S.: *Embedding insects and other specimens in clear plastic.* Australian Entomological Press, Sydney 1975.
Romaniak, T. H.: The use of unsaturated polyester resins for embedding material. *Science* 104/1946: 601–602.
Roukes, N.: *Sculpture in plastics.* Watson-Guptill, New York 1968.
Scarfe, H.: *Introducing resin craft.* Batsford, London 1973.
Sills, B.: Mechanics of plastic embedding. *Bulletin of the Medical Museum of the Armed Forces Institute of Pathology* Washington D. C. 1953.
——, and S. Couzin: Dry preservation of biologic specimens by plastic infiltration. *Curator* 4/1958: 72–75.
Spolyar, L. W., R. N. Harger, J. F. Keppler and H. E. Bumsted: Generation of phosgene during operation of trichloroethylene degreaser. *AMA Archives of Industrial Hygiene* 1951: 156–160.
Stevenson, W. T., and K. L. Hughes: *Synopsis of zoonoses in Australia.* AGPS, Canberra 1980.
Winsor, L.: Safety aspects of solvent use. *Kalori* 45/1973: 3–10.

12
Modelmaking

IN THE COURSE of museum work it is often necessary to represent some forms of life through lifelike models. For example, many invertebrate animals cannot be preserved without causing drastic changes to their appearance. Although such specimens are useful for scientific study, they do not properly represent the live animal, and therefore cannot be used for exhibition purposes. A well-sculpted model of the animal can take the place of a preserved one in an exhibit. Scale models, either smaller or larger than life, are produced by sculpting methods. The scope of work for a model-maker–preparator is enormous and it is only limited by the designer of the exhibitions, and by the ability of the preparator.

The range of techniques, materials and equipment is equally enormous; in this chapter we discuss only those which are relevant to exhibition preparation in a museum of natural history.

SCULPTING

SCULPTING IN PLASTICINE

This synthetic clay, Plastiline or Plasticine, contains oil, wax and fine clay, and other additives such as pigment and glycerine, depending on the formulae of the various manufacturers. Plasticine is available in many colours and in

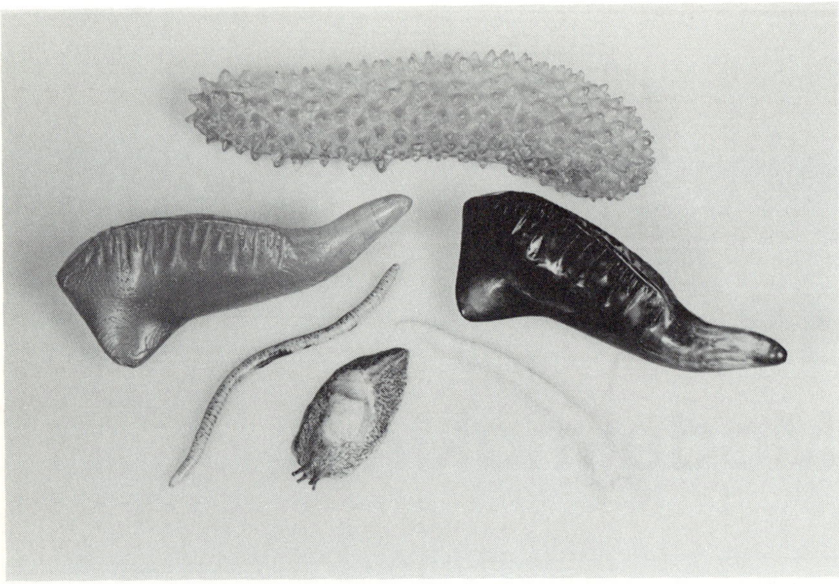

Fig. 12.1. Polyester casts made of Plasticine models. Model by J. Hood. Photo A. Pitman. Courtesy of The Australian Museum.

various hardnesses. The degree of hardness mostly depends on the melting point of the wax used in the compound. It is useful for making small to medium size models. It is not a permanent medium since it never hardens, and can be distorted by heat and careless handling. Plasticine models are always reproduced by moulding and casting into a more permanent material such as plaster, polyester (Fig. 12.1) or epoxy (further information is contained in Volume 1, Chapter 6).

If the model is of a very simple shape, such as an earthworm or slug, there is no need for an armature. The Plasticine is simply worked between the fingers, becoming malleable from the heat of the hands. It can be shaped as required. Surface detail is modelled with fine clay modelling tools, dental instruments or special modelling instruments made for achieving special effects. Many such tools can be improvised by adapting nails of various gauges, pieces of wire, glass rods, or even match sticks.

In hot weather the Plasticine may become too soft for convenient work. Smaller models can be placed in a refrigerator until the Plasticine hardens (Fig. 12.2). This may be necessary even in colder weather if very fine detail is to be worked into the surface.

If the model is more complex, a wire armature becomes necessary. Samples of armatures are shown in Fig. 12.3. The armature should be made of soft wire and securely attached to a timber base.

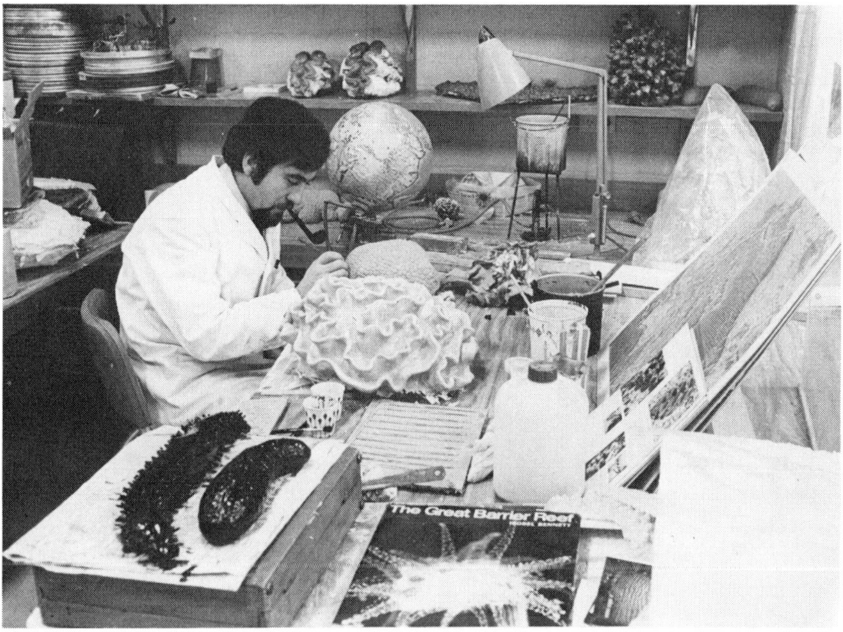

Fig. 12.2. These shapes were modelled by using chilled Plasticine. Model by J. Cases. Photo G. Millen. Courtesy of The Australian Museum.

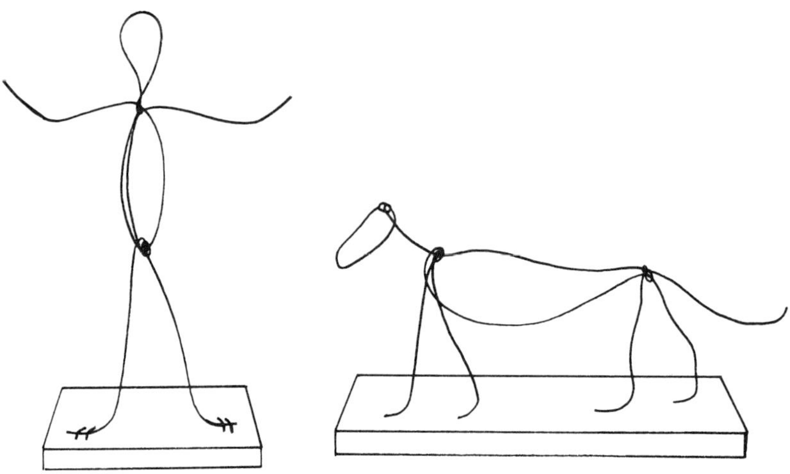

Fig. 12.3. Simple armatures for small models. Drawing by G. Hangay.

Larger models need strong armatures. These armatures can be made up from appropriately bent steel pipes and rods welded together. The bulk of the model can be made with chickenwire, wrapped on the armature, or by blocks of rigid foam (either polyurethane or polystyrene) wired on the armature. This not only saves large quantities of expensive Plasticine, but also reduces the weight of the model. The model's surface can be worked with scrapers of various sizes. If a very smooth, high gloss surface is required, the Plasticine can be 'glazed' with a flame.

Flaming the surface needs practice. It is advisable to try the method first on lumps of Plasticine and not on the model. A welding torch can be used by an experienced person, but a hot air gun on high temperature may be more suitable for the beginner. The surface can also be smoothed by 'washing' with mineral turpentine or turpentine oil. Tools can also be cleaned with these substances.

The greatest advantage of Plasticine is that the material remains malleable and it can be reused again and again. It is very important that Plasticine is kept clean. Particles of foreign materials, such as plaster, fibreglass and foam, render the plasticine almost useless for fine modelling. If the material becomes contaminated with small debris it must be melted and strained. The Plasticine is placed in a pot, and this vessel is placed in another, larger pot containing water. The water is brought to the boil and kept simmering until the Plasticine melts. The melted substance is strained through a fine metal mesh and decanted into disposable plastic containers. After cooling, the containers are cut away and the Plasticine stored in plastic bags.

SCULPTING IN CLAY

Water clay is the traditional material for sculptors. However, it is usually suitable for large models only. The method of building a large animal model in clay over an armature is described in Volume 1, Chapter 6.

SCULPTING IN WAX

Wax is a most suitable medium for direct modelling of small sculptures or for preparing models which are to be cast in a more permanent material. Wax, like Plasticine, is affected by heat and improper handling, but the range of properties of the various types of waxes is far greater than that of Plasticine.

For modelling purposes, one needs a wax which becomes malleable from the heat of the hands, and which can be worked conveniently. Beeswax meets this requirement, but in recent times it has become rather expensive. Microcrystalline waxes, which are petroleum products, have replaced beeswax in

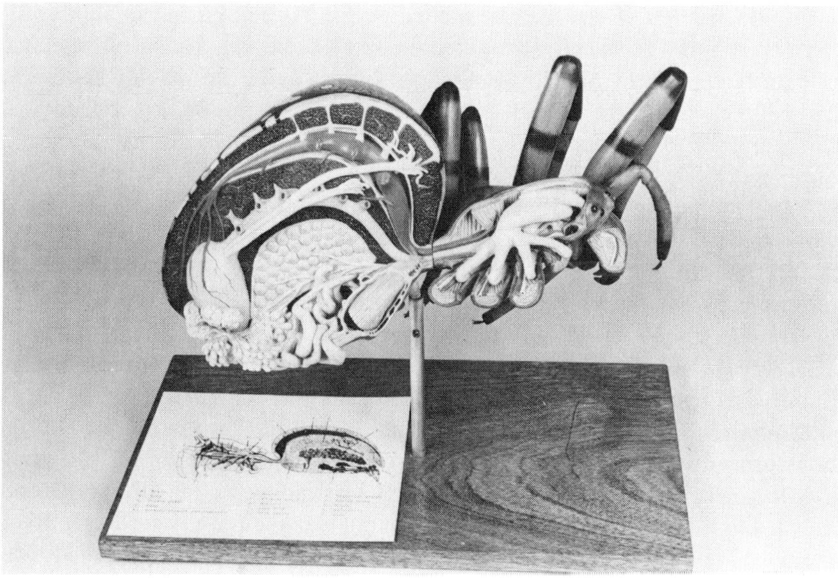

Fig. 12.4. Model made from hard wax. Photo A. Pitman. Courtesy of The Australian Museum.

many fields. These are available in various blends and marketed under many different brand names.

Preparing models with very fine, small detail can be difficult if a very soft low melting point wax is used. For work such as this, a harder wax is more suitable (Fig. 12.4). Wax with a higher melting point may not become soft from the warmth of one's hand. If working with such wax, heated modelling tools should be used. The tools are heated either electrically or over a flame. Electric spatulas can be purchased from dental and jewellers' suppliers. The spatula, which can be fitted with various tips, is thermostatically controlled. If an electric spatula cannot be purchased one can be made from a small soldering iron with a replaceable tip. The regular tip is replaced with a short length of copper wire. The protruding end of the wire is flattened and shaped and the edges filed round. The iron is plugged into a small power rheostat and, while testing it on a lump of wax, the required temperature adjusted. The spatula is used to work fine detail on the model or to apply small, softened amounts of wax to it.

The same work can be carried out with a flame heated metal spatula. Care must be taken not to heat the tool end of the instrument directly in the flame. The flame should be applied on the shaft only, letting the heat travel to the tool end.

Small amounts of wax can be melted over a bunsen burner in small pots (porcelain ladles are usually favoured by wax modellers). The liquid wax can be transferred to the model with a brush. Larger quantities of wax should be melted in a water bath (like Plasticine) and not overcooked. Prolonged heat can change certain properties of the wax and render it useless for some aspects of the work.

Basic modelling techniques are somewhat similar to those used when working with Plasticine. Smaller models of simple shape can be built without an internal armature, but those with a more complex design or larger in size require the same kind of armatures as depicted on Fig. 12.3.

To build up the initial bulk of the model one can start off by dipping the armature in melted wax, then in cold water. Repeat this procedure until the required mass is obtained.

Models without armatures can be hand built and sculpted in one continuous procedure as follows: The wax is melted and poured on a lightly and evenly oiled slab of marble or plate glass. After a few seconds the wax cools down, and forms a leathery layer. The layer of wax is carefully separated from the marble or glass with a broad spatula or painter's knife and turned over so that the other side can also cool against the marble. The hands should be lightly oiled and the leathery wax kneaded into a ball from which the basic shape of the model can be formed. Details are modelled in, first with the fingers, and then with tools. The wax gradually cools and becomes more and more rigid. Final detail can be worked on the surface with heated tools. If necessary, heated wax can be added to the model. The surface can be flame or heat finished if so required.

A small flame polishing torch can be made by fitting a hypodermic syringe and needle to a Bunsen burner (Fig. 12.5).

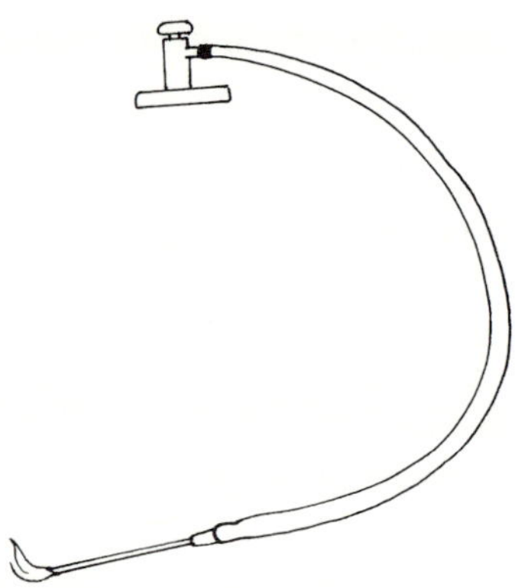

Fig. 12.5. Hypodermic needle fitted to a Bunsen burner. Drawing by G. Hangay.

A syringe without the needle can also be used to achieve special surface texture. The syringe can be filled with molten wax, swiftly rotated over the flame a few times and the wax extruded on the model's surface. An eye-dropper can be used the same way.

SCULPTING IN PLASTIC COMPOUNDS

If the nature of the model permits, it is practical to prepare the model directly in a permanent material. Epoxy resins are the most practical for the purpose. The readymade compounds are usually too thin for modelling purposes, and therefore must be modified. The most practical approach is to obtain a gap-filling compound and fill both components (resin and hardener) with kaolin or powdered pipeclay until the compound becomes thick enough for convenient modelling. Once the two components are mixed, the compound must be used up before it hardens. Compounds filled with microballoons are not suitable for modelling, since the balloons act as ball bearings and the material flows no matter how thick it is.

Mr Daniel Oppliger of the Basel Museum of Natural History, Basel, Switzerland, told us how he prepared life size models of various prehistoric animals using epoxy putty as a direct modelling medium over a metal–timber armature and wire mesh (Fig. 12.6).

SCULPTING IN PAPIER MACHE

Papier mache is a cheap and practical modelling compound. Good quality mache is available from craft suppliers. However, one must make sure that the mache contains no asbestos fibre. In the past it was common practice to use this dangerous substance in papier mache. Suppliers sell papier mache in dry form. It comes a light, grey, fluffy dust which is mixed with water to form a paste or putty-like substance. If commercial mache is not available, a good modelling quality compound can be made as follows. Paper (such as old newspaper) is macerated in water for a few days until the fibres separate and the paper turns into a grey pulp. Churning and breaking of the paper can speed up the process. Boiling also accelerates the breaking up of the fibres. Glue is then heated with about one third its weight of water. Carpenters' gelatine glue (also called animal glue) is most suitable, but casein glue can be substituted. The latter needs no heating since it dissolves in cold water. Excess water is then squeezed from the paper and the glue (carpenters' glue is added hot) is added. The mixture is stirred into a soft paste. It is important that the mix is very even. Whiting is added and also mixed in very well. Finally, a few drops of glycerine are added along with some carbolic acid. These two ingredients are required to make the mache smoother and to preserve it from mould if it is to be stored for some time before use.

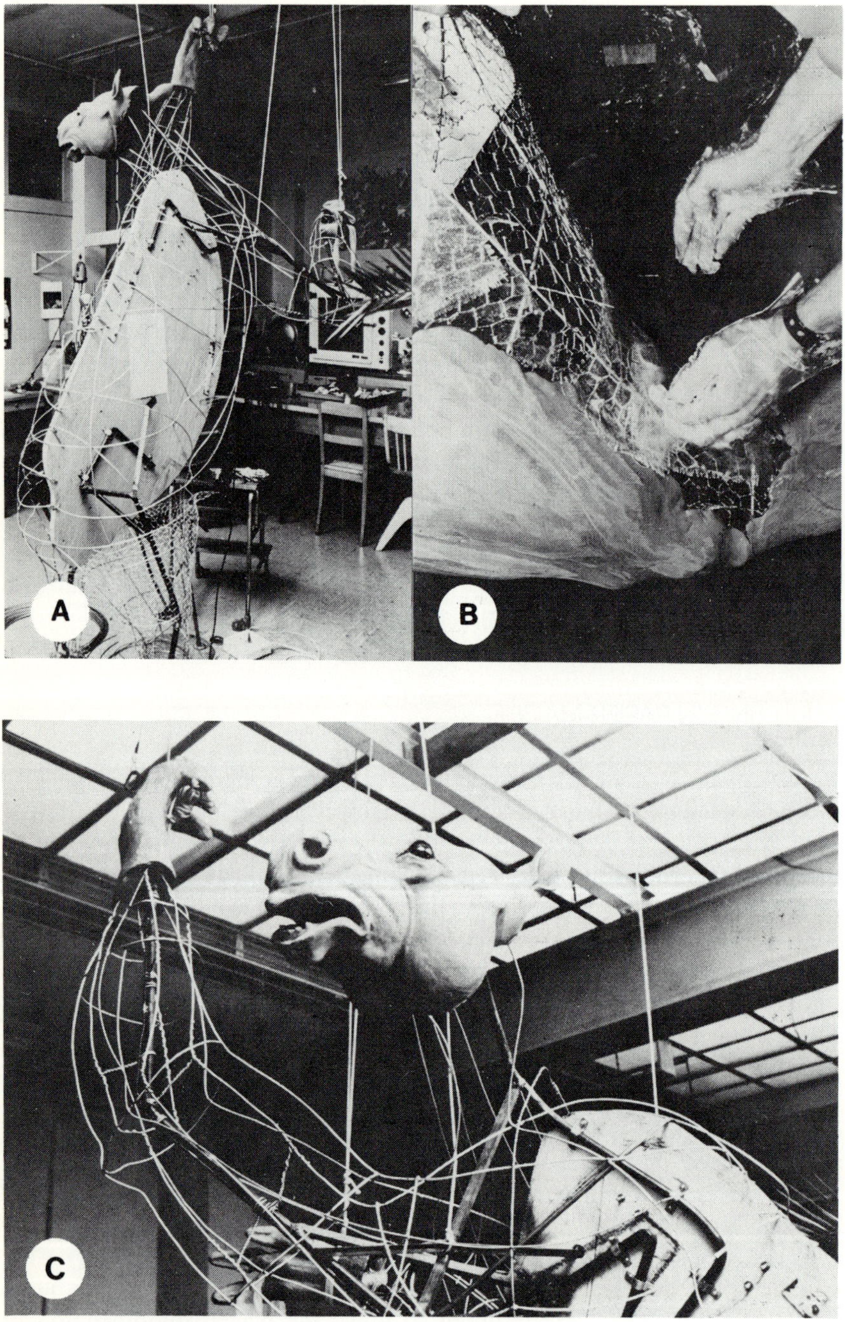
A
B
C

Just before the mache is to be used, the plaster is mixed in. The compound is then kneaded into a stiff, putty-like substance which sets hard in approximately three hours. The setting time can be prolonged by adding a little borax to the mix, or speeded up by adding a little salt. An increase in the amount of plaster may speed up setting also. The basic mix (without the plaster) can be stored in tightly sealed jars.

An alternative method for preparing papier mache is described in Volume 1, Chapter 6.

Papier mache is used to prepare models by direct sculpting. The compound is somewhat inferior to modern plastic compounds. It is not translucent, as is wax, and its preparation is time-consuming. However, it is a safe substance to use. It is also cheap and durable. The formula for papier mache is:

Wet paper pulp	1 part by weight
water	3 parts by weight
whiting	3 parts by weight
plaster of Paris	8 parts by weight
glue (hot or casein)	2 parts by weight

CARVING

Models with certain shapes, such as those with long, slender extremities, are better carved than modelled in a pliable medium (Fig. 12.7).

Wood is a common carving medium. Soft wood, with a dense, even grain, is the most suitable. Balsa is the classic modelmaking wood for making small to medium sized objects. Sugar pine, which is considerably cheaper than balsa is also suitable, but mostly used for larger models. Dense, rigid plastic foams, such as polyurethane, can also be used for carving. Polyurethane foam of various density is used to carve large models. Because of the porous nature of the material it is often necessary to coat the finished carving with some other material such as rubber latex, polyester or epoxy resin.

Polyurethane carving

A carved model of a Minke whale was prepared under the instructions of Michael Traynor, the senior preparator of the Museum of Victoria, Melbourne.

Fig. 12.6. Large model of a prehistoric animal. **A.** The armature was made of timber and metal. **B.** Two part epoxy putty was applied to the chicken-wire base. **C.** Detail of the head. Model by D. Opplinger. Courtesy of the Natural History Museum, Basel, Switzerland.

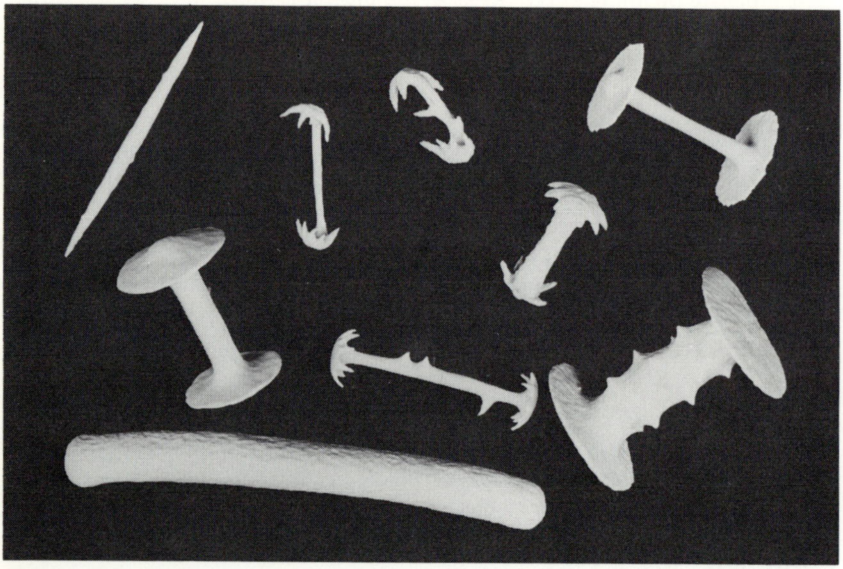

Fig. 12.7. Carved shapes. Models by J. Hood. Photo A. Pitman. Courtesy of The Australian Museum.

PETER SWINKELS, preparator of the same museum describes the process: "A stranded whale was reported at the mouth of Spout Creek, Victoria. It was determined to be a young Minke whale *(Balaenoptera acute restrata).* As it was in good condition, it was decided that it should be prepared as a skeleton as well as a display model. The specimen was taken to the museum. It was measured and plaster moulds made of all its features; that is, the tail, fins, vents, throat furrows, and skull. It was decided to make an adult Minke whale using the information gathered from this specimen. Adults average from 7.3 to 7.6 metres in length, while the stranded specimen was 3.61 metres. Working from the measurements and moulds, a half-scale drawing was made of the plan and side elevations with the shape required. A graph was then drawn over this to work out the size of the polyurethane blocks needed for the body. The body height and width did not exceed 1.22 m, so ten 1.8 m × 1.22 m × 0.51 m polyurethane blocks of sufficient density were ordered. Nine were to be used for the body and fins, while one was needed for the tail. The polyurethane blocks were laid out and housing made for timber planks between the blocks for reinforcement through the joins. These were fixed using general purpose polyester resin (GPR) with fibreglass reinforcement (Fig. 12.8). Some heavy duty contact adhesive was applied over all other exposed urethane surfaces to be joined. After all the blocks had been glued

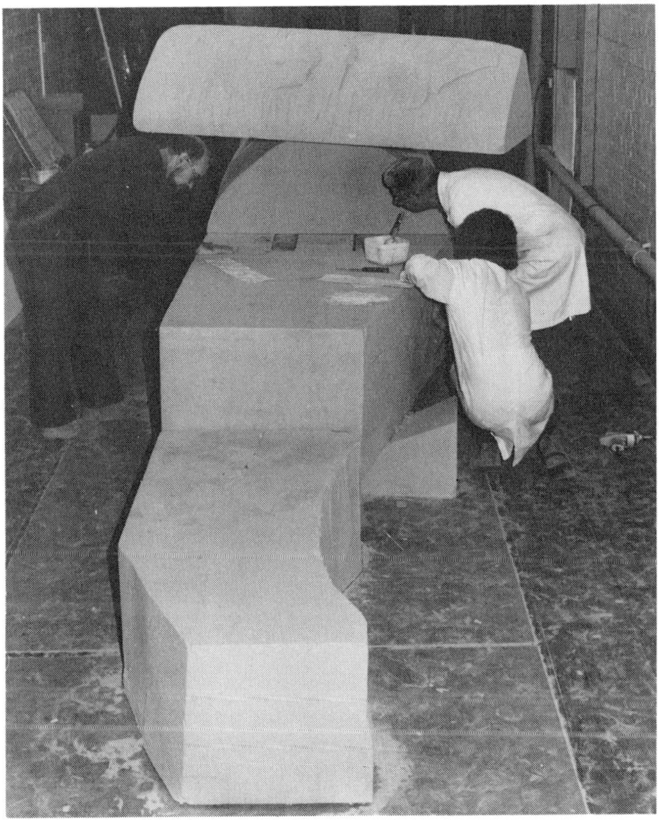

Fig. 12.8. The shape of the whale is carved from blocks of polyurethane foam. Courtesy of the National Museum of Victoria, Melbourne.

and roughly carved into shape, two hanging points were worked out. Housings were carved out on the top and underside of these points for timber blocks to be set using GPR and fibreglass. A 9 mm steel rod was bent and welded at one end to form the hanging point and threaded on the other end for bolting to the timber. It was then possible to hang the model up and get a better idea of the shape and angle that we were aiming for. Casts had been taken from the moulds using GPR and, working from these, the tail, dorsal and pectoral fins were shaped. Positions for the fins were then measured and joints cut and shaped on the body. The pectoral fins were held in place and a 9 mm piece of dowel hammered through from the top of the fin into the body, leaving some protruding for ease of removal. The dorsal fin was fixed with hidden dowels. The musculature running into the tail was left on the

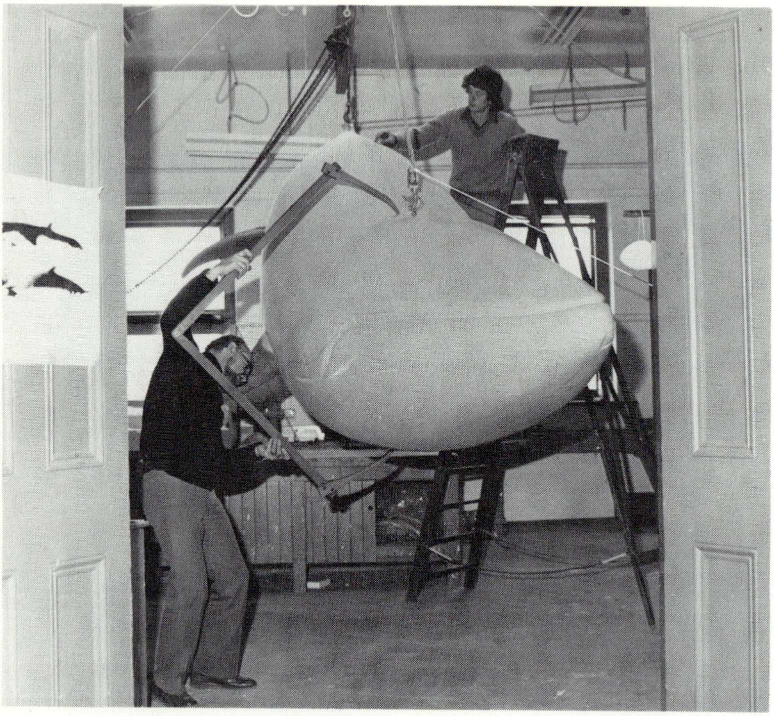

Fig. 12.9. The carved model. Courtesy of the National Museum of Victoria, Melbourne.

body with the tail flukes having a large V-shape, giving a maximum glueing area. Dowels were hammered in from the leading edge of the flukes through the body from both sides to secure it. The eyes were made with clear polyester embedding resin poured into a round embedding mould. These were then painted, sealed, and waxed into position. Using large calipers, measurements were taken regularly and, when close to the desired size, the glue lines were cut out. To stop ridges along the glue lines, we experimented with various soft glues but were unable to find one that would be strong enough. Some heavy duty wallboard contact adhesive was used, which dried hard. Thus, using a sharp knife, glue lines were cut out to a depth of 3 cm–5 cm, keeping the sides as straight as possible. Thin, flat strips of urethane were then cut, and, using a flat sanding pad, tapered to suit the groove cut into the model. These were hammered into the groove, cut off and sanded flat. It was then possible to carry out final shaping with surforms and sandpaper, without interference of glue lines (Fig. 12.9). The front of the model was then raised and the longitudinal throat furrows measured and marked with chalk. Next, the fins and tail were removed, the model lowered to the floor and rolled on to its back. A small dental drill fitted with several flat sanding discs to increase the width of the groove, was used to cut the furrows accurately. The model was rolled again, raised, the fins and tail replaced, then the final shaping done. Photographs, slides and measurements, along with casts of whales, were used constantly at this stage for detailing around the eyes, vents, blowhole and mouth. The inside of the mouth line was deepened with a fashioned modelling tool. Fine sandpaper was used for all the smoothing down and highlighting of musculature. When the finished shape was obtained, the dust was blown off and the fins and tail removed. A sealing coat of GPR was applied using a roller and brush over the whole of the model, except where the fins and tail had to be attached and where the furrows occurred on the underside. Realizing that the model would occasionally have to rest on its belly while it was being moved around, it was decided to use a flexible resin to prevent it cracking under the weight. Reinforcing would have been hard to apply while leaving the furrows open. So a mixture of 75% GPR, 20% flexible polyester resin and 5% cobalt octoate was catalysed and applied to the undersurface. This proved an adequate shield.

Once sealed, a second coat of GPR was applied and some 230 g m^{-2} fibreglass matting reinforcement, torn into as large a working piece as possible, was applied with a roller. Tearing the fibreglass matting left frayed edged which were overlapped the width of the fraying to disguise the joins. Pressure rollers were used on the fibreglass after it had been soaked with resin to flatten it and remove any air. Brushes were needed occasionally to stipple the resin in areas inaccessible to a roller. The nonreinforced underside was given three coats of flexible resin. Water paints were selected for painting because of the ease with which they can be touched up. The surface was lightly sanded, then washed with acetone to remove any grease or wax, thus allowing a good bond. The eye was then covered with Vaseline and a white undercoat sprayed over the whole surface. The second coat was gloss white. The grey was mixed using white, black, yellow, and brown. This was sprayed on all the upper parts and blended into the white by reducing the amount of paint from the gun (Fig. 12.10). An airbrush was used for finishing touches. A sealing coat of clear satin varnish was then applied to even the sheen".

Polyurethane spraying

Large, complex models can also be carved from polyurethane foam. Weak sections, like the legs of a model animal, need reinforcing, preferably with an internal metal armature. To construct such a model, the armature should be made first, the bulk of the body formed of chicken wire, and the whole object sprayed with polyurethane foam. To spray foam, a special foam spraying outfit is needed. The equipment consists of a compressor and spraying unit. The two components of the foam are attached to the spraying unit in two

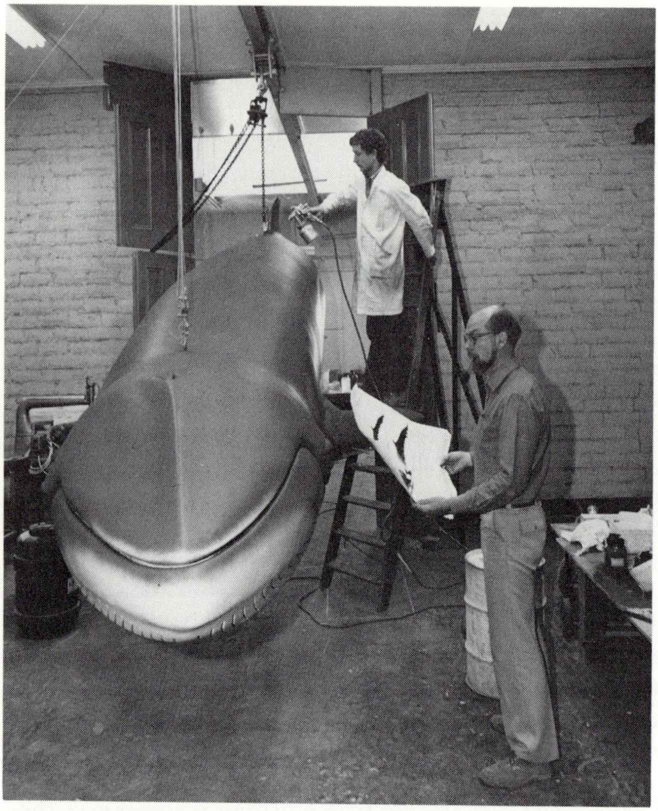

Fig. 12.10. Lifelike colouring is sprayed on the finished model. Modelling by M. Traynor and P. Swinkels. Courtesy of the National Museum of Victoria, Melbourne.

separate containers. When the equipment is in operation the foam components are pushed by air into the spray gun through two hoses. The components leave the spray gun in two fine jets and mix in the air. They are thoroughly mixed by the time they reach the model and the foam starts to develop. After it has hardened the model can be carved to the required shape.

MOULDING

Until the 1950s, mould making techniques and materials had remained the same for centuries. In the second half of this century a number of synthetic materials have been developed which have become popular, and most practical, moulding media. Moulding materials can be divided into two main groups: rigid and flexible.

RIGID MOULDS

In museum work the most frequently used rigid moulding materials are plaster of Paris, fibreglass reinforced resins (either polyester or epoxy), and wax. These materials are suitable for making moulds of soft models (for example, Plasticine sculptures) or flexible objects, such as fish or reptile specimens. Rigid objects, such as relief type models, topographical relief maps, or any other object without undercuts, can also be moulded with rigid materials. (Undercuts are areas of a model or object which prevent the easy separation or removal of the mould).

Waste mould

A rigid mould prepared over a soft model with undercuts can be removed for the cost of the model, which gets destroyed during the process. A flexible object can be pulled out of the mould, or if it becomes necessary, can even be cut out and removed in pieces. However, when a rigid, or semirigid cast is made in the mould, it can only be removed if the mould is destroyed. Plaster is the usual material from which this type of mould is made. A detailed description of the method of preparing plaster moulds of fishes is given in Volume 1, Chapter 2.

Piece mould

Piece moulds must be prepared for models with deep undercuts. The preparation of such a mould is explained in Volume 1, Chapter 6. The essence of the technique is that the model is sectioned by clay walls or brass shim and the moulds of each section are prepared one after the other. The sectioning is done in such a way that the undercuts are eliminated. After the moulding is finished, the sections are removed and prepared for casting.

Moulding and casting human hands: It is often necessary to use human hand reproductions in museum displays, either to demonstrate the use of certain tools and impliments or as parts of a human figure. One practical method has been developed for moulding and casting hands in plaster of Paris by Mr RAY WITCHARD, of the Australian Museum, Sydney. His description of the method follows:

"An experienced moulder will have little difficulty casting a human hand in an extended a relaxed position, but the preparation of these hands, with flexed muscles holding actual specimens, presented difficulties which the casual observer might not fully realize. Several methods have been attempted, but, for our purposes, the most successful was for preparators to cast one of their own hands. Unless completely relaxed, living flesh is subject to constant motion (pulse vibration, involuntary muscular twitching, and fatigue). For ths reason a piece mould of a living, working hand in a rigid material like plaster of Paris is impracticable unless all the pieces can be poured at the one time. A low table or bench should be selected for the work. The bench should just about reach the hand as it hangs freely by the side. Place a 25 cm^2 piece of plate glass on the table on which is spread a circular pad of clay about 200 mm in diameter and 20 mm in thickness. A thin bladed knife, blunt at the tip, and a piece of sheet celluloid about 50 mm × 25 mm are the required tools. Moulding of the hands is made easier and less tiring if a moderately quick setting plaster is used; some practice will show just what margin of working time is necessary. Roll the celluloid into a cylinder and tie with string, in a bow knot, at both ends. The bow is important because it has to be undone quickly with one hand. This cylinder of celluloid must have a large enough diameter to circle the hand and whatever it is holding, leaving a margin of at least 15 mm. The implement in the hand may touch the cylinder wall; the operator can recognize its position in the mould in this way. In the case of a club or spear or similar object, make a cast of the section that touches the hand and use that instead of the whole object. Now, shave the forearm to the elbow; failure to do this is disastrous. If a porous object is to be held in the hand it must be thoroughly soaked in water for some time. With a relatively quick setting plaster, the following is the order of procedure. Dry the surface of the object, apply a coating of separator, and place it in a clean spot. Leave the tin of separator handy. Should it be the left hand that it to be moulded, place the cylinder of celluloid upright in the clay with the knots to the right, and press down till the edge rests on the glass. Place the knife in a small bowl of water, also on the right hand side. Depending upon how much of the arm is required to be moulded, in addition to the hand, decide on the amount of plaster to be mixed as there can only be one batch. Sift into the bowl of water sufficient dry plaster to make a thinnish mix. Wash all dry plaster off the hands, wipe them dry, and apply some separator, especially to the back. Sir up the soaking plaster with the left hand (in this case) and ladle it into the cylinder. If any time has been lost and the plaster is warming, or showing signs of setting, discard it and make another mix. Do not fill the cylinder right to the top. With a layer of set plaster applied over the left hand, pick up the object and adjust it to the hold required. Carefully and slowly push the hand down into the wet plaster, flexing the wrist to remove bubbles from under the fingers. This must be done slowly or the clay will be forced open at the bottom. Should the plaster flow out at the bottom it can be staunched by pressing down on the top end of the celluloid, but do not cause the celluloid to buckle. Then place some wet cotton wool over the leak. If too much plaster is forced out, the job must be begun again. When the hand reaches the

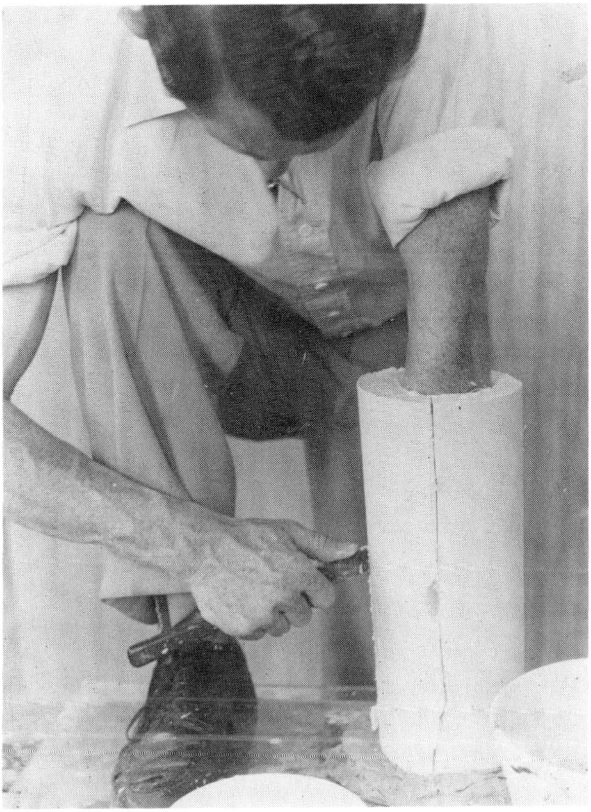

Fig. 12.11. The mould is cut with a knife before the plaster sets hard. Photo H. Hughes. Courtesy of The Australian Museum.

bottom, retract it a little, and, if the cylinder is a rather close fit, deliberately touch the wall, then retract the hand. This should serve to locate the hand in the centre. The final position of the hand must be maintained without movement until the plaster has set to the stage where you can feel the weight of the object being taken off the hand. As soon as a knife can make a cut in the plaster which does not fill up, the string is loosened and the celluloid unwound and dropped to one side. During this cutting operation a strong conscious effort must be made to maintain the embedded arm upright and in the same aspect, or the mould around the forearm will be distorted (Fig. 12.11). For most positions of the hand, it will be found necessary to make three cuts in the mould, but, for a clenched fist or flat extended position, two may be enough. As the pressure of the knife can be felt through 15 mm of the partially set plaster, it is not difficult to make a cut in such a way that the knife passes about 5 mm clear of the skin. The first and most important line is along the outside of the thumb over the base of the radius and straight up the arm. A second line is made opposite, down the inside margin of the hand, and across the ends or backs of the fingers. When the thumb is widely separated from the forefinger, a third line will be necessary down the back of the wrist and over the first finger knuckle to link with either of the other two cuts. Dig a wedge shaped slice out of the cut lines above the deepest parts to give leverage for a chisel. Now compose yourself and wait till the plaster sets; the heat generated does not become uncomfortable. When the mould can be felt cooling off, lay it on a table and place a wide bladed chisel in the leverage notches. Apply very gentle but firm pressure until the mould fractures. This will be felt rather than seen, and, if possible, all cut lines should be fractured before the removal of any part. It will be easier to work by touch in the removal of the pieces as the delicate manipulation required to preserve the sharp edges is often too slow and minute to be seen. Watch the stability of the mould to prevent it collapsing as the pieces are taken away. Extract the finger tips singly, using the other hand if they are trapped. Lay the large pieces aside and carefully gather up all the small chips and fragments which contain some of the skin detail; they will have to be glued back onto the main pieces before shellacking. Endeavour to keep whatever object the hand was holding fastened in the mould as much as possible. It is not necessary to leave the mould to dry completely before going on with the casting. After glueing all worthwhile chips back onto the main pieces, apply two coats of very thin shellac and then a coat of separator. These must be liberally applied to the palm area if the hand is half closed. Now, soak the mould for a few minutes in water, then carefully fit the pieces together. Peer inside and make sure that the edges have come properly together before strapping the mould securely for casting. Unless the plaster is mixed to a perfect consistency, simply pouring it into the standing mould will not give a good result. There will be deterioration of the surface around the wrist, or bubbles will be trapped in the finger tips. These difficulties can be overcome by tumbling the mould. Make a flat Plasticine plug to fit tightly into the armhole, but make it with a rebate so that it cannot be pushed in. It is wise to hasten the setting time of the plaster by adding some salt to the water, especially if the mould is a heavy one. Tilt the mould and pour a small quantity of plaster in and work it around into the finger tips before filling the mould almost to the top. A large bubble should be deliberately left at the top to collect the excess water. Clap on the plug and swing the mould steadily until the plaster ceases to move inside. Remove the plug and fill to the top with a small quantity of new plaster using the bubble as a key. Let it stand for a few days before attempting to chip it out. In chipping out the cast, begin by freeing the finger tips and working back to the wrist. If the object has been allowed to touch the side

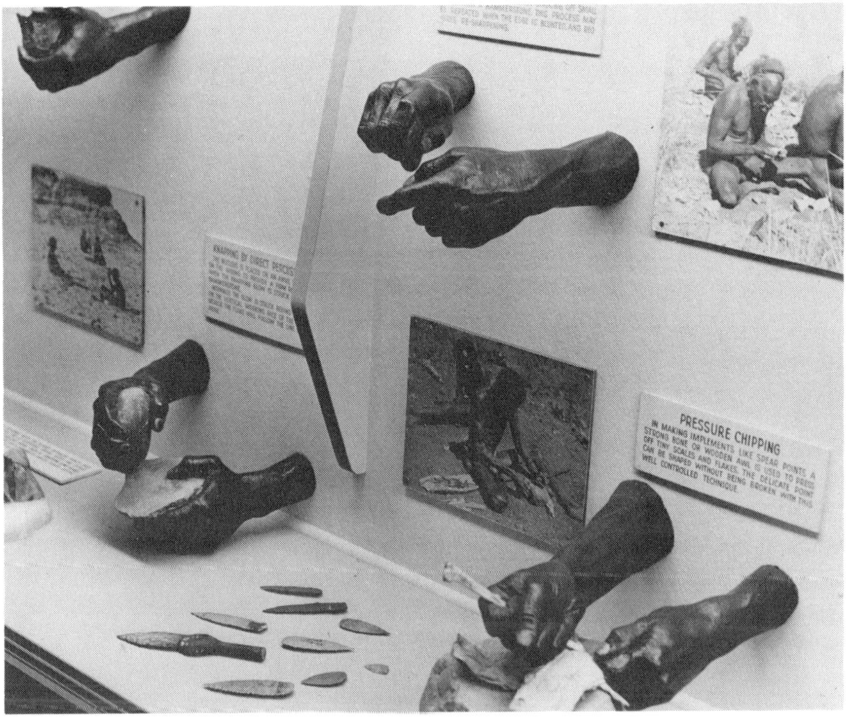

Fig. 12.12. The finished hands in the exhibition. Models by R. Witchard. Photo H. Hughes. Courtesy of The Australian Museum.

of the celluloid cylinder it will now serve as a useful guide to the work. When the fingers, thumb and undercuts present around the object are clear, the arm pieces of the mould will come off after some gentle tapping. Since the fingers of the cast hand do not exert pressure in their grip on the object in the same way as the real hand did in the moulding process, some thought should be given to arranging the fingers to get the maximum security allowed by the design. For example, for the sake of comfort in moulding do not have the hard sharp edges of stones pressing into the inside of the finger joints. Endeavour to keep a pad of muscle over them. In the exhibit the hands shown in Fig. 12.12 were required to be fixed to a vertical board. The wrist of each was sawn across in a flat plane to give it the right aspect when held against the board. A 12 mm hole was drilled about 75 mm deep at right angles to the flat end of the wrist and a 3 mm bolt with a head removed was cemented into it with artificial stone. A test hand supported in this fashion held 75 kg downward pull on the fingers at a distance of 230 mm from the board without fracturing".

FLEXIBLE MOULDS

There is a great variety of materials which can be used to produce flexible moulds. Some are more suitable for one particular job than the others.

SILICONE RUBBER MOULDS

In order to produce a flexible and durable mould of this material, one must choose carefully from the available rubbers. If in doubt, it is best to consult the local supplier. There is still much experimentation in silicone rubber manufacture and product characteristics sometimes change. The coding of the various compounds also varies with each manufacturer.

Small models of rigid materials, such as woodcarvings, plaster, or hard wax models, are usually moulded the following way:

The object is embedded in Plasticine and a Plasticine wall is formed around it. The catalysed rubber is poured on until the walled in area is completely filled. The rubber is allowed to set and after that the Plasticine is removed. While the model remains in the mould, it is turned upside-down and the now exposed flange is painted or sprayed with a release wax solution. A Plasticine wall is again erected around the model, and the walled in area is filled with another batch of catalysed rubber. After this batch has set, the Plasticine is again removed, and the mould is opened.

This is the simplest method of making a solid rubber mould, but it is only suitable for reproducing small to medium sized objects. Silicone rubber is expensive; therefore, a different method should be used for larger objects. From these, a silicone rubber skin mould can be made. The method is more labour and time consuming, but required only a relatively small amount of material. The technique is described below.

The object is embedded in Plasticine, or, if it is too large for that, it is sectioned with Plasticine walls, brass shims, or pieces of celluloid sheeting. The rubber is not poured, but painted on the surface of the model and allowed to set. After setting, another layer is painted on. Before this coat sets, pieces of cheesecloth should be laminated on with more rubber. Once this layer has set, one or two more layers (depending on the size of the model) of cloth are applied with rubber. Finally, a layer of heavier cloth, such as stockingette, is laminated on with more rubber. A mould constructed in this way cannot retain its shape once it is removed from the model. To help retain its shape, a rigid or semirigid jacket or mother mould has to be built over it. This can be made of fibreglass and polyester resin. First, a batch of catalysed resin is painted on the rubber mould and pieces of fibreglass cloth are laminated onto it. Depending on the size of the model, another layer of fibreglass cloth may also be applied to make the jacket strong enough. After sufficient curing time the jacket is removed, then the skin mould is peeled off the model and placed into the jacket, which will retain the shape of the model.

Complex shapes require more complicated techniques. In 1983, a massive silicone rubber moulding job, perhaps the largest ever in a museum, was undertaken in the Museum of Victoria, Melbourne. The work, which took several months, resulted in a full set of moulds (Dow Corning silicone rubber in plaster jackets) of a Memenchisaurus skeleton. Kelly (1983) describes the process in detail.

RUBBER LATEX MOULDS

Natural rubber latex is also a popular moulding compound. This material is cheaper than silicone rubber. It requires no catalyst, it is water based, and, as the water evaporates from the compounds, the rubber solidifies and sets to a very flexible consistency. Volume 2, Chapter 8 has a description of the method used to create a rubber latex mould of a termite mound. This method can be adopted for smaller objects, the latex being applied with a brush rather than a spraygun. Brushes may be preserved by being dipped first into liquid soap, then wiped with a rag.

Rubber moulds often need release agents before casting. They are not self-releasing like silicone rubbers, although some casting compounds can be cast from latex moulds without releases. Plaster releases well from a latex mould, but polyester resin does not. A polyester cast can also be removed with a lot of physical effort from an unreleased rubber mould.

Rubber latex is sensitive to most commercially made release agents. Those which contain any sort of oil, kerosene, trichloroethane, turpentine, or white spirit may blister and deteriorate the rubber on contact. One particular substance seems to be most suitable for releasing rubber moulds; it is a low-kilojoule cooking spray marketed under various brand names. This spray, sold in a spray can, contains vegetable extracts and lecithin. A light spray usually assures excellent release from a latex mould, regardless of which casting material is used. Glycerine and silicone oil are suitable releases, too. PVA release agents also can be used, provided they are water or alcohol based. Rubber can be released from rubber by using talcum powder or plumbago (graphite powder) as release agents. These substances should be dusted on the surface of the set rubber before wet rubber is applied to it.

Rubber moulds should always be stored either with a cast inside of them or filled with dry sand to prevent them from sagging and distorting. Rubber usually deteriorates over 15 to 20 years and moulds become brittle and weak.

POLYURETHANE RUBBER MOULDS

Polyurethane rubbers are marketed under the trade names of Desmocap (a Bayer product) and Ureol (a Ciba-Geigy product). These are viscous liquids.

Desmocap is clear while Ureol is milky in colour. Both rubbers are supplied as two component kits. The components are mixed in the described proportions and used to make moulds in a similar manner to silicone rubber. By adding dibutyl pthalate, the compound can be made very flexible. Without added flexibilizer, the rubber sets rigid, almost to the consistency of hard rubber, such as car tyres.

Mr Kurt Küng (Küng, 1982) of the Bern Museum of Natural History Bern, Switzerland, pioneered the use of this material for museum purposes. He prepared a life size model of a hippopotamus by sculpting it in plaster over a metal armature (Ter-Meer method). Final detail was achieved by coating the model with Plasticine and casts taken from various parts of a live hippopotamus. The model was painted with polyurethane rubber, without partitioning. After this coat of rubber had set a new batch of polyurethane rubber was catalysed and thickened with a filler, to form a paste. The paste was applied to the model 10 mm–15 mm thick to form a strong, one piece skin. This mould was further reinforced at strategic points with plastic tubings and strips. A 14 piece plaster jacket was constructed to support the rubber mould. After the pieces of the jacket were removed, the rubber skin was slit open at suitable points and peeled off the model. The mould proved to be very strong, more so than a similar silicone rubber mould, and retained its shape without shrinkage when placed in the plaster jacket. The rubber mould, without the jacket, weighed 120 kg.

ALGINATE

This material is frequently used by dentists to take moulds of sections of the mouth's interior. Dental alginate is available under various trade names. It is a dry powder, usually pink, light green or white with peppermint scent. When mixed with water, it quickly sets into a rubber-like substance. It is an ideal moulding material as it penetrates every crevice, thus preserving minute detail of the object being moulded. Its quick setting time renders it most suitable for live-moulding parts of the human body (Fig. 12.13).

Live moulding with alginate

Mr Brian Bertram contributed the following. It is for use with a particular alginate.

"Materials and equipment required consist of: CA37 alginate (made by Keur and Sneltjes, Holland) available from dental supply houses; fumed silica dioxide (Aerosil or Cabosil); trisodium orthophosphate in 5 g lots; measuring jug, 1 L or larger; thermometer for 30°C; medium sized plastic hot food containers; round measuring sticks, fairly heavy (18 mm dowels about

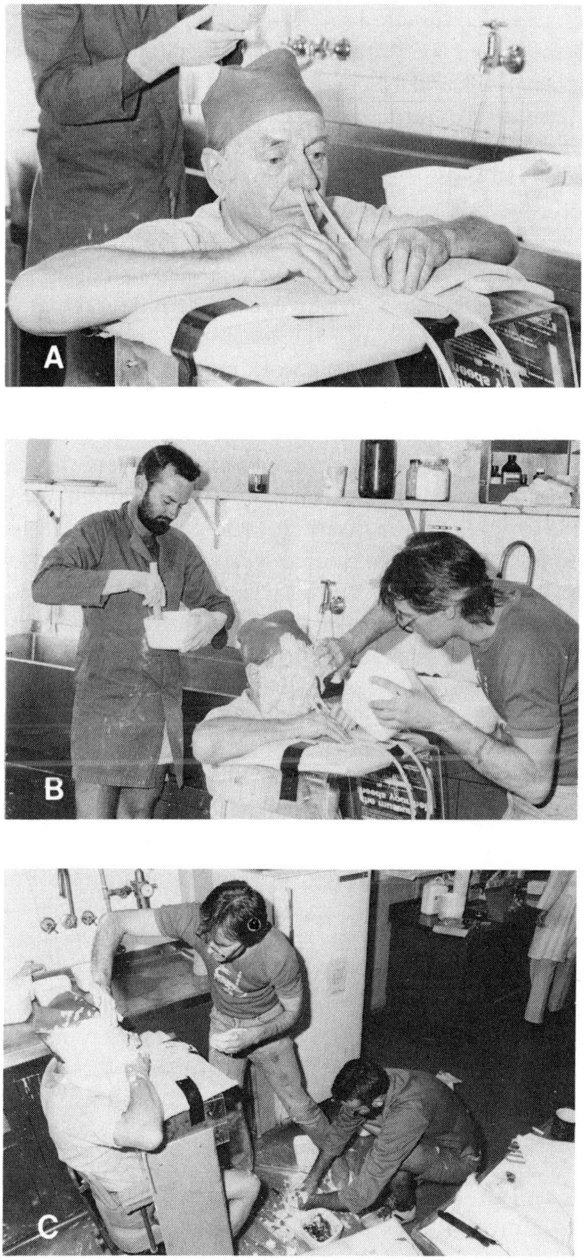

Fig. 12.13. A. The subject is prepared for moulding. **B.** The alginate is applied to the face. **C.** The plaster jacket is made over the alginate mask. Moulding and casting by O. Keywan and P. Townsend. Photo A. Pitman. Courtesy of The Australian Museum.

200 mm long); wooden tongue depressors; small measuring cups cut from any suitable polythene or PVC bottles, of 100 mL, 110 mL and 120 mL capacity, to be known as 10, 11 and 12 part measures; spray dispenser of very strong washing soda solution; starched and stretched stockingette, cut into squares of 30 mm–40 mm, and about 60 mm, plus some strips about 20 mm × 80 mm; plaster of Paris, fast setting; salt; plaster bowls (ice cream containers); spatula or large spoon for mixing plaster; open-weave hessian, cut into squares of mainly 75 mm with some of 125 mm; and, for taking face masks, a hair binder, such as elastic netting, a bathing cap or bald-cap, is required, plus mineral oil (baby oil). Shake the alginate powder in its tin to aerate it, then dispense it with the '10 part' measure into the hot food containers, one lot per tub. Prepare sufficient containers for the work planned. Five containers are more than enough for a face. When dispensing the powder, scoop it up in the measure and then screed it off with a tongue depressor to make it level. Add a '10 part' lot of Aerosil to each container and stir them together dry. Place 5 g of trisodium orthophosphate in the water jug and add a little hot water to dissolve the crystals. When dissolved, add more water until there is a litre of solution at 30°C. The water is used warm for the comfort of the subject. Warm water, however, speeds up the setting of alginate, hence a retarder is added. The orthophosphate is the retarder, but must be added in precise quantities to achieve consistent result. The CA37 mix will set in about four minutes, but vastly different results will be achieved with some other alginates. When using other brands, therefore, careful prior testing must be undertaken. Arrange the subject at a comfortable working height and do any necessary hair preparation. Eyelashes, eyebrows and temple hair should be lightly oiled when moulding faces and hairy skin on certain parts of the body may require Vaseline or other treatment. One person can both mix and apply the mould, but it is easier if an assistant mixes and cleans up. When all is ready, add a brimming '12 part' measure of orthophosphate solution to a container of powder and stir it in. At first the mix will appear too stiff, but vigorous beating will produce a smooth, just flowing paste. Apply the paste with a smearing action, using a tongue depressor, or the hand if large areas of skin are involved. The first mix will set in about four minutes, so fast work is called for. If the first mix has not set by the time the second batch is mixed, work the second straight on from the first. As soon as the first mix sets, however, the second will no longer adhere to it, so, if this has happened, spray the first mix lightly with the washing soda. This will ensure good adhesion of subsequent mixes. The first alginate coat will be thin in places, so must be reinforced. Spray it, then add another layer where necessary, this time using only 11 parts of liquid to get a stiffer mix. When the second layer is completed, spray it again and quickly add another thin overall layer. Into this layer lightly press squares of stockingette, more or less all over the surface. This is an interface which will provide a physical bond between the weak alginate and the stiff plaster jacket which is applied next. Use barrier cream

on your hands and then place about 0.5 L of warm water in an ice cream container. Do *not* use leftover orthophosphate solution. Dissolve about a teaspoon of salt in it and quickly make a creamy mix of plaster. Smear a little of this into the stockingette with your fingers, then start dipping hessian into the plaster, removing excess plaster by running the pieces through your fingers, and applying the hessian to the stockingette. Work fast as the plaster will 'go off' very rapidly. Apply two layers only of hessian, as the mould must be weak enough to break up with the fingers later. When the plaster has set, manipulate the subject's skin around the edges of the mould to loosen it, and ask the subject to work his muscles very gently until he can feel that the skin is free. The mould can then be easily removed. It is important to cast the positive immediately, before it starts to deteriorate. If it cannot be cast immediately then immerse it in water, wrap it in wet cloth or seal it in a plastic bag until it can be done. Alginate is easy to remove from utensils when it has fairly recently set. Therefore, if there is an assistant, cleaning up should proceed progressively. Any alginate that sets on clothing will have to be broken down with washing soda solution before it can be washed out".

POLYVINYL CHLORIDE RUBBER MOULD

Polyvinyl chloride (PVC) rubber is a hot melt compound popular with sculptors, mostly because moulds made of it can be recycled. To make a simple PVC rubber mould, the procedure is as follows.

The rubber, which can be purchased in blocks, is cut up into 2.5 cm cubes and melted. Small quantities can be melted in a heavy saucepan but larger amounts must be heated in a double boiler, such as a gluepot. For best results, oil should be used instead of water in the double boiler. When heated PVC gives out harmful fumes; therefore, melting should take place in a fume cabinet, or at least in a room with adequate ventilation. The PVC rubber melts at approximately 120°C. The object, which must be of a material which can withstand 120°C, is placed on a nonporous surface, such as sheet metal or a marble table top. If the object is round, it is practical to support it on a small, cylindrical prop. A wall is made around the object; if it is small, a clay wall is adequate; if it is a large object, a strip of linoleum or plastic coated plywood can be used, supported by clay buttresses on the outside.

The melted PVC rubber is poured in to fill the walled in area. The rubber must form a 20 mm thick layer around the object. After while, the compound cools and sets. The retaining walls can be removed, the mould turned upside down, and the object removed from it. If the object was supported on a prop it will be necessary to slit the mould to release the model from it. The mould is ready to be used as soon as the object is removed from it.

Polyvinyl chloride rubber moulds can be repaired or altered with hot tools. A heated spatula can be used to correct small blemishes, in the same fashion as wax.

CASTING

POLYESTER RESINS

The various polyester resins can be considered the most versatile casting media in present use in a museum's preparation department. Polyester resins are suitable for solid casting, providing the object is not too large, for preparing hollow laminated casts, and for the making of flexible casts.

Polyester resin can be applied to almost any mould; however, the moulds must be prepared to ensure good release. Plaster moulds must be thoroughly dry, then sealed (shellac is one of the most widely used sealants), and coated with a release agent. A detailed description of polyester casting in plaster moulds is given in Volume 1, Chapter 2. This method can be used to prepare casts of sculpted models as well as polyester replicas of animal specimens.

Silicone rubber and PVC rubber moulds require no sealing or release agents; the polyester resin can be applied to the mould surface and after setting it can be easily removed (Fig. 12.14). In PVC moulds, only low exotherm setting resins should be used, otherwise the rubber may melt.

Epoxy resins are also often used, mostly when special effects are required which are difficult to achieve with polyester resin. Moulds can be prepared for epoxy casting more or less in the same manner as for polyester casting. Epoxy tends to inhibit silicone rubber moulds; therefore, in order to protect the rubber, it is a good idea to use a release agent before casting. If repeated epoxy casting is required from a silicone rubber mould, the mould should be heated to approximately 250°C for 10 minutes after each 10 to 15 casts. This drives epoxy residues out of the rubber and prolongs the life of the mould.

Both epoxy and polyester resins can be filled with inert powders, such as whiting, talcum powder, kaolin and microballoons, in order to thicken them and render them thixotropic. Some of these fillers are more suitable for some applications than for others. For example, if the resin should be filled in order to form a thick gel coat to be applied to a horizontal mould surface, talcum powder or whiting is suitable. Whiting filled resin is also excellent for solid castings.

Powdered fillers tend to dissipate the heat formed during the setting of the resin, prevent cracking, and strengthen the cast. The more fibrous the filler, the stronger the cast will be. In the past, asbestos fibre was widely used to fill resins. It is an excellent filler, especially blue asbestos mixed with resin which forms an almost indestructable substance. However, asbestos proved to be an extremely dangerous substance and its use is banned in most workshops.

If the resins are to be painted on vertical surfaces, they must be filled with a thixotropic filler. Here, again, asbestos is an excellent thixotropic agent.

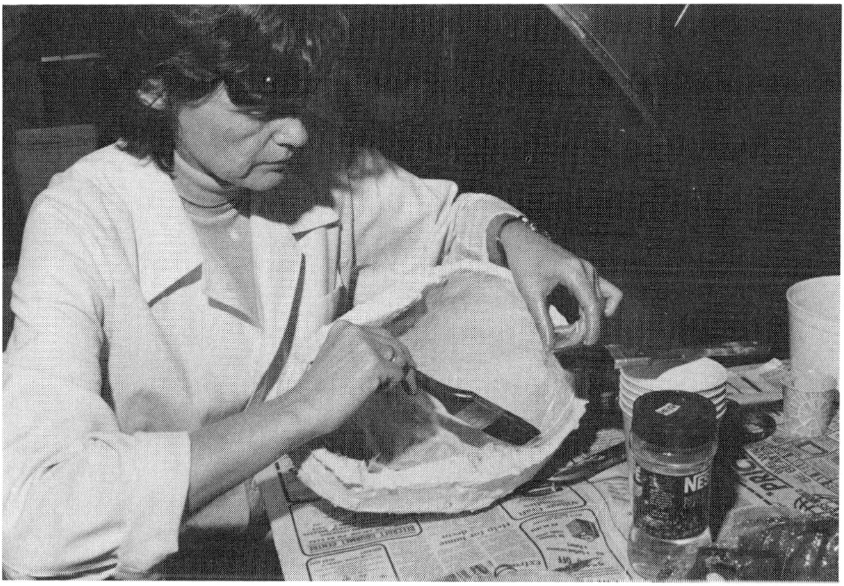

Fig. 12.14. Polyester resin is brushed on a silicone rubber mould. Photo A. Pitman. Courtesy of The Australian Museum.

Special grades of silica sand or silicone dioxide sold under various trade names (Aerosil and Cabosil) are also excellent. The advantage of silicone dioxide is that, if mixed with a clear resin, it forms a very translucent substance. Most other fillers turn resins opaque. Unfortunately, no clear filler has been discovered yet.

All fillers are dangerous for workers. The main dangers occur when the filler is mixed into the resin and when the cast material is further formed by sanding, drilling or in any other way which creates dust. The inhalation of the particles can cause severe illness. Aspirators and protective clothing must be worn when working with fillers and filled plastics (for safety procedures, see Volume 2, Chapter 11).

Resins can be reinforced with fibreglass or acrylic fibre matting. Ropes and tapes are available for reinforcing certain areas of a cast, and woven or chopped strand cloth can be used for making laminated casts. The woven cloth is stronger, and neater to work with, than the chopped strand material. However, it must be cut precisely to conform with the mould's surface, otherwise it is very difficult, if not impossible, to work it in smoothly without forming air pockets. Chopped strand cloth, especially lightweight mats, are more suitable for sharp corners and complex shapes in the mould. Very fine grade chopped strand mat, called the surfacing mat, is specifically designed

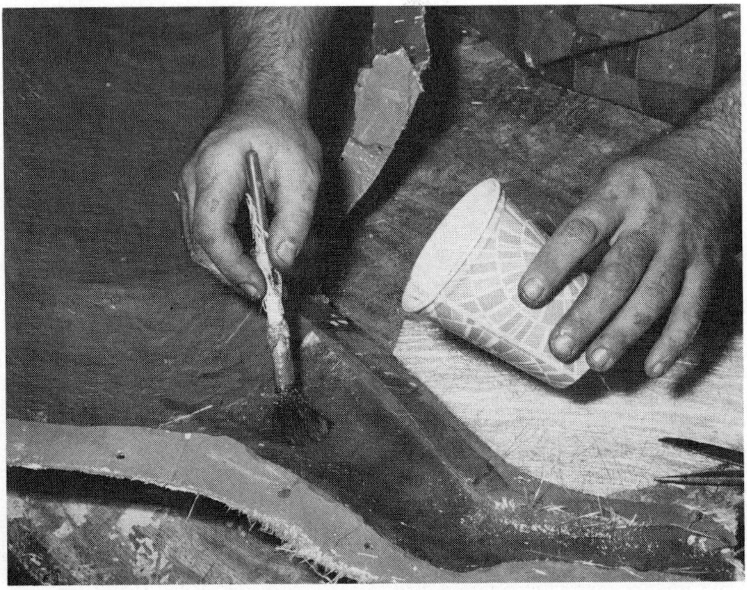

Fig. 12.15. Laminating fibreglass and polyester resin into a mould.

to be laid in the mould first, impregnated with resin, and worked in intricate detail on the mould's surface. The heavier the mat, the more difficult it is to impregnate and work into depressions in the mould. Most chopped strand mats can be peeled apart to give thinner layers.

A standard laminating method is as follows: Catalyse (as recommended by the manufacturer of the resin) a batch of filled resin gel coat and paint it in the mould. Make sure that the resin forms an even layer all over the mould's surface and penetrates all detail.

Air trapped in the crevices of the mould, bubbles, and other imperfections will show on the surface of the finished cast. Allow this coat to set. Cut chopped strand mat into convenient pieces (the more sharp angles and complex shape the mould has, the smaller the pieces should be). Catalyse a batch of laminating resin and paint it on part of the mould. While it is wet, place a piece of mat on it, daub it with resin until impregnated, and work it into the curves of the mould with the brush (Fig. 12.15). Place another piece of mat next to it, slightly overlapping at the edge, and work that into the mould with more resin. Continue this process until the entire surface of the mould is covered with resin saturated cloth. If the mould is large, apply another layer of heavier cloth. While applying cloth to the mould make sure that there are no air bubbles trapped under the fibreglass. These can weaken the cast as well as ruin its appearance. The gel coat, which has little strength

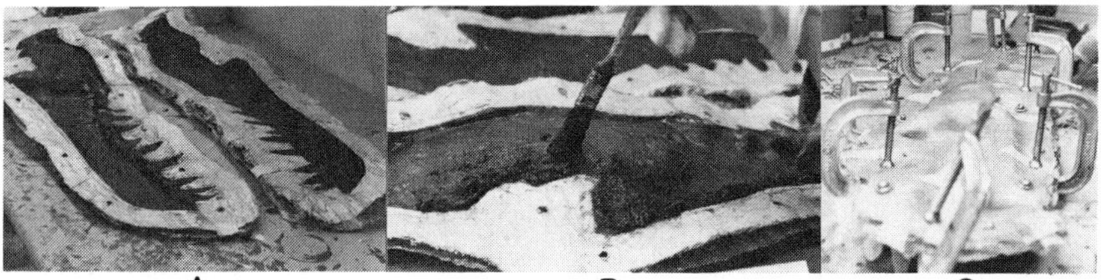

Fig. 12.16. A. The two halves of the mould with the laminated cast inside. **B.** A thick paste made of resin and filler is applied to the edges of the casts. **C.** The two halves are joined before the paste sets. Casting by A. McLaughlin. Courtesy of the Provincial Museum of Alberta, Alberta, Canada.

by itself, may cave in where the cloth is not attached to it, forming unsightly holes on the surface.

Multipiece moulds which are laminated in separate sections must be given special attention. The edges of the mould pieces, where the seams of the casts are formed, must be worked very neatly. Fibreglass and deposits of resin must not protrude over the edges,otherwise the surfaces will not meet properly. The casting process is shown in Fig. 12.16. Smaller multipiece moulds can be tied together securely before the resin is applied to them, and the seams sealed with Plasticine or potter's clay. The catalysed resin is poured into the mould through a hole, the hole is sealed, and the mould is slowly rotated. Rotation can be by hand or by a motor driven rotator. Once the resin has gelled (but not set completely) a quantity of chopped short strand fibreglass is added. The mould is rotated again and the strands of fibreglass form an even coating on the interior surface of the cast. Allow the resin to set completely. Once the strands of glass have properly adhered to the surface, add more catalysed resin to strengthen the cast.

Resins can be pigmented to any colour. Opaque colours can be achieved by mixing the resin with pigments or with artist's oil paints. Translucent colours can be achieved by mixing in dyes dissolved in acetone. Any dye or paint which dissolves in acetone can be mixed with resins. Polyesters are more tolerant than epoxies. The latter are more sensitive to additives and some properties of the resins may be ruined by acetone or oil paint.

Most casts require some sort of finishing. Cured resin can be machined almost like hardwood. Small protruding imperfections (usually caused by pinholes on the mould's surface), and burrs at the seams can be carefully carved or sanded off. A dental drill, with various burrs, is an indispensable tool for finishing resin casts. Various small, handheld electric tools are also very good for the purpose, especially if they work at variable speeds.

Holes, cracks and the like can be filled with thickened resin of the same kind of which the cast was made.

Resin casts are often painted (Fig. 12.17). Before painting, all residues of release agents must be removed from the cast's surface. Polyvinyl acetate release can be scrubbed off with warm water; waxy compounds are usually washed off with mineral turpentine. Once the surface is clean it can be painted with a turpentine or lacquer based undercoat. Compatible paint can be applied to this to achieve the required colour effect. If translucency is required, the cast should be prepared in clear resin and translucent lacquer sprayed on it.

Techniques for colouring plastic replicas of fishes are discussed in Volume 1, Chapter 2. These techniques can be applied to any other resin cast. Epoxy surfaces can be difficult to paint, since paint often does not adhere to them. A light wash or spray of acetone may render the surface slightly tacky, and this helps the paint to adhere.

POLYURETHANE FOAM

Polyurethane foam produces very good casts of medium complexity. Objects with long, slender, protruding parts (an octopus, for example) are less suitable for polyurethane reproduction. All moulds except silicone rubber moulds must be very well sealed and released before casting because polyurethane foam adheres to almost any surface. Silicone moulds can be used without release agents, but the application of a sealer and release paint is recommended. This paint forms a tough, skin like surface in the mould and ensures good detail in the cast. Polyurethane may have bubbles and blemishes on the surface after curing, but, if the sealer–releaser paint is applied in the mould first, these will be hidden underneath it. The authors have prepared various casts in polyurethane foam. Some fossil replicas with very fine detail were produced from one piece silicone rubber moulds, closed with a glass plate, using the following technique:

First, silicone rubber moulds were prepared of fossils. Dow Corning Silastic J was chosen for this purpose because of its durability. The rubber was put in a vacuum before pouring to eliminate air bubbles. The mould was housed in a box made of timber and angle aluminium to prevent warping.

Once the silicone rubber had cured properly (about 48 hours) a coat of sealer–releaser paint was sprayed into the mould. The paint dried in approximately 10 minutes; then the thoroughly measured and mixed liquid components of polyurethane foam were poured. The flat back plate (also sprayed with sealer–releaser paint) was immediately placed on top of the moulds and kept in position with weights. The foam used was a medium density formula. This foam had a 40 second cream time on +20°C. It expanded 12 to 15 times its liquid volume and weighed 60 kg m^{-3}. Any other polyurethane foam with the same properties would be just as suitable.

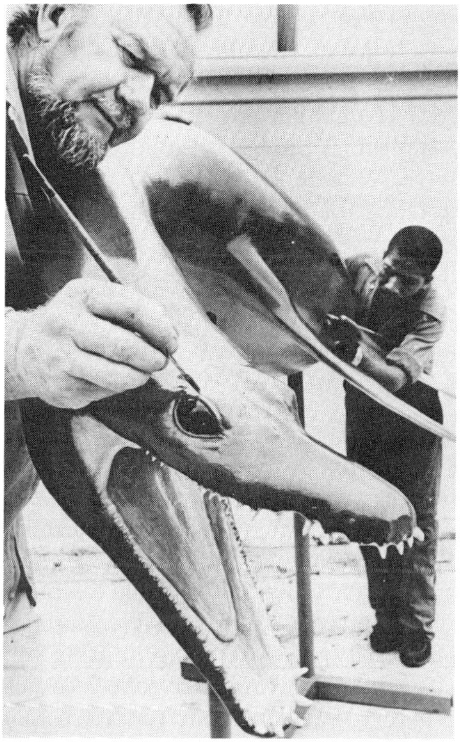

Fig. 12.17. The finished model is painted. Model by B. Newman. Courtesy of the Port Elizabeth Museum. Port Elizabeth, Republic of South Africa.

When the foam reached its final rigidity (approximately 20–25 minutes after pouring), the casts were removed from the moulds. The moulds needed very little cleaning before a new coat of sealer–releaser paint could be sprayed into them, thus preparing them for another pour. The casts themselves needed practically no trimming, apart from a very thin lip around the edges where the back plate pressed against the silicone rubber. The detail of each cast was very true to the original. Most replicas had to be painted to achieve the right effect. Generally, paint was sprayed on lightly to avoid clogging minute detail. Different washes were applied to this base colour. Final colour detail was added with a fine brush.

The above technique is very useful if repetitious work must be carried out in a reasonably short time. The more moulds that can be worked with at the same time, the less time needed for preparing the casts.

Multipiece moulds can be rather difficult to handle unless they are specificially designed for polyurethane casting. Such moulds should have very efficient locking devices and a pour hole which can be closed quickly once the foam is poured into the mould. Strong fibreglass jackets are also necessary for multipiece rubber moulds.

Where fine detail is unimportant, another polyurethane and another technique can be applied. A fibreglass (multipiece) mould can be used for taxidermy models (see Volume 1, Chapter 6).

Flexible polyurethane foam is generally used to produce small animal mannikins. Strong wires are incorporated in the legs of the models and, once the skin is mounted, the position can be altered by bending or repositioning the model's limbs. These mannikins are also available from many taxidermist suppliers.

POLYURETHANE RUBBER

The flexibility of polyurethane rubber can be increased by adding di-butyl pthalate. The substance can be poured into moulds made of any material, provided they are properly sealed and released. Silicone rubber moulds are ideal for this material as they do not need to be released before casting. Polyurethane rubber is mostly suitable for solid casting. It can be pigmented with the usual colouring agent or painted after curing with lacquer or oil based paints. It is especially practical for reproducing human figures or other hairless animals. Touchable polyurethane rubber casts are especially effective. The material is very tough and almost indestructible, as was proven to one of the authors (Hangay) in Mr Küng's studio in the Bern Museum of Natural History. A polyurethane rubber test cast of a gavial is kept on the floor in one corner of the studio where visitors (and Mr Küng himself) can kick it, stand on it, or abuse it in any other way. The cast, after years of such treatment, shows no signs of deterioration.

Mr Küng also prepared a very convincing cast of a model hippopotamus (Fig. 12.18) from this material. The polyurethane rubber mould was painted with a coat of catalysed polyurethane rubber. After this coat had set, another batch of rubber was mixed and thickened with a filler. This thickened rubber was applied to the entire inner surface of the mould to build up a 5 mm thick layer. While this batch was still wet, the mould was assembled and flexible polyurethane foam was poured in, filling the mould completely. In order to save material some empty drums and plastic containers were also included. Thirty millimetre pipes were also foamed in the legs to provide fittings by which the model could be fastened to a base later on. The cast was finally spray painted with flexible acrylic colours.

RUBBER LATEX AND SILICONE RUBBER

Although rubber latex cannot be considered as a permanent material, it is still a practical casting medium. Plaster moulds are most suitable for rubber

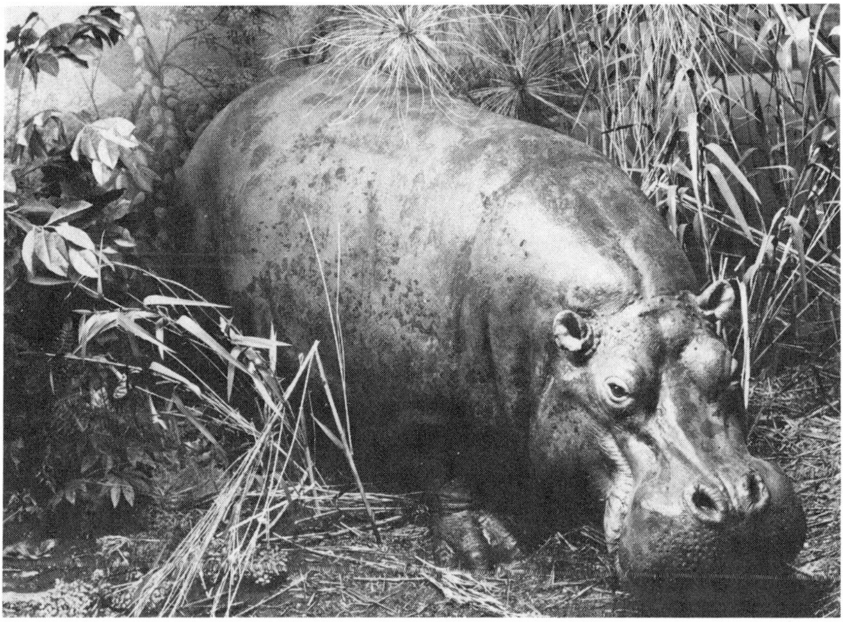

Fig. 12.18. A polyurethane rubber cast of a sculptured hippopotamus. Model by K. Küng. Photo K. Grossenbacher. Courtesy of the Natural History Museum, Bern, Switzerland.

casting. The mould surfaces must be clean and not treated with any sealer or release agent. If the mould consists of several pieces, they should be securely tied together. Liquid rubber latex is poured in through a hole and the mould is filled three quarters full. The latex is swirled around in order to coat the entire interior of the mould, then poured back into its container. The dry plaster of the mould absorbs the water from the latex which, upon drying, forms a skin on the mould's inner surface. The interior must be completely dry before a new batch of latex is poured in the mould. Air from a hair dryer or compressor can speed up drying, or the whole mould can be placed in a hot air drying cabinet. If the interior is accessible, a very thin layer of cotton wool can be spread over the entire surface and saturated with latex after two or three coats have been applied. This helps to build up the thickness of the cast quicker, and the wool serves as reinforcement. In Volume 1, Chapter 4, the technique for casting rubber replicas of reptiles is described.

Silicone rubber replicas can be made of any moulds, although some silicones do not set on rubber latex. Latex moulds must be either sealed with polyvinyl acetate or shellac before silicones are used in them. Moulds made of plaster, fibreglass, wax, or polyvinyl acetate need no sealing or releasing agents. Because of its water content, alginate is not suitable for silicone casting. Silicone moulds, of course, need to have a release agent applied before liquid silicone rubber is poured into them.

WAX

Wax casts, especially those with a high melting point, hard waxes, can be used as display specimens or models in an intermediate stage in sculpting. It is a sculpting technique to make an initial model of a complex object in water or oil based clay first, then mould it in plaster, before preparing a wax cast from the mould. The wax cast can be modelled then in all details and moulded again so that the final cast can be produced in a permanent material. It is easier to form the bulk of the model in clay, but better detail can be modelled in wax.

Plaster moulds with minimal undercuts are suitable for wax casting. Multipiece moulds with no undercuts, or waste moulds without reinforcements and thin walls are also suitable. It is important that the waste mould can be cracked away with minimum effort, otherwise the wax cast within may suffer damage.

The plaster mould should be saturated in water before casting wax in it. In a one piece, relief type mould, the molten wax can be brushed in. Wax should be applied evenly, and hot, until a thickness of 6 mm–8 mm is built up. If necessary, the cast can be reinforced with burlap, saturated with hot wax.

Multipiece moulds made of rigid or flexible materials must be tied together securely and sealed at the seams with water clay. Fill the mould with molten wax, making sure that the entire inner surface of the mould is coated with it, and pour it back into the melting pot. Allow the wax in the mould to cool and harden, then repeat the procedure until the wax is 6 mm–12 mm thick. The larger the cast, the thicker the wall.

If the interior of the cast is accessible, pour some plaster in after the wax has solidified, swirl it around until the interior of the cast is coated with plaster. Allow this to set. Mix some more plaster and add some wood wool to it. Mix the wood wool thoroughly with the plaster and pack it loosely in the cast. When this mixture sets, the cast will be solid and sturdy. The cast can then be demoulded and cleaned up.

It is not safe to fill moulds with wax in one pour and let the wax solidify. Shrinkage usually occurs, which will distort the cast.

ALGINATE MOULDS

Alginate contains water; therefore, it is not compatible with a number of casting media. Some epoxy and polyester resins do not react favourably in the presence of water and it is impossible to produce a cast with good detail from an alginate mould with these materials. Those which are formulated to accept water (for example, polyester resins to be used in a water extended

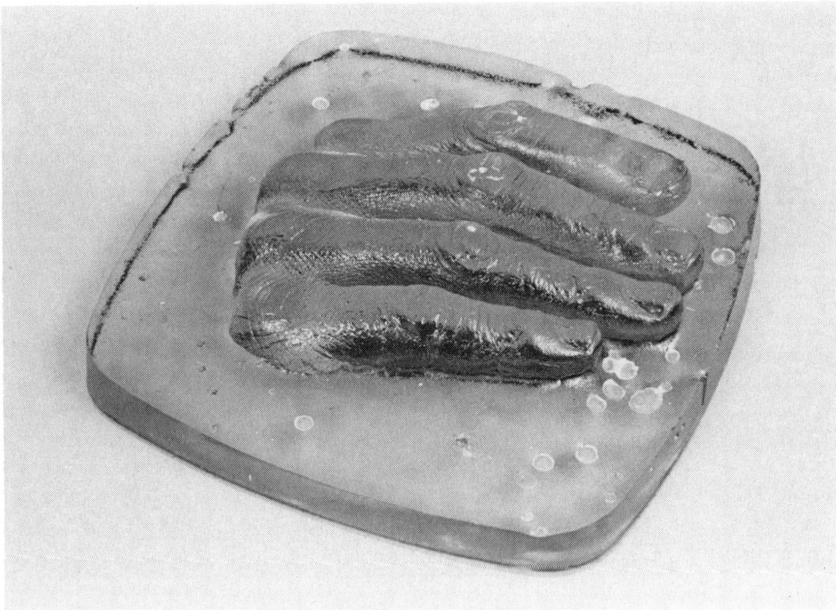

Fig. 12.19. A cast, made in a fast setting polyester resin from an alginate mould. Model by O. Keywan. Photo A. Pitman. Courtesy of The Australian Museum.

form) may be used, but perfect detail cannot be achieved with certainty. Some fast setting polyesters give excellent results (Fig. 12.19), while a very good material for casting from alginate moulds is dental acrylic.

Casts of wax, hard plaster and plaster of Paris can be made in an alginate mould. Plaster of Paris is most commonly used, mostly because of its availability and low cost.

Bertram's method

Spray the inside of the mould with detergent and drain out any excess. This will prevent surface bubbles on the cast. Pour a creamy mix of plaster into the mould and move it about to coat the inside fully. As it thickens, gently spatulate it up the sides of the mould so there is a good, full cover. When it is reasonably well set, add another layer, reinforcing with hessian. Be very careful, particularly around the edges, because the just set plaster is extremely fragile in its flexible mould. Do not completely fill the mould as the expansion of the setting plaster will probably crack the cast's already set surface.

After the plaster cast has exothermed, remove it with as little damage as possible to the mould. This cast is the master cast.

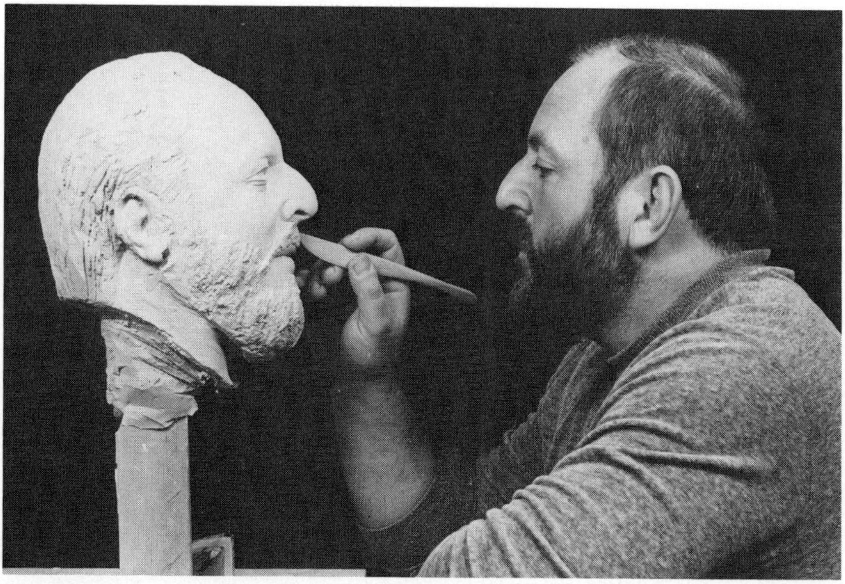

Fig. 12.20. A plaster cast, made from an alginate mould, is being cleaned up. Cast by O. Keywan. Photo A. Pitman. Courtesy of The Australian Museum.

If the cast was made very soon after the mould was completed, the mould should still be in good condition. It is often useful to take a second cast from the alginate mould. This can be kept as a record during the final modelling.

When the cast has cooled down, its surface should be carefully freed of the small plaster protrusions and any other blemishes that form in the alginate mould (Fig. 12.20).

VACUUM FORMING

Sheets of rigid plastic, such as acrylic, toughened polystryrene and acetate, can be vacuum formed to produce artificial flowers, leaves, fish fins, and so on (Fig. 2.21).

The basic principle of vacuum forming is that a sheet of plastic is heated until it becomes flexible, then drawn by vacuum on a rigid mould and held tight against it until the plastic cools and becomes rigid again.

Vacuum formers are available commercially, but one can be made easily for a relatively small cost. Moulds made for vacuum forming must not have undercuts or sharp, protruding parts. In most cases, the difference between the highest and the lowest part of the mould should not be greater than

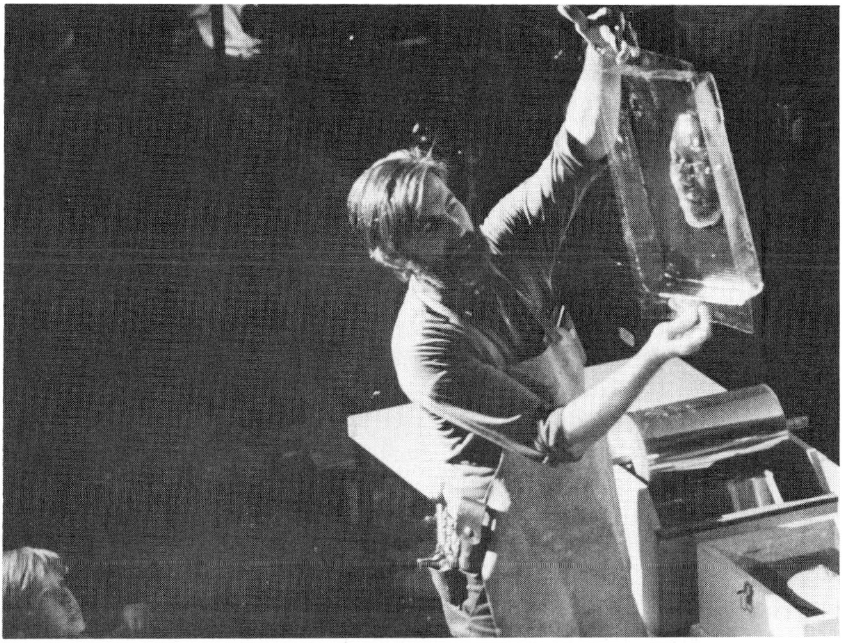

Fig. 12.21. A vinyl acetate mask reproduced by the vacuum former during a lecture. Model by W. A. Bell. Photo John Field. Courtesy of The Australian Museum.

40 mm–50 mm. The mould should be perforated to facilitate suction. Plaster of Paris is a suitable material for the mould, although it may deteriorate if a great number of replicas are made of it. For large scale reproduction work, epoxy moulds are best. Dry plaster can also be reinforced with epoxy impregnation.

The formed shapes are cut manually from the cooled sheets after vacuum forming. Dental drills, rotating engravers or scissors can be used for cutting, depending on the nature of the material. Large numbers of the same models can be clamped together and cut on a bandsaw. Certain models (fish) are formed in two halves. After trimming, the halves are glued together. It is usually very difficult to match the two halves perfectly. Vacuum formed models are generally used when they cannot be closely scrutinized.

SURFACE FINISHING OF MODELS

Unless models are cast in a pigmented resin, they usually need some sort of surface finish. Since most models used in natural history museums depict live plants or animals in a realistic manner, a close resemblance to the original is desirable. Painting is usually employed to achieve lifelike appearance. Before painting, the surface must be cleaned of all release agents and dust. Blemishes of the cast are removed or filled and, if the model should receive an opaque finish, it is undercoated. The various colours can then be applied until the desired effect is achieved.

In most cases it is very difficult, if not impossible, to achieve lifelike colouring with hand painting. An airbrush and a small spray gun (shading gun) is a great asset for the modelmaker. Often, a combination of hand painting and spraying gives the best results.

Patterns with sharp edges can be sprayed by blocking the various areas out with liquid rubber latex. After the latex has dried it is sprayed over with the colour. When the latex is peeled away, it reveals the pattern which was protected from the colour.

Spray painting techniques vary a great deal and constitute a craft of their own which can be mastered only through experimentation.

ARTIFICIAL HAIR

To reproduce the effect of mammalian hair, paint alone is not enough. Mr Michael Traynor developed a technique to apply coloured floc to the surface of mammal models in order to achieve a lifelike appearance. Mr Traynor contributed the following description of his technique:

"With the body modelled and cast to the preparator's satisfaction, the eyes, mouth, lips and toenails are masked. The body is then sprayed with a coat of flat, oil-based paint coloured as closely as possible to the colour of the animal being reconstructed. This gives a good base on which to work. A tongue can be made and inserted before or after floccing, but painting of all soft parts is done on completion. Floccing is the application of false hair to a modelled body to represent as closely as possible the true coat of the animal being reconstructed. The floc is a vegetable fibre, dyed and dried, then finely chopped with a guillotine. The colour used depends on the animal being flocced, but no one pigment will achieve a natural looking colour texture. As many as half a dozen colours may be used to achieve a desirable finish and close study of the fur of the animal will reveal these colours. In the case of extinct or fossil animals, a good imagination is helpful. Dylon dyes have been used successfully for dyeing the fibre, but the colours marketed are limited, so mixing is usually necessary. The dyed fibre is then chopped with a guillotine to the length required, which varies depending on: (i) the stage of floccing — (short floc is used early because of a higher degree of adhesion); (ii) the part of the animal being flocced (short floc around nose, and so on);

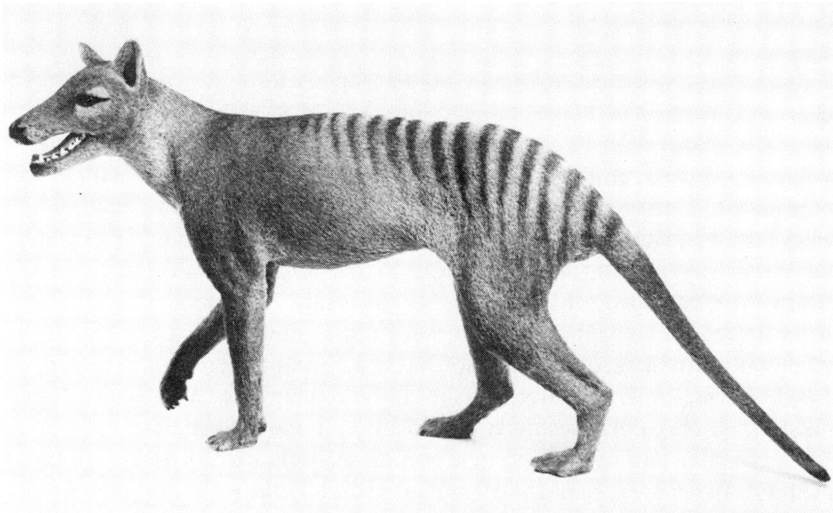

Fig. 12.22. A polyester model, covered by sprayed-on floc. Model by M. Traynor. Photo F. Coffa. Courtesy of the National Museum of Victoria, Melbourne.

and (iii) the final applications (true length of hair used — the maximum length of floc that can be sprayed feasibly at the moment is one centimetre). Two spray units, one of which is a floc gun, and an adhesive satin finish varnish are required for floccing. The varnish should be sprayed on to the body liberally, but not enough to run. The hair tracts of the animal should be considered carefully and the floc applied through the floc gun in the same direction. It is advisable to spray the animal in sections. The body can then be moved into different positions so that correct direction of floc is possible. After initial floccing, the body must dry and then can be brushed through until all excess floc is removed and all remaining floc has the correct direction. Only 20%–30% of floc will adhere at first. Many applications will be needed to cover the body, and each application will require drying and brushing. As the floc forms a coat of hair, colour markings need to be considered. In the case of the Thylacine (Fig. 12.22), the stripes were applied by spraying only those areas with varnish, and then applying the darker floc. Floccing is a very slow process. It cannot be rushed, as the varnish must be allowed to dry completely before the floc can be brushed. Once a satisfactory coat has been achieved, long hair areas, such as whiskers and those around the feet, can be added by hand. Soft parts are then coloured, and the animal is complete. It should be stressed that the models prepared in this way will not stand close scientific scrutiny. The technique is geared towards public display, to be viewed at a given distance".

SCALE MODELS

A good three dimensional scale model can capture the imagination of museum visitors. Volcanoes, sea bed trenches, or models of the local region can allow features to be viewed from many angles. Also, scale models or archaeological or anthropological areas can be extremely revealing and informative. Cut-away models revealing the inside of museum exhibits are always crowd pleasers. Most people know a ship or a building from outside, but there is something very satisfying to the museum visitor to have the exterior shell stripped off so they can see the inside. Scale models do not have to be static. Movement can make them more interesting. Care should be taken with moving scale models so that they do not become a gimmick, or toy-like. This spoils the effect. It should be remembered that scale models made for display purposes must be educational and informative.

Scale models are often used in museums as tools for the design of major displays. An accurate scale model will immediately show the designer his / her concept in three dimensions. Any change in design can then be made. It is much more economical to change a scale model than the real thing.

TOPOGRAPHICAL MODELS

Topographical models, such as three dimensional maps or detailed representations of particular sections of the earth's crust, are very easily understood presentation methods in exhibitions. These, as opposed to graphic or photographic renderings of the same subject, generate more interest amongst visitors. Younger visitors tend to spend more time studying three dimensional models than graphic panels and doubtlessly retain more information from them.

Mr Jim Hood, modelmaker and preparator at The Australian Museum, Sydney, contributed the following considerations and techniques regarding topographical models (from Hood, 1980):

"Once it has been decided to construct a topographical model, one of the first considerations is that of scale. The actual area to be depicted, whether it be a whole continent or a detail of a mountain peak, and the amount of detail required, usually suggests a scale. With museum work, the physical limitations of the display area may dictate the actual size of the model. A technique other than true-to-scale may be considered. Horizontal scale will decide the length and breadth of the model. The same scale may be used vertically, but this may require additional thought, particularly with large land masses at a small scale. An instance would be a contour model of a continent. If built true-to-scale, mountains and valleys would be barely distinguishable from plains. Therefore, a considerable vertical exaggeration in scale would be desirable. Once the scale has been decided upon, a map of the site showing contours, represented by continuous lines, should be reproduced to the scale of the model.

"One of the simplest ways to construct ground contours is to laminate them from sheet material such as card, balsa wood, plywood acrylic, or cork. Such materials can be cut and glued with ease, and need a minimum of finishing. The natural colour and texture of cork is aesthetically very pleasing. For economy of material and weight, the model can be constructed from only two sheets of material. The thickness of the material should reflect the vertical scale. Now, transfer all the contour lines to the sheets of material using carbon paper. As a datum, establish the lowest contour level on sheet 1 and cut around it. Locate two contour levels above the first and cut along this also. This thickness may now be stuck with a polyvinyl acetate glue to a Pyneboard or plywood baseboard which has been faced with the contour material.

"On sheet 2, cut along the contour, one level above the datum level, and two levels above the first cut. This can now be glued into position on the previous sheet; each layer of materials represents a contour increment on the map. This cutting and glueing method is carried out using alternate sheets until the highest level is reached. If the assembly is sagging, it can be supported with offcut pieces glued underneath. The edges of the base can be finished with strips of the material cut to the correct profile. The exposed contour model, with its series of stepped levels is, of course, a stylized form, but the same system can be used for a much more accurate and natural looking model. Layers of material, preferably plywood, are cut with a jigsaw or router and built up in exactly the same way, and edge glued in place (Fig. 12.23). Plaster can then be carefully applied with a palette knife into the stepped areas, and sanded to produce a continuous contour.

"A quick and very light method of construction may be employed if the site is fairly level. Three millimetre plywood is an ideal material to use. Divide the contour map into grid squares. Using carbon paper, transfer all the grid and contour lines of the site onto a sheet of plywood, and also mark on the actual heights. Make a height gauge from a strip of ply or cardboard, marking off, to scale, all the graduations from the lowest to the highest. Using a circular saw, cut one strip of ply for each grid line. Each strip will be cut to become a former for the grades that appear along the grid line, so it will need to be wide enough for the highest grade that will be encountered. On each strip, draft each contour line as the grid line passes through it, using a square to mark up from the contour intersections. Using the gauge, mark the various heights with a dot and connect them. Also mark on the strip lines representing the intersections of the strip, with all the strips perpendicular to it. Using a band saw or jigsaw, cut along the contour lines of each strip. Cut 3 mm wide

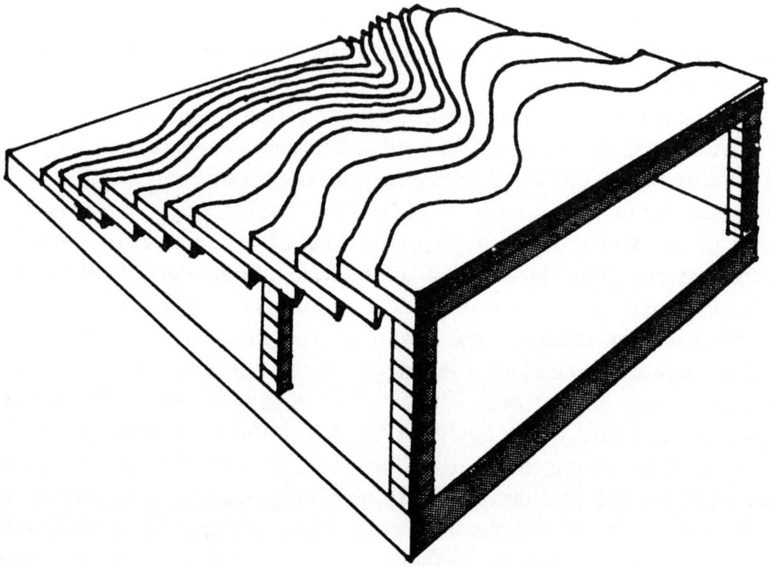

Fig. 12.23. Exposed contour map.

slots half way into each strip at the grid intersection line; the slots should run from the top on all the strips which run in one direction, and from the bottom on all the others. Glue the strips to the play base to form an egg crate. Cut strips to form the edges of the base and glue in place. Glue an oversize sheet of ply onto the formers with the aid of weights to hold it in position, and staple the ply to the formers (Fig. 12.24). When dry, plane the oversize ply flush with the edging."

Polyurethane foam lends itself well to scale landscape contour models. It can be built up in sheets or blocks and can be easily carved with a serrated-edged knife or a rasp. Block foam can be incorporated in models made from other materials. For instance, a dramatic mountain rising from a plain could be sculpted from foam and stuck on a plywood base. All of this could then be covered with a plaster skin. Polystyrene foam can be used for open contour models, but it does not carve well. It is best cut with a hot wire. Alternatively, models can be moulded, either because multiple copies are required, or because elaborate electrical work is to be incorporated. A fibreglass model can be made from the mould (Fig. 12.25). Other sculpting methods combine wire mesh, plaster and polyurethane foam.

It is often necessary to represent a particular three dimensional picture. Houses and roads, as well as other human constructions, are parts of the landscape in most parts of the world. To create small scale models of these

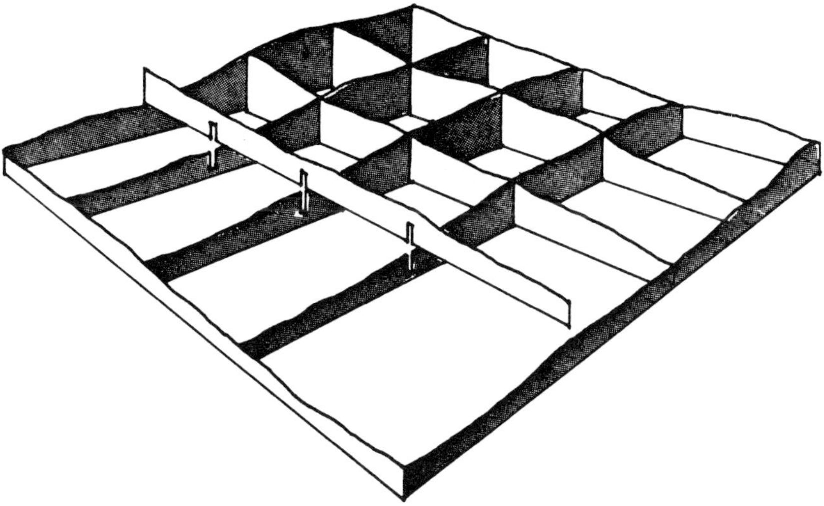

Fig. 12.24. Egg crate construction.

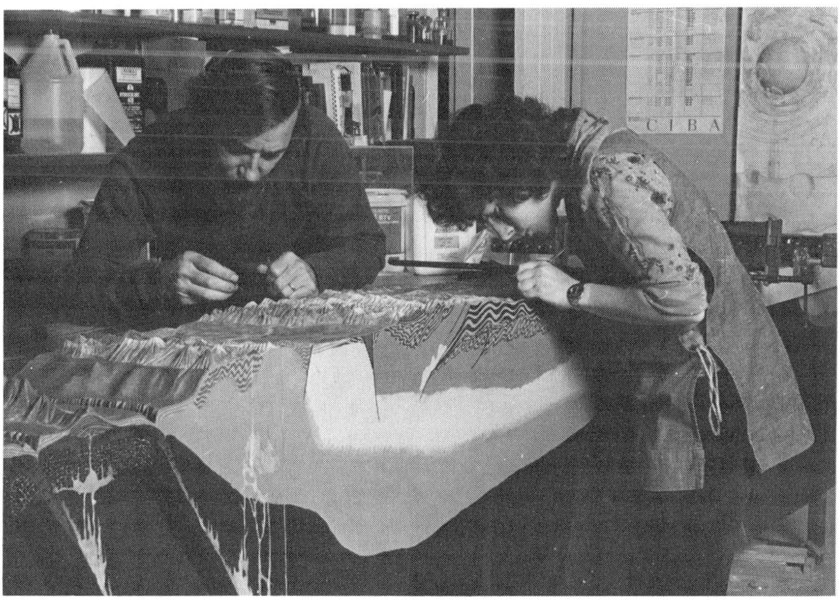

Fig. 12.25. An epoxy resin cast of a Plasticine model of the Earth's crust. Model by J. Heller. Photo G. Millen. Courtesy of The Australian Museum.

and skilfully install them requires the knowledge and dexterity of trained modelmakers. Mr RICHARD SCOTT-CHILD, a model maker and preparator at the Australian Museum, Sydney, detailed some techniques to create highly detailed models, and his work follows.

Miniature masonry

Scaled down masonry can be extremely convincing if made correctly. Fine gauze or bandage type material can be laminated to a piece of plywood (any thickness) with a contact type adhesive (spray pack type). The adhesive should not be water based. The laminated panel is then sprayed with a fine mist of water on both sides. Plaster is spread on to the gauze side of the panel with a palate knife, or similar tool, to a depth of one to two millimetres. When the plaster is approximately 50% cured, the stone or brick relief effect is added. This is done by dabbing a sponge, piece of synthetic foam, wad of course fabric, or the end of a thick paintbrush on to the half-cured plaster. A similar, but rather uniform, effect can also be achieved by adding fine sand to the plaster. When the plaster if fully cured, the horizontal and vertical brick joints can be created. Deliberate chipping of the joints adds to the realism. The colouring of the brickwork or stonework is very important. Pigment can be added to the plaster, or it can be brushed or sprayed on. Four or five different shades of colour should then be mixed up and applied, either by hand or by spraying. Spraying is best done by using a stencil the same measurement as the bricks. The bricks should be painted in a random fashion to avoid a uniform appearance. For repetitive miniature masonry work, a mould can be made of the stone or brickwork panel so that any number of walls or elevations can be made. Polyester resin, epoxy resin, or polyurethane rubber can be used as a casting material. A polyurethane rubber facade has the advantage of being very flexible. By controlling the amount of di-butyl pthalate added to the rubber, various degrees of rigidity can be achieved. Latex rubber is not recommended as a casting material for models that are going to be on display for a long time as latex is prone to disintegration. The miniature masonry technique does not necessarily have to be done on a single panel. The scale model of the building or wall can be made first, and the plaster applied later. Plaster will cause plywood to warp. To combat this, either the plaster can be applied to both sides, or suitable bracing must be applied to the back of the panel. With small masonry panels, the shape can be maintained by spraying a fine mist of water on both sides of the panel (when it is finished) and then putting it under a heavy weight or press for a short period.

On small scale models, other techniques and materials can be used to achieve realistic masonry work. Acrylic sheet or polystyrene sheet can be scored horizontally to show the brick or stone course. A suitable base colour can be painted with either a spray gun or a brush. The final effect is achieved by speckling different colour tones on to the panel with a spray gun on very low pressure. As with the plaster technique, the scale model could be built first and painted afterwards. Windows and doors have to be masked out. The miniature masonry technique can be used on many different types of scale model structures, from ancient Egyptian temples to farmyard walls to paved roadways to modern houses.

ACRYLIC AND STYRENE SHEET CONSTRUCTION

Scale model buildings can be built out of acrylic and styrene sheet very successfully. An elevation should be drawn on a panel of styrene in pencil. The horizontal brick course can then be scored, if required. All windows and doorways should be cut out. Each panel of the model is now placed over a similar size panel of acrylic, and all windows and door openings marked on the acrylic with a light score line. Window bars and frames should be ruled or scored into the acrylic and the score lines filled in with paint. The acrylic part of the model, now with windows and doorways painted in, should be glued at the corners to make a box. The styrene elevations can be laminated to the acrylic. Windows and door frames should line up with the window and door cut-outs in the styrene. It is also possible to laminate the styrene to the acrylic before sticking the building together.

The floor levels and roof can now be completed. Interior masking is important so that the viewer cannot see right through the building or see from one floor to another. All windows and doors should be masked out with clear tape (not masking) before painting.

Styrene sheet should not be confused with polystyrene foam. Styrene sheet is compressed flat, and resembles card. The adhesive used for this type of construction is dichloromethane. It is applied with an artist's water colour brush, or a hypodermic syringe. Care should be taken to avoid breathing in the fumes.

Miniature roof surfaces

Ceramic tile type: In many cases, styrene sheet can be used to model roofing tiles. It comes in many thicknesses. The styrene sheet is scored vertically to represent the individual tiles. A tool may have to be specially made for this purpose. The scored sheet is then cut into strips at 90° to the scorelines. These strips should be wide enough for an overlap. The strips are overlapped from top to bottom, like a real tiled roof, on an acrylic panel.

Thatch and palm roofs: Raffia, shredded into fine strips, is applied to card or thin plywood to the desired thickness. The card or plywood panel is cut to shape, braced from behind, and placed on the model. The raffia can be coloured with a wood stain. It may be advantageous to spray the stain once the basic colour has been applied.

Palm: Raffia is shredded into fine strips and applied to thin card or heavy paper similar to the thatch. The paper raffia panel is then cut into strips at right angles to the raffia allowing for an appropriate overlap. The strips are then stuck to heavier card or thin ply and applied to the building.

Timber effects with styrene

A realistic wood effect can also be achieved with styrene sheet. The styrene can be scored to resemble wooden boards, or cut into strips for fences. A stiff wire brush or coarse file should be used on the styrene surface to make it resemble wood grain. Creative and skilful painting will give the model a realistic timber finish.

Styrene should be used for wood only on small scale models. On large scale models, the real thing should be used. On large models, a close-grain softwood should be used to keep the real wood in scale with the miniature building or structure.

Balsa wood

Balsa wood is known to every modelmaker, and is an extremely useful material. However, balsa wood does have its limitations. Used as scaled down timber, it can look out of scale or odd, as some balsa wood can have a rather coarse grain. When painted, it can become rather furry and coarse looking. When it is used in an exterior or prominent position, it should be used with caution.

LANDSCAPE SURFACE FINISHING AND PAINTING TECHNIQUES

The finished surface on any model is extremely important, for this is what the viewer will see.

MATERIALS FOR GRASS

Sand, sawdust and floc can be used. Grassed areas should be painted first. Diluted polyvinyl acetate glue and a drop of dishwashing liquid should be mixed and stirred well. This mixture can then be applied to the grassed area. Coloured fine sand or sawdust should then be sprinkled on through a sieve. Various colour tones should be used, as grass is never a solid colour. Not even a well kept lawn or a green field is the same colour throughout.

Colouring sand or sawdust is very simple. A small drop of paint is mixed in and this very quickly colours all the grains. The sand or sawdust is then spread out on a piece of newspaper and allowed to dry.

PREPIGMENTED PLASTER METHOD

The surface is prepared in a similar way to the sand or sawdust method. Dry plaster is mixed with powdered paint pigments and sprinkled on the surface. A fine mist of water, mixed with a small quantity of polyvinyl acetate glue, is sprayed on to the pigmented plaster and allowed to dry. This method can be repeated until the right colour and shading is reached.

The surface of the landscape can have a final clear mat spray to lock the surface materials together, although this is not mandatory.

GRASS FLOC METHOD

Floc can be used to model grass. Different colours may be mixed to achieve the right shade. To apply the floc, paint the surface with a slow drying paint approximately the same colour as the floc, then sieve the floc on to the paint. Brush off the excess when the paint has dried.

Subtle spray painting will give the model added detail and realism. Successive layers of various colours can be speckled on a base coat with a spray gun set at low pressure. Colours can be blended in this way to simulate natural areas. Always use surface colours in a subtle manner, as this will be nearer to reality than bright colours.

ROUGH SURFACES AND EMBANKMENTS

Various materials are used to create rough or craggy surfaces. Miniature rocks and cliff faces can be carved out of polyurethane foam. The carved foam should be covered with a light layer of plaster or diluted polyvinyl acetate glue, and the appropriate finish applied.

Real rocks and stones can be incorporated into the simulated material. The right rock or stone should be used to fit into the scale of the landscape. The texture of the rock or stone should be fine, and the colouring in harmony with the rest of the landscape, although the colour of the rock can be altered with subtle spraying. Cork or pinebark chips, if painted correctly, can pass as

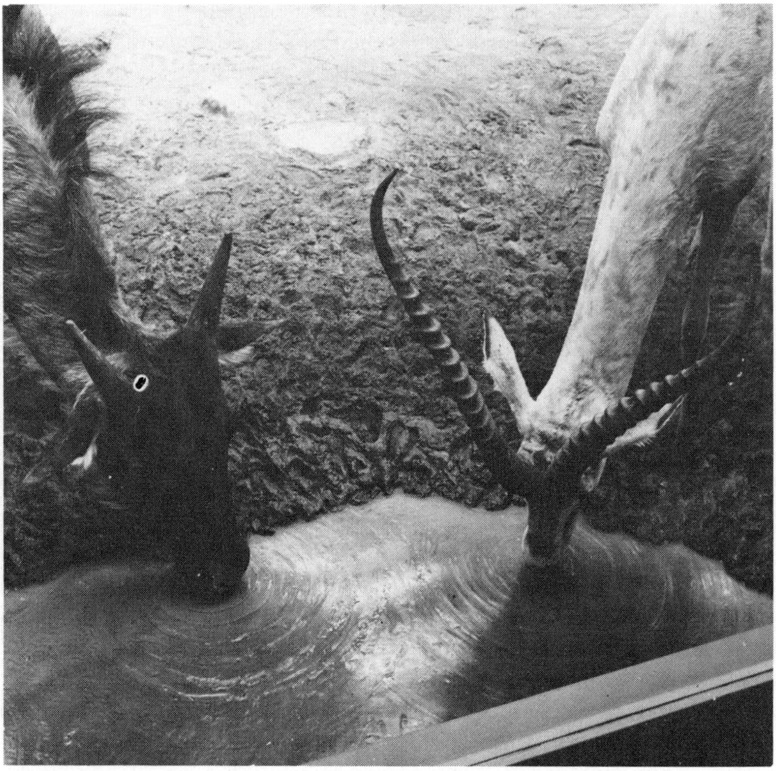

Fig. 12.26. Water, made of polyester resin. Courtesy of the Naturhistorisk Museum, Arhus, Denmark.

rocks or stone surfaces. Very small stones and sand can also have a place on the scale landscape. Horizontal areas may be finished in a stippled plaster effect, similar to the miniature masonry technique, only much coarser.

Paths and roads can be carved or scraped into the plaster surface by using a sanding block the thickness of the path or road, or an old chisel or broken hacksaw blade. When cured, the plaster will be quite easy to carve or scrape in this manner. Painting and surface techniques may then be used on the surface of the plaster.

WATER EFFECTS

Polyester resin poured onto an acrylic sheet will give the effect of water. To produce an impression of small waves or ripples on the surface, an electric fan can be directed on to the curing resin (Fig. 12.26).

If the acrylic sheet is painted on the underside it will give a realistic effect. Clear polyester resin can be poured directly into the model for river and small pool effects. A slow curing resin should be used.

REFERENCES

Hood, J.: Topographical models. *Proceedings of the Conference of Museum Preparators.* Australian Museum, Sydney 1980.

Kelly, K.: *The moulding and casting of dinosaurs: A workshop manual.* Museum of Victoria, Melbourne 1983.

Küng, K.: Bemerkungen zum Aufbau eines Flusspferdes aus Kunst off. *Der Präparator* 28/4 1982.

SOURCE MATERIALS

Bertram, B.: Miniature foliage from brass. *Curator* 7/3 1965: 206–210

Blechschmidt, E.: Rekonstruktionsverfahren mit Verwendung von Kunststoffen. *Zeitschr. für Anat.* 118/1954: 170.

Blunrechli, H.: Bemahlte Gipsmodelle. *Anat. Anz.* 78/1926.

Clemens, H. J.: Zur Technik der Leimform. *Anat. Anz.* 100/1953/54.

Eichner, D. and B. Ahlers: Zur Verwendung von Kunstharzen in der anatomischer Museumstechnik. *Anat. Anz.* 103/1956.

Gutebier, Th.: Die nahtlose Textilbeschichtung (Beflockung). *Der Präparator* 26/1980.

Kaden, J.: Über die Anfertigung von Lehrmodellen einiger Invertebraten. *Der Präparator* 21/3 1975.

Katona, J.: Neue Technik der Modellierung anatomischer Präparate. *Anat. Anz.* 88/1939.

Kolbe, W. and E. Schicha: Die Stechmucke. *Der Präparator* 21/2 1975.

Kolligs, D.: Plastikmasses für Bildhauer. *Der Präparator* 20/1–2 1974.

La Fortes-Pinheire, L. C.: Le moulage capillaire. *Archives d'Anatomie* 82/1956.

Lunk, W. A.: Making and finishing small models. *Curator* 11/2 1969: 165–173.

Orem, H. G.: Lightweight hollow models of reinforced plaster. *Curator* 7/4 1965: 282–286.

Schicha, E.: Lehrmodell einer Mücke. *Der Präparator* 25/3 1979.

Schuldes, W. K. F.: Magnified models of microscopic mites. *Curator* 9/1 1967: 36–40.

Sheffield, R. E.: Microcrystalline wax as a medium for model preparation. *Curator* 7/3 1965: 244–254.

Spanner, R.: Die Bearbeitung des Celluloids zum Montieren makroskopischer Sammlungspräparate. *Anat. Anz.* 62/1926.

Swinkles, P.: Building a minke whale. National Conference of Preparators and Technicians. Museum of Victoria, Melbourne. 1983.

Traynor, M., and K. Kelly: A method of reconstructing furred mammals both living and extinct. *Proceedings of the Conference of Museum Preparators.* Australian Museum, Sydney, 1980.

Veit, O.: Zur Konservierung von Wachsplatten-Modellen. *Anat. Anz.* 73/1931/32.

Wall, C.: Preparation of articulatory models for instruction and display. *Curator* 18/4 1976: 256–262.

Weidemann, G.: Kleben von Wachsmodellen mit Hausenblase-(Ichtiocola electa). *Der Präparator* 22/1 1976.

Wick, K.: *Die Maler-und Lackiererpraxis.* Dummler Verlag, Bonn 1970.

13
Dioramas

THE EFFECT of even the most skilfully prepared plant or animal specimen can be lessened by wrong display techniques. Background, surrounding, light effects and the entire design of a display should not only complement and feature the specimens, but fuse them into one with the whole exhibition. Only in this way can a museum exhibit fulfil its true obligation: the dissemination of knowledge.

DISPLAYS

Most people can never see a wild animal going about its usual business in its natural environment. Even common animals, such as small rodents, foxes or many of the bird species, can only be seen fleetingly by those who do not possess the skill of the hunter or bird watcher. Yet people are fascinated by the activities of animals and their relationship to the environment. Habitat displays, which show plants, animals and part of the landscape they can be found in, can cater for this need. Such displays, if well made and presented, can have immense educational value. Aesthetic exhibitions convey the educational message in a pleasant way and the information is easily absorbed and retained by the viewer.

IDENTIFICATION DISPLAYS

Large exhibitions, which show a great number of species in systematic order, are very useful for those who wish to identify a particular plant or animal. However, these displays, which often take up considerable space in the public areas of museums, can be made more interesting for the general public. The exhibited species can be divided into groups, according to their scientific classification, and the groups broken up by small habitat displays which depict an interesting characteristic of one or two species from the nearest group.

SMALL HABITAT DISPLAYS

The simplest form of habitat display is the so called natural stand. Instead of mounting a taxidermic specimen (a duck, for example) on a piece of polished wood, it can be installed on an artificial sand mound, depicting a section of a sand bank, where ducks are often found. If some preserved vegetation is added, the stand becomes more interesting, and more educational, since it starts to tell a story about the duck's environment. The idea can be developed further. Instead of a small stand, a larger area can be designated within the exhibit for the purpose of representing a part of the environment.

Part of a marsupial exhibition is shown in Fig 13.1. The information about the animals is written on the panel, in which circular holes have been cut. Through these holes, the viewer can see a small part of the animal's environment. The animal in Fig. 13.1 is a spotted cuscus. The viewer of this exhibition can perceive that the animal is arboreal and nocturnal, has a prehensile tail, and robs bird's nests. This is enough to stimulate enough interest for the viewer to obtain more information from the written text displayed next to the animal. Small habitat displays are relatively cheap to build since they can be installed front of a plain, flat background, and require small amounts of materials and little time to construct. The display can be constructed as an island exhibit. This design allows the viewer to see the display from all sides, but it should be protected by glass.

Small habitat groups can be installed in specially designed, diorama-like showcase units (Fig. 13.2).

In building a small diorama, the first thing is to establish its concept. That is, what should it explain in a visual form, and how. An initial design should be drawn up, following the principles of composing a picture. The animal specimens can be mounted according to the design, or, if they are already mounted, the diorama may have to be designed around them. Foreground

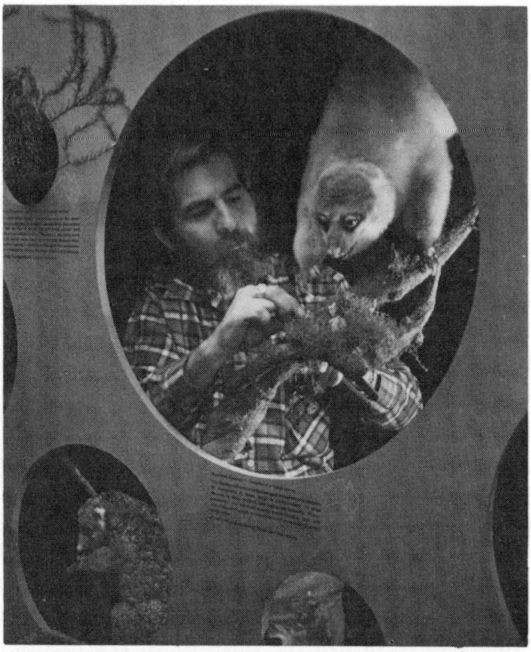

Fig. 13.1. Small habitat displays can be very effective. Foreground by W. Bell. Taxidermy by G. Hangay. Photo J. Fields. Courtesy of The Australian Museum.

material, such as vegetation, soil and small rocks, can be collected now and prepared.

In the mean time, the wooden shell of the diorama is constructed and light fittings are installed. If the diorama depicts a night scene, the back wall should be painted matt black. Many small dioramas show scenes where the viewer cannot see a distant view in the background. These small displays are the best and easiest to construct. For example, it is easier to build a reproduction of a corner of an attic and show an owl family in the nest than recreate the distant views and atmosphere of the semidesert with some ground squirrels speeding across it. However, if a distant view is required in a small diorama, it is possible to use a large colour photograph mounted on the back wall. But it is difficult to match foreground to a photographic background; the transition, especially, is problematic. A painted background is still the best choice, but landscape artists who can paint in a realistic style are very difficult to find. If the design permits it, the background can be one colour only (Hohorst, 1980). This could provide a semiabstract impression, whereby a reallistically detailed foreground and lifelike mounted specimens are matched with a plain background.

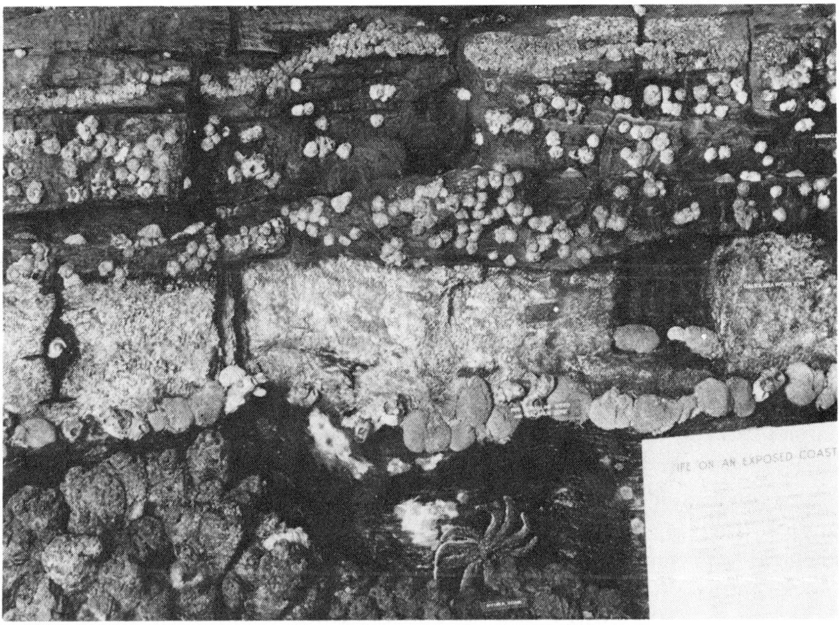

Fig. 13.2. Habitat group, depicting life on seashore rocks. Diorama by R. Witchard and K. Gregg. Photo H. Hughes. Courtesy of The Australian Museum.

Special light effects can be employed to simulate certain colours of the sky. Bindel and Rasmussen (1975) have described how the light of the rising sun was simulated by special illumination in a diorama built for the Zoological Museum in Copenhagen, Denmark. In small dioramas, the foreground can be made of a rigid plastic foam. Polyurethane is the most suitable, as, unlike polystyrene foam, it can be poured as a liquid on the floor of the diorama. Once the foam has risen and set hard, it can be shaped with a saw, knife or wood rasp to correspond with the design. Preserved or artificial plants, together with other props, such as logs or rocks, can be placed now. Large objects, such as tree trunks or fallen logs, can be bolted to the timber frame of the diorama shell. These large props should be sunk into the foam and glued in position with small batches of freshly mixed foam. Rocks can be carved out of foam and installed the same way. A coat of rubber latex improves the appearance of such foam rocks. The foam which forms the basis for the foreground is painted with a coat of polyvinyl acetate glue and covered with dry soil. Once the glue has dried, the excess soil can be removed. Small plants should be installed after the soil has been positioned. Most of these can simply be pushed into the foam, while others, like moss or lichen, will need to be carefully glued on at appropriate spots.

The animal specimens can be installed at this stage. The leg wires of small specimens should be pushed into the foam at an angle to provide satisfactory anchorage. Specimens to be installed on the vegetation can be wired to suitable branches and tree trunks in the usual manner.

The atmosphere of a diorama greatly improves with skilfully installed and directed lights. Broad daylight can be easily simulated with daylight fluorescent tubes; and the light of the rising or setting sun can usually be imitated with spot lights fitted with colour filters. Small spotlights can be used to recreate filtered light.

LARGE DIORAMAS

A large diorama which shows a group of animals in their typical environment comes close to the essence of a museum.

The first part of creating one is to construct the shell. Although compromises can be made, the best shape for a diorama is a half dome. This is usually built by forming the basic shape of steel rods and wire mesh which are then covered with cement (in the same way that ferrocement boats are built). The most important facet of the work is that the interior of the dome must be perfectly even and smooth. This is not easy to achieve. The inner surface of the dome must be coated with plaster and finished with the usual plasterer's methods. Usually, professional plasterers are employed for this work. The size and shape of the dome are dictated by the availability of space, design, and, of course, money.

The design of a large diorama is a task for a skilled artist. Simplicity is not always a merit here, since the more elaborate a display, the more interesting the story it conveys. A large diorama could have several focal points, depicting a number of stories. These may be almost independent of each other, yet still work towards the same goal: to show the viewer a particular place with its plant and animal life. As an example, a diorama can picture life at a waterhole of the African savanna. A dozen or more antelopes of various species can be depicted around the waterhole, some drinking, and some nervously looking at a stalking lioness in the nearby thicket. Closer to the front of the display, a bird's nest may be installed on a tree showing a bird feeding its young. Closer to the viewer, an insect hunting chameleon can be shown capturing its prey with its long, prehensile tongue. The entire diorama can be designed along these lines, in order to provide the viewer with a long lasting, educational experience, as shown in Fig. 13.3.

Once the shell has been constructed and the content designed, a three dimensional small scale model should be made. This model should be finished with a fair amount of detail and in colour to demonstrate how the finished diorama will look like. In a true-to-life diorama, the background should be

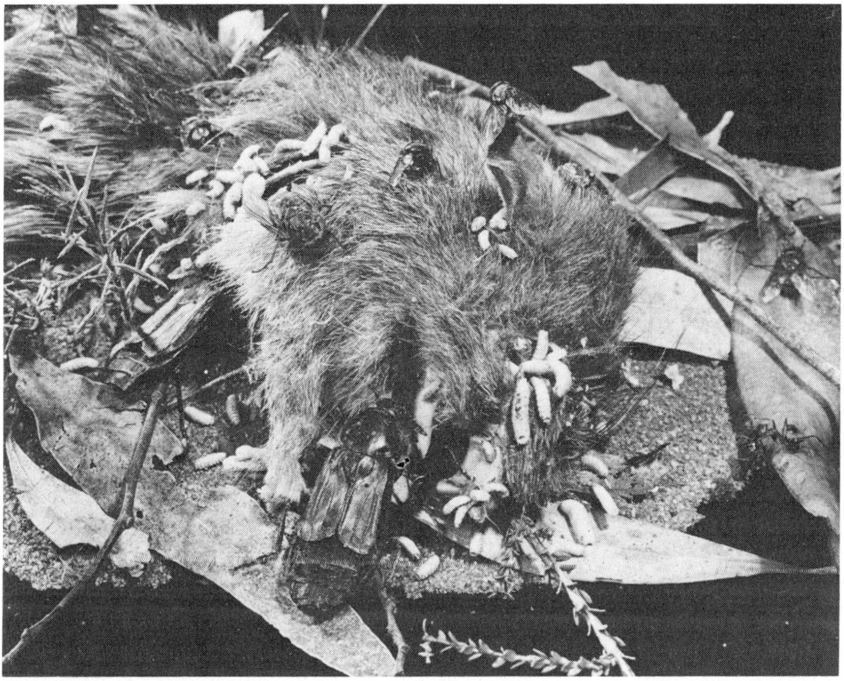

Fig. 13.3. Detail of a small diorama, showing the work of insects on the carcass of a bush-rat. Foreground and taxidermy by M. Robinson. Photo A. Pitman. Courtesy of The Australian Museum.

always painted. First, large size colour photographs, which would be needed for the purpose can be extremely expensive. Furthermore, the detail of such huge enlargements is usually not good enough for a realistic presentation. Mounting the photograph on the curved surface of the dome's interior presents another problem. The main objection is, however, that a photograph cannot achieve the same effect as a skilful artist. Photography can aid the artist; a transparency of the required landscape can be projected on the back of the diorama and the outlines of it sketched on the wall.

The horizon should usually be placed at the eye level of an average person. This assists realism. In real life, we always see the horizon at eye level. The landscape should be painted by conventional methods with the accent on realism. An airbrush is often used to paint realistic clouds, drifting haze, or fog, but detailing is carried out with conventional brushes (Fig. 13.4).

After the background has been painted, but not necessarily finished, work on the foreground can start. The floor understructure of the diorama should be built from timber (Fig. 13.5). The floors of dioramas are usually higher than the floor of the rooms they are installed in. The understructure can be

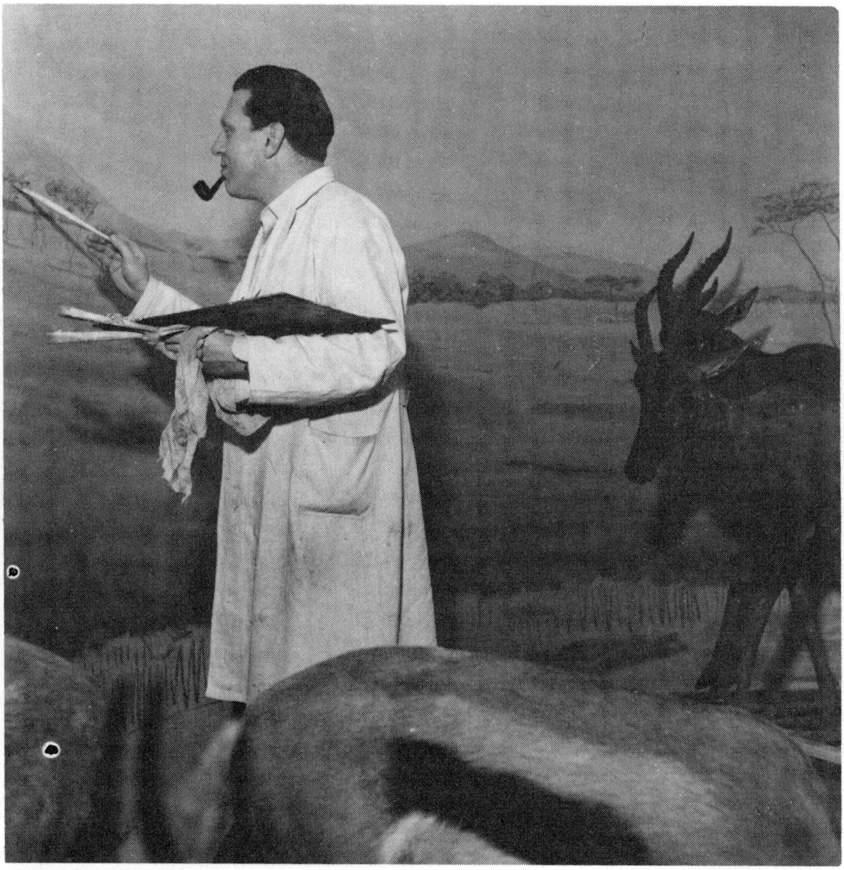

Fig. 13.4. The background is sketched on the wall of the diorama. Courtesy of the Naturhistorisk Museum, Arhus, Denmark.

a strong timber frame with wire mesh stapled on. It is vital that the floor does not run up to the back wall. A gap must be left between the wall and the floor of the diorama to act as a service trench (Fig. 13.6). This trench can be used during the construction of the exhibit and later on when maintenance is carried out. The trench must be accessible, preferably from a passageway behind the diorama; otherwise, access will have to be through the crawlway from the front or from any other, convenient, concealed spot.

Timber frames or platforms to accommodate the leg fittings of the specimens should be built where large specimens are to be placed. Large props, such as artificial or real trees and rocks can be installed at this stage.

Large rocks can be made as hollow casts of plaster of Paris or resin laminated fibreglass. The casts are made from real rocks. Making rock

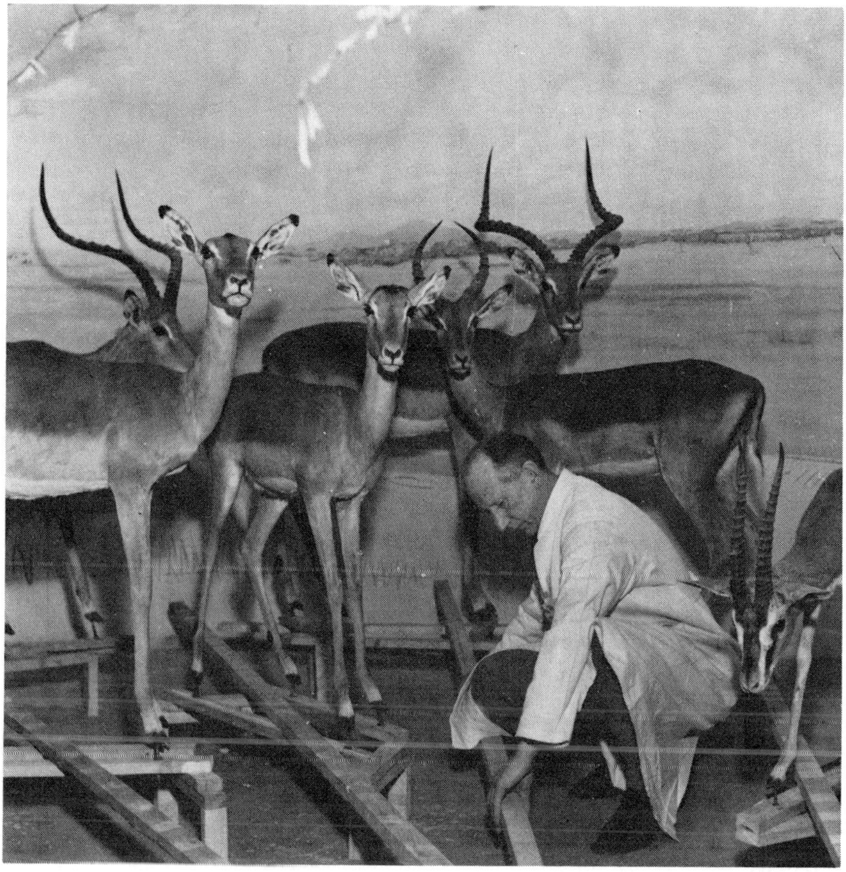

Fig. 13.5. The floor structure is built of timber. Courtesy of the Naturhistorisk Museum, Arhus, Denmark.

replicas is very similar to the method described to reproduce termite mounds (Volume 2, Chapter 12). Tree trunks can be treated the same way. Rubber latex moulds of living trees are used to make casts which can be painted later to true-to-life colours. After all the large objects have been installed, the foreground can be laid on the wire mesh.

First, pieces of burlap soaked in plaster of Paris are laid on the mesh (Fig. 13.7A). Small bushes, then scrub is mounted on pieces of plywood and placed in the desired positions. Tufts of grass or other vegetation can be set in paper cups filled with plaster. After the plaster has set, the cups can be removed and the clusters of vegetation arranged on the ground (Fig. 13.7B). Another layer of plaster soaked burlap can then be applied to lock the vegetation in place. Soil mixed with taxidermist's body paste should now be applied to the surface

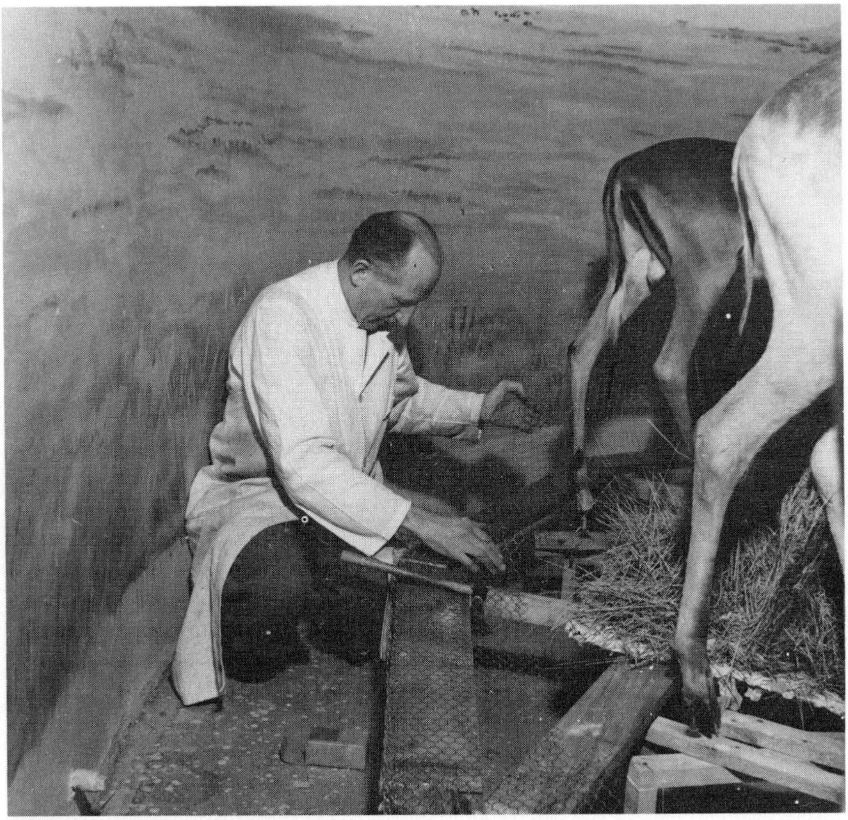

Fig. 13.6. Between the wall and the foreground a service trench is formed. Courtesy of the Naturhistorisk Museum, Arhus, Denmark.

to form the ground. After drying, this should be painted with polyvinyl acetate glue and covered with dry soil. Once the glue has dried, most of the excess soil should be removed. Leaf and other appropriate litter can then be applied to make the ground realistic.

After this, the animals are installed (Fig. 13.8). In some institutions, the animals are installed before the first layer of burlap and plaster is applied (Evans-Olsen, 1980). In many instances, this is the best method to follow; however, the animals need to be protected from splashes of plaster, dust, paint and dirt.

An alternative method of forming the ground is to use a polyurethane spraying unit. After carefully masking the background and everything else not to be covered by foam, the liquid components of polyurethane foam were

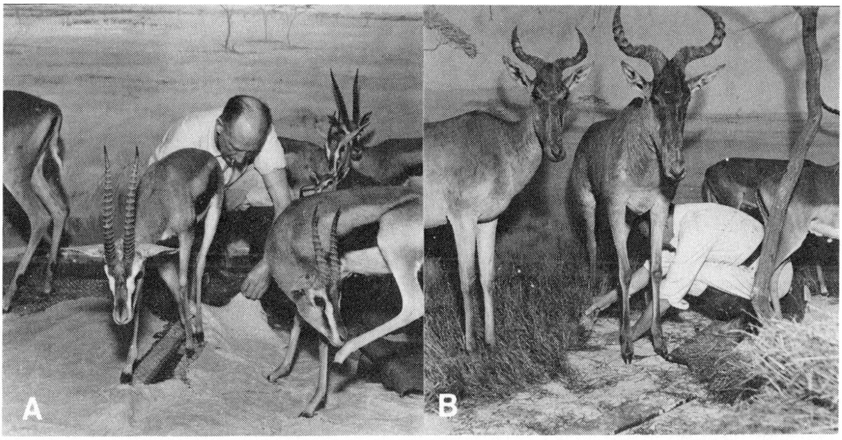

Fig. 13.7. A. Burlap, dipped in plaster, is draped over the wire mesh. **B.** The preserved vegetation is installed. Courtesy of the Naturhistorisk Museum, Arhus, Denmark.

sprayed on the wire mesh. After the foam had formed and set, the surface was carved to the required shape. Plants and other props were set in position in impressions carved into the foam; these were fastened with fresh batches of foam. Polyurethane foam emits cyanide containing gases which are extremely dangerous. Masks should be worn when working with foam (Fig. 13.9).

Vegetation can cause problems in diorama building. Volume 2, Chapter 2, gives some insight into the intricacies of this work. Some of the problems can be eliminated at the beginning if the diorama is designed with this difficulty in mind. For instance, a spring scene in a temperate climate forest can be extremely difficult because of the multitude of flowers, fresh leaves, and budding plants necessary. In addition, a flower bedecked clearing in the forest may be breathtakingly beautiful in real life, but the same scene reproduced with plastic flowers and leaves can turn into kitsch. Autumn or winter scenes of the same forest are more rewarding. Most of the foreground material can be simply air dried and there may be no need for artificial plants. This does not mean that a spring scene cannot be created successfully (there are many fine examples in many museums), but some settings are more difficult than others.

Diorama building requires a great sense for innovation in a preparator. Almost every project calls for a different approach. The scope of work is almost limitless. There are excellent examples depicting scenes in equatorial jungles, deserts, the polar regions, coral reefs, or on the bottom of the sea in many museums.

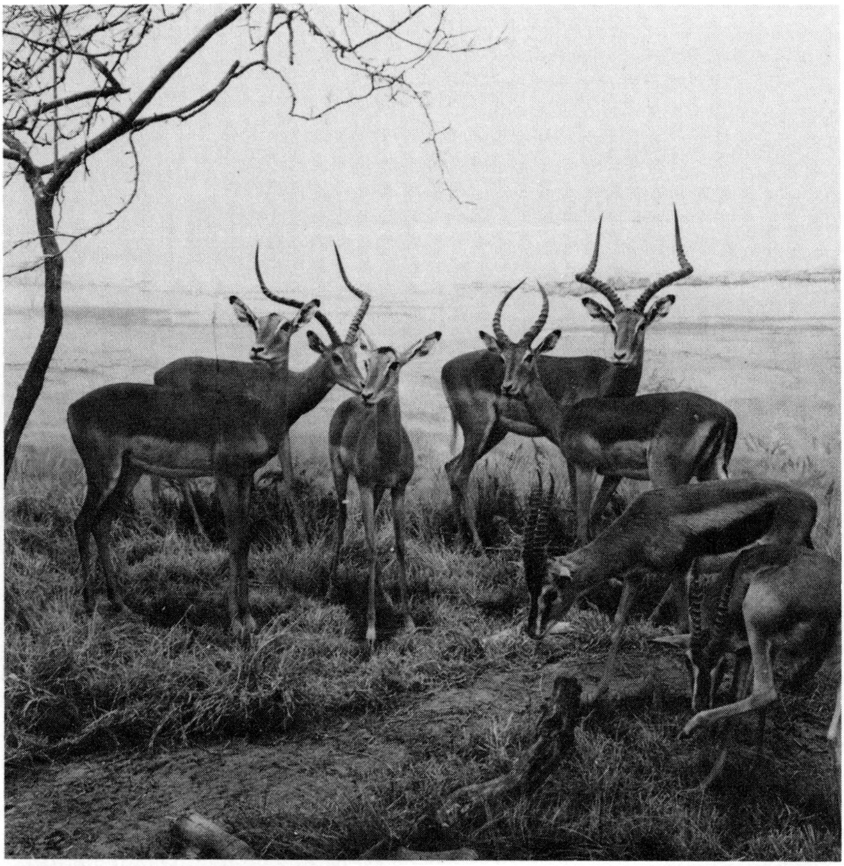

Fig. 13.8. The finished diorama. Courtesy of the Naturhistorisk Museum, Arhus, Denmark.

SPECIAL FOREGROUND EFFECTS

WATER

Small puddles of water can be made from clear, slow setting polyester resin. The catalysed resin is poured into a suitably formed depression and allowed to set. Clear embedding resin is most suitable for this application. Larger bodies of water cannot be made with this method, as big quantities of polyester resin become too hot while setting and discolour, crack and shrink. Resin representing large volumes of water should be poured in several layers, allowing enough time for the resin to cool down between pours. Less resin

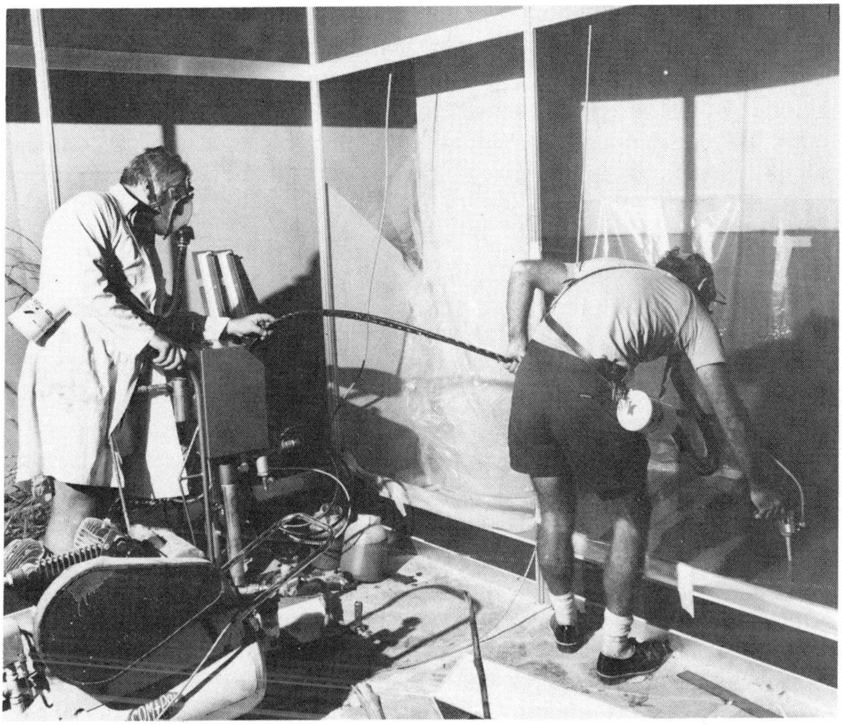

Fig. 13.9. Polyurethane foam is applyed by spraying equipment to form the ground in a diorama. Photo J. Fields. Courtesy of The Australian Museum.

can be used by fitting a sheet of acrylic over the depression which is supposed to hold water and pouring resin on top. Ripples can be formed by pouring slightly overcatalysed clear resin onto a sheet of transparent acrylic. To get the right ripples, one must experiment with small amounts of resin and various quantities of catalyst. Small waves can be formed by playing a fan on the surface of setting resin. Distinctive waves must be modelled in clay, moulded, and finally cast in clear resin. Water can also be imitated by sheet glass, acrylic sheet, or several layers of fine muslin. Evans-Olsen (1980) mentions the use of layers of bridal veil for water effects.

ICE

The surface of frozen ponds or creeks can be simulated with glass or acrylic sheets. The underside of the glass may be lightly airbrushed white to increase effect. Lumps of ice, with a melting surface, can be made of clear polyester resin. The resin should be catalysed and allowed to gel. While in the gel stage, it should be poured on a sheet of polyethylene (or into a plastic bag) and gently agitated until set. With some practice, one can make beautiful icycles the same way. The gelled resin should be allowed to flow out of the cup in which it was mixed and carefully laid on polyethylene. It is also possible to catch the resin at the right time, and let it set in mid-air while hanging off the rim of a cup.

SNOW

Snow can be made in a number of ways. The oldest method uses cotton wool sprayed with wax. While the wax is still hot, it is sprinkled with lucite. Plaster of Paris can be used instead of cotton wool, but the former is more effective. O. Hoegh Post (1983) describes a method where a bronze hand-sprayer (the kind used to spray insecticides in a garden or in a house) was used to spray hot paraffin wax in order to form snow. In this method a base is formed of cotton wool or polystyrene foam and the hot wax is sprayed on. Commercially made snow crystals can be added to the wax snow.

Imitation snow can be made over a white base (cotton wool, polystyrene, or plaster) by adding white polyvinyl acetate glue and sprinkling fine glass beads mixed with borax on top. The procedure must be repeated until the snow is the required thickness. Glass beads are used by road sign makers and occasionally can be bought from them.

DRIPSTONE

Stalactites can be made by carving the basic shape out of polyurethane foam and repeatedly dipping this into molten paraffin wax. The wax can be coloured the required hue with small amounts of oil paint.

Stalagmites are made the same way, except that the molten wax is poured on the erect foam core in several batches. If necessary, the artificial stalagmite is flame polished with a bunsen burner.

Small amounts of fine mica or lucite can be added to the surface to imitate the glittery effect of stalactites and stalagmites.

ROCKS

There are many ways to make artificial rocks. The best results are gained from the most labour consuming technique; moulding and casting real rocks. Rubber latex moulds are prepared from selected rocks, rock faces and walls and casts are made. Polyester resin casts are light weight but the making of large casts prepared from this material can be a health hazard. Such work should be carried out in a well ventilated area which is equipped with adequate exhaust systems.

Plaster of Paris is a more agreeable material to work with and the end result can be just as good as polyester. The casts may turn out to be a little heavier but, if a large rock formation has to be built and various elements must be fitted together, plaster casts are more practical. To trim and join plaster casts, a hand saw or a power jigsaw is an adequate tool. The casts are hung on timber frames and the gaps between ill-fitting elements are modelled in by hand. For this modelling and filling, freshly mixed plaster can be used. The parts to be joined should be wetted first to ensure good adherence. The setting of the plaster can be slowed by adding borax to the mix. The freshly modelled areas can be given texture by patting the surface with a coarse brush. Experience will show the preparator when to start the texturing. It is best done when the plaster is still in a cheesy stage; set solid, but not absolutely hard.

Plaster rocks accessible to the public should be hardened. Epoxy resins used for impregnating lesser quality timbers and plywood are ideal for this purpose. One application of these resins can render the plaster rocks 'rock-hard'. Resin hardened rocks should be painted after the resin treatment. The colouring of artificial rocks should start during the casting operation. The batches of plaster should be pigmented with oxides. The finished casts can be painted with a darker wash. When the wash starts to dry, it is wiped off with rags. This allows the darker paint to stay in the depression, while high spots are wiped almost clean, allowing the base colour to show through. Here again, experience will show the worker how to achieve the best effect. The use of spraying equipment can make the work easier.

CLOUDS OF DUST

A splendid example of a small cloud of dust can be seen in the Denver Museum of Natural History, Denver, Colorado, USA. The dust is shown at the foot of a speeding cheetah in one of the African dioramas. It was made by airbrushing a faint image of a small cloud on a piece of acetate sheeting. The sheet was installed front of the foot of the animal and its sheen was neutralized with skilful lighting. The viewer can see only the small puff of dust created by the pouncing animal, without a hint of the acetate sheet.

There are many more interesting and useful techniques for creating lifelike accessories or props for dioramas. The reader can gather more information from the literature listed at the end of this chapter.

REFERENCES

Bindel, F. P., and G. M. Rasmussen: Neue Diorameneffekte im Kranichdiorama. *Der Präparator* 21/1 1975.

Evans-Olsen, S.: Details, details, detail. Preparation of exhibits with particular reference to foregrounds and habitat. Part 1. *Taxidermy Review* 9/2 1980; Part 2 *Taxidermy Review* 9/3 1980.

Hohorst, B.: Dioramen in Naturkundlichen Heimatmuseum Benrath. *Der Präparator* 26/3 1980.

Hoegh-Post, O.: Snow in dioramas. *Taxidermy Review* 11/5 1983.

SOURCE MATERIAL

Allan, D. A.: Dioramas in the Zoological Museum of Berlin University. *Museum Journal* 36/1936:

Blakely, D. M.: Suggestions for increasing the effectiveness of habitat groups. *Museum News* 17/1939: 7–8.

Breiner, E.: Neue Wege der musealen Darstellung und ihre Problematik. *Der Präparator* 4/1958: 99–104.

Burns, N. J.: Group lighting. *Museum News* 10/1932: 5–7.

——. The history of dioramas. *Museum News* 17/1940: 8–12.

Cappel, L.: *Guide to modelmaking and taxidermy.* A. H. & A. W. Reed, Sydney 1973.

Dummel, R. G.: Lighting and the habitat group. *Museum News* 10/1932: 6–8.

Fuehrer, O. F.v.: Das Karbondiorama. *Der Präparator* 1/1955: 3–7.

Gottschalk, R.: Was erwarten wir vom Diorama? *Neue Museumskunde* 7/1964: 120–125.

Lanham, F. M.: Modern trends in exhibition work in museums. *Bulletin of the South African Museum Association* 2/1941: 261–268.

——. The display of material. *Bulletin of the South African Museum Association* 3/1943: 1–4.

Legge, E. M.: Glass for dioramas. *Museum Journal* 33/1934: 352.

McNeely, P.: Problems in the painting of a large background. *Museum News* 8/1931: 11–72.

Mielke, C. H. Th.: Aufbau einer Schausammlung der stammesund kulturgeschichtlichen Entwicklung des Menschen im Museum Nacional, Rio de Janeiro. *Der Präparator* 3/1957: 3–7.

Miner, N. A.: Artificial water for habitat groups. *Museum News* 12/1934: 8.

Perry, R. K.: Atmosphere and colour in simple habitat groups. *Museum Journal* 37/1938: 533–537.

Prater, S. H.: Modern museum methods: Part 3. *Journal of the Bombay Natural History Society* 33/1929: 149–162.

Richardson, J.: Artificial water and its limitations. *Museum News* 11/1934: 7–8.

Stansfield, H.: Coloured backgrounds for display cases. *Museum Journal* 32/1932: 96.

Tose, F.: Hints on museum preparation. *Museum News* 12/1934: 7–8.

Wallace, H. M.: Display methods. *Museum Journal* 33/1933: 100.

Formulae

THE following formulae are used by preparators and commercial taxidermists all over the world. However, many of them are often modified to suit the needs of an individual worker. These modifications are sometimes published, sometimes not. In this selection, the authors have endeavoured to collate a number of formulae which suit most purposes.

FORMULA 1: AMMANN'S LACTOPHENOL

This formula is used for mounting oligochaets.

phenol	100 g
lactic acid	100 mL
glycerine	200 mL
water	100 mL

FORMULA 2: BOUIN'S FLUID

picric acid (saturated aqueous solution	75 mL
formalin	25 mL
glacial acetic acid	5 mL

FORMULA 3: CHROME-ACETIC MIXTURE NO 1

chromic acid 1%	100 mL
concentrated acetic acid	5 mL

FORMULA 4: FORMOL-ACETIC-ALCOHOL (FAA)

absolute alcohol	90 mL
formalin	5 mL
glacial acetic acid	5 mL

FORMULA 5: FORMALDEHYDE SOLUTION (NEUTRAL)

To 40% formaldehyde solution add calcium carbonate or magnesium carbonate to make a saturated solution. To make neutral formalin take the clear supernatant fluid and dilute it with water.

FORMULA 6: ISOTONIC SALINE SOLUTION

Wash worms in the isotonic solution that closely resembles that of the host. For example; for frogs, 0.65 g sodium chloride dissolved in 100 mL distilled water; for mammals, 0.9 g sodium chloride dissolved in 100 mL distilled water.

FORMULA 7: CORROSIVE SUBLIMATE

This is a saturated solution of mercuric chloride in water (sea water for marine specimens). After fixation remove all traces of mercuric chloride by soaking the material in iodized alcohol for 12 hours.

FORMULA 8: CORROSIVE ACETIC

Add acetic acid to corrosive sublimate to make a 5%–10% solution.

FORMULA 9: IODIZED ALCOHOL

iodine	3 g
potassium iodide	6 g
70% alcohol	300 mL

Add enough of this solution to 70% alcohol to produce a sherry brown colour. Specimens are washed in this until the brown colouration no longer disappears. When this is done transfer the specimens into fresh 70% alcohol to remove any traces of existing iodine.

FORMULA 10: FORMOL ALCOHOL

70% alcohol	50 mL
water	45 mL
formalin	5 mL

INDEX

acanthocephala, 45
acid Lugol's solution, 18
acrylic, 187–189, 220–221
  annealing, 189
  antistatic compounds, 203
  cements, 189
  cutting, 199
  glueing, 201–203
  glues, 189
  lugs, 205
  nibs, 204
  polishing, 203
  properties, 188
  sanding, 203
  spacers, 204
alcohol, see ethanol
algae, 17–19
alginate moulds, 273–276, 285–286
Ammann's lactophenol, 316
ammonia, 84
amphioxus, 75
amphipods, 70
annelida, 47
  see oligochaeta, earthworms, polychaeta, hirudinea
aquatic
  insects, 104
  molluscs, 59
arachnids, 168–174
  collecting, 168
  freeze drying, 185
  webs,
    collecting, 169
    making, 172
aschelminthes, 50
  see gastrotricha
    kynorhyncha
    nematodes
    nematomorpha
    rotifera
ascidiaecea, 74
aspirator, 93
attracting insects with females, 118

baiting insects, 113
benzol, 82
Berlese funnel, 96
bitumen, 213
black light, 107
bladders, 213
blattodea, 155
Blés fluid, 186
Bouin's fluid, 316
Boyle, Robert, 1
brittle stars, 40
buildings, models, 295–297

calcium cyanide, 82
Canada balsam, 147
carbon tetrachloride, 82
carcasses to attract insects, 115
carding, 130
Carnoy's fluid, 84
carving, 260–265
casting
  ant nest, 165–168
  gel coat, 218
  laminating, 279
  latex, 283–284
  plaster, 286–287
  polyester resin, 277–281
  polyurethane foam, 281–283
  polyurethane rubber, 283
  silicone rubber, 283–284
  wax, 285
catalogues, 4
cephalochordates, 75
cephalopods, 62
  freeze drying of, 183
cestodes, 52
chilopoda, 173
chitons, 59
chloroform, 83
chordates, 74
chrome-acetic mixture No 1, 317
cirripedia, 70
coelenterates, 33–39
  freeze drying of, 183
coleoptera, 159
collembola, 153
coloured flowers, 21, 242–243
corals, 37
corrosive
  acetic mixture, 317
  sublimate, 317
crab, 70
critical point drying, 144
crustaceans, 67–73
  freeze drying of, 184
cuttlefish, 60
cyanide killing jar, 78

degreasing insects, 152
de-pinner, 160
dermaptera, 155
dioramas, 301–314
diplopoda, 173
diplura, 153
diptera, 158
display, 301–314
  dripstone, 313
  dust clouds, 313
  habitat, 302
  ice, 312
  rocks, 313
  snow, 313
  water, 311
dissecting, 210
Dornisthorpe's fluid, 85
Dowicil, 7
dredging for insects, 104
dried specimens
  rejuvenation of, 212
drift fence, 116
dung traps, 117
dust cloud making, 314

earthworms, 48
echinoderms, 40–44
  collecting, 40
  display, 43
  drying, 43
  killing, 42
  fixing, 42
  preserving, 42
echiura, 51
eggs of insects, 140
embedding, 223–244
  calculating volumes, 227
  coloured backgrounds, 229
  cylindrical blocks, 239
  dehydrating, 240
  dry specimens, 235–237
  flowers, 242
  hazards, 243–244
  labels, 231
  mirror under block, 234
  mould release, 225
  moulds, 224
  polishing, 237
  procedure, 235
  silvering, 231
  wet specimens, 240–242
embioptera, 155
emergence box, 103
envelope for packing insects, 121
ephemeroptera, 153
epoxies, 219
ethanol, 7
ethyl acetate, 80, 82
ethylene glycol, 116
euparal, 147
  essence, 147
eviscerating large insects, 122

FAA *see* formol acetic acid
feather stars, 40
ferns, 19
figs, 15
fixation, 5
flatworms, 52
flowers
  colour restoration, 242–243
  embedding, 242
flukes, 53
formaldehyde, 5
  buffering, 6
  neutralizing, 5, 317
formalin, 6
formol acetic acid, 17, 317
formol alcohol, 318
formulae, 316–318
freeze drying, 177–186
  foliage, 180
  fruit, 180
  fungi, 180
  insect proofing, 180
  invertebrates, 178, 181–186
  plants, 179
  procedure, 178
  reconstituting, 181
  succulents, 180
fruit, 21
  freeze drying, 180
fungi, 21
  freeze drying, 180

gastritricha, 50
genitalia preparation of, 144
glyceel, 148
green plants, 19
grylloblattodea, 154
gum chloral, 147

hair artificial, 289–290
hair worms, 50
hemiptera, 156
hirudinea, 49
Hofmann, A. W. von, 1
hot melt compounds, 222–223
Huffaker's XA mixture, 85
Hunter, J., 2
hymenoptera, 159

ice display, 312
insect attack prevention, 151
insecticide
  spraying light sheet, 112
  spraying trees, 93
insects, 77–168
  aquatic, 104
  baiting, 113
  carding, 130
  casting ant nest, 165–168
  cleaning, 151
  degreasing, 151
  display, 160

eviscerating, 122
field preservation, 119
freeze drying, 185
genitalia, 144
habitat display, 163–165
killing
agents, 82–86
bottles, 78–81
heat, 86
injection, 86
labelling, 144
larvae, 140–143
microscope preparations, 146–148
moulding ant nest, 161–163
pinning, 128
preservatives, 84–86
relaxing, 125
setting, 134–140
sorting, 127–128
storage, 148–150
terrestrial, 77
invertebrate freeze drying, 181–186
iodized alcohol, 318
isopods, 70
isoptera, 156
isotonic saline, 317

jellyfish, 33

Kahle's fluid, 85
kynorhyncha, 50

labels, 3
lacto alcohol, 84
lactophenol (Amann's), 316
lanceolates, 75
landscape models, 297–299
larvaceae, 74
larvae of insects, 140
inflating, 141
degreasing, 143
dehydrating, 143
wax filling, 143
latex, 121
moulds, 272
layering, 121
leaves, 23–26
leeches, 49
Leonhard forceps, 132
lepidoptera, 157
lichens, 21
light
black, 107
collecting, 110
mercury vapour, 107
trap, 107
limpets, 59
lobster, 70

maintenance, 9
of insects, 150–151
malacostraca, 70
malaise trap, 118
mallophaga, 156
mantodea, 156
mecoptera, 157
mercury vapour lamp, 107
methyl ethyl ketone peroxide, 217
micromolluscs, 60
minuten pins, 128
mistletoe, 15
mites, 169–171
Moczarsky-Winkler selector, 99
modelmaking, 252–300
artificial hair, 289–290
buildings, 296
carving, 260–265
landscapes, 297–299
masonry, 295
roof surfaces, 296–297
scale models, 291–295
sculpting, 252–260
surface finish, 288–289
topographical, 291–295
vacuum forming, 287–288
water effects, 299
models
leaf, 23, 25, 26
plant, 23
surface finish in plastic, 288
molluscs, 57–66
Montagu Brown preservative, 18
mosquitoes, 106
mosses, 21
moulding, 265–276
flexible moulds, 271–276
human hands, 266–270
latex, 221–222
termite mounds, 161–163
mould prevention of insects, 150
mould release, 225
moulds
alginate, 273
flexible, 271
piece, 266
polyurethane, 273
polyvinyl chloride, 276
rigid, 266
rubber latex, 272
silicone rubber, 271
waste, 266
mounting media, 147
museum pots, 187–209
sealing lids, 213–214

nemathelminthes, 50
nematodes, 50
nematomorpha, 50
nemertines, 51
nets
aerial, 87
beat net, 91
butterfly, 87
dip, 104

nets—*continued*
mobilized, 90
plankton, 67
seine, 104
sweep, 90
tow, 67
neuroptera, 157
neutral formaldehyde, 5
nicotine water, 42

octopus, 60
odonata, 154
oligochaeta, 47
onychophora, 173
orthoptera, 154
ostracods, 70
freeze drying, 182
Oudeman's fluid, 85, 124

Pampl's fluid, 86
paper packing for insects, 121
leaves, 34
mache, 258
paraformaldehyde, 7
paraloid B72, 163
pauropoda, 173
peanut worms, 54
pentastomides, 52
Peterson's KAA, 85
phenoxyethanol see propylene phenoxytol
phasmida, 155
phoronides, 52
pinning insects, 119, 128
pins for insects, 128
pit traps, 116
plants, 13–27
algae, 17
artificial, 23–26
coloured flowers, 21
display, 22–23
dry preservation, 13–15
ferns, 19
freeze drying, 177
fruit, 21
fungi, 21
green plants, 19
herbarium sheet, 15
lichens, 21
mosses, 21
press, 14
seed plants, 19
wet preservation, 15–22
plastic display boxes, 187–209
acrylic properties, 188
annealing, 194
backing plate, 204
cements, 189
equipment, 194–198
glueing, 201–203
heat formed, 207–209
polishing, 203
sanding, 203
special considerations, 210–214
plastic leaves, 24
plastics, 216–223
elastomers, 217
epoxies, 219
fillers, 221
flexibilizers, 218
hazards, 246–249
hot-melt compounds, 222–223
laminating, 217
latex, 221
polyester, 217
polyurethane, 220
silicone rubber, 221
thermoplastic, 217
thermosetting, 217
urethane, 220
platyhelminthes, 52
see also, cestodes, trematodes, turbellarians
plecoptera, 154
pogonophora, 53
polychaeta, 48
polyester, 217
embedding, 223
flexibilizers, 218
hazards, 218
laminating, 279
polyurethane, 220
casting, 281–283
moulds, 272–273
rubber, 283
spraying, 264–265
polyvinyl chloride moulds, 276
potassium cyanide, 78, 82
prawn, 70
preservation
alcohol, 7
definition, 2
Dowicil, 7
fixation, 5
formaldehyde, 5
formalin, 6
methods, 1–10
narcotization, 5
paraformaldehyde, 7
propylene phenoxytol, 8
purpose, 2
preservations
acid Lugol's, 18
Bouin's, 316
Blés, 86
Carnoy's, 84
chrome-acetic no. 1, 317
corrosive acetic, 317
corrosive sublimate, 317
Dornisthorpe's, 85
Dowicil, 7
ethanol, 7, 84
formaldehyde, 5
formalin, 6
formol acetic acid FAA, 17
formal acetic alcohol, 317
formol alcohol, 318

Huffaker's XA, 85
Kahle's, 85
lacto alcohol, 84
Montagu Brown, 18
Oudeman's, 85, 124
Pampl's, 86
paraformaldehyde, 7
Peterson's KAA, 85
propylene phenoxytol, 8
Smart's, 123
Transeau's algal, 18
Van Emden's, 86
press for plants, 14
priapulida, 54
propylene phenoxytol, 8
protura, 153
psocoptera, 156
pupae, 140

Reitter sieve, 96
relaxing insects, 125–127
ribbon worms, 51
rock display, 313
rotifera, 50
roundworms, 50
RTV rubber, 221
Ruysch, F., 1

safety, 244–249
sculpting
clay, 255
papier mache, 258
plastic, 258
plasticine, 253
wax, 255
sea
anemones, 35
cucumbers, 40
fans, 38
pens, 37
squirts, 74
urchins, 40
seed plants, 19
segmented worms, 47
setting boards, 136–140
shore washing, 102
silica gel, 123
silicone rubber, 221
moulds, 271
siphonaptera, 159
siphunculata, 156
sipuncula, 54
slurp gun, see yabbie pump
Smart's embalming fluid, 123
snails, 59
snow display, 313
sodium cyanide, 82
spiders, 168–174
webs
collecting, 169
making, 172
spiny headed worms, 45
sponges, 31–32
spoon worms, 51
squid, 60
beak preparation, 64
pen extracting, 64
stalagmite, 313
stalagtite, 313
starfish, 40
steamer, 210
stems, 23
Stockholm tar, 213
storage of museum specimens, 8
storage boxes, 148
strepsiptera, 159
symphyla, 173

tapeworms, 52
termite mounds, 161–163
terrestrial,
insects, 77
molluscs, 57
thaliacea, 74
thysanoptera, 157
thysanura, 153
ticks, 169–171
tongue worms, 52
topographical models, 291–295
Transeau's algal fixative, 18
trematodes, 53
trichoptera, 158
Tullgren funnel, 98
turbellarians, 53

urethane, 220
urochordates, 74

vacuum forming, 287–288
leaves, 24, 26
Van Emdens fluid, 86
vapour killing jar, 80
velvet worms,173

wax
casts, 285
leaves, 24
sculpting, 255
webs
collecting, 169
making, 172
wires, tapering, 26
wood naptha, 126
worms, 45–56

yabbie pump, 68

zoonoses, 248
zoraptera, 156

5 6 7 8 9 0 1 2 3 4
A B C D E F G H I J